59th International Scientific Conference on Experimental Stress Analysis (EAN 2021)

Litomysl, Czech Republic
29 September – 1 October 2021

Editors:

Petr Bittnar
Pavel Padevet

ISBN: 978-1-7138-4520-1

Printed from e-media with permission by:

Curran Associates, Inc.
57 Morehouse Lane
Red Hook, NY 12571

Some format issues inherent in the e-media version may also appear in this print version.

Czech Society for Mechanics (CSM)
Dolejskova 5
Praha 8
182 00
Czech Republic

Phone: +420 266 053 045
Fax: +420 286 587 784

csm@csm.cz

Additional copies of this publication are available from:

Curran Associates, Inc.
57 Morehouse Lane
Red Hook, NY 12571 USA
Phone: 845-758-0400
Fax: 845-758-2633
Email: curran@proceedings.com
Web: www.proceedings.com

EXPERIMENTAL STRESS ANALYSIS
59th International Scientific Conference
September 29 – October 1, 2021
Litomyšl, Czech Republic

EAN 2021

59th conference on experimental stress analysis

BOOK OF FULL PAPERS

Editors: Petr Bittnar, Pavel Padevět

Czech Technical University in Prague, Faculty of Civil Engineering, Department of Mechanics
&
Czech Society for Mechanics, Expert Group for Experimental Mechanics

Title:	Experimental Stress Analysis 2021 Book of Full Papers
Editors:	Petr Bittnar Pavel Padevět
Published by:	Czech Technical University – CTU in Prague Faculty of Civil Engineering Department of Mechanics

The Contents and Index of Authors are sorted in Czech alphabetical order. Typeset in LaTeX.

The complete set of EAN 2021 Conference Proceedings consists of two parts:

- Book of Extended Abstracts (printed)
- Full Papers (on-line, web)

Prague, September 2021

Conference organized by:

Department of Mechanics, Faculty of Civil Engineering, Czech Technical University in Prague
https://mech.fsv.cvut.cz

in cooperation with:

Conference Partners Prague, s.r.o. – namely Mrs. Alexandra Sternberg and Mr. Martin Zein
https://www.conferencepartners.cz

and partners:

HBP měřící technika, s.r.o.
https://www.hbm.cz

LABORTECH, s.r.o.
https://www.labortech.cz

LABTECH, s.r.o.
https://www.labtech.eu

Webpages
https://mech.fsv.cvut.cz/ean2021/

Scientific Program

- Development of experimental methods in mechanics
- Development of experimental methods in biomechanics
- New methods and applications serving to analyse deformation and stress in mechanical constructions
- New methods and applications serving to analyse deformation and stress in building and other constructions
- Experiment as a tool for verification of analytical and numerical methods
- Experimental research and prediction of strength, durability, and operating reliability of constructions and devices
- Tracking and monitoring of operating loads and operating conditions of structures and devices
- Mechanics of advanced materials, composites and 3D print materials
- Methods and means of teaching experimental methods

Scientific Committee
prof. Ing. Nikolaj Ganev, CSc.
prof. Ing. Radim Halama, Ph.D.
prof. RNDr. Miroslav Hrabovský, DrSc.
prof. Dr. Ing. František Holešovský
prof. Ing Stanislav Holý, CSc.
prof. Ing. Josef Jíra, CSc.
prof. Ing. Milan Růžička, CSc.
doc. Ing. Peter Frankovský, PhD.
doc. Ing. Róbert Huňady, PhD.
doc. Ing. Tomáš Návrat, Ph.D.
doc. Ing. Iva Petríková, Ph.D.
doc. Ing. Pavel Padevět, Ph.D.
doc. Ing. Miloš Vlk, CSc.
doc. Ing. Robert Zemčík, Ph.D.
Ing. Karel Doubrava, Ph.D.
Ing. Lubomír Houfek, Ph.D.
Ing. Kamil Kolařík, Ph.D.
Ing. Aleš Lufinka, Ph.D.
Dr. Ing. Roman Růžek
Ing. Jaroslav Václavík

Organizing Committee
doc. Ing. Pavel Padevět, Ph.D.
Ing. Petr Bittnar, Ph.D.
Ing. Ondřej Zobal, Ph.D.
Mrs. Alexandra Sternberg (Conference Partners Prague)
Mr. Martin Zein (Conference Partners Prague)

Sponsors
HBP měřicí technika, s.r.o.
LABORTECH, s.r.o.
LABTECH, s.r.o

The 59th International Conference "Experimental Stress Analysis 2021" ("Experimentální Analýza Napětí" in Czech) is organized under auspices of prof. Ing. Jiří Máca, CSc. – the Dean of Faculty of Civil Engineering, Czech Technical University in Prague.

This year's conference is dedicated to the memory of prof. Ing. František Plánička, CSc. – longtime member of Czech Society for Mechanics, Expert Group of Experimental Mechanics.

Pavel Padevět

Contents

Influence of Overlapping on Surface Integrity Parameters of Laser-hardened Railway Axles

J. Čapek[1,*], K. Trojan[1], I. Černý[2], J. Kec[2], N. Ganev[1], S. Němeček[3]

[1] *Department of Solid State Engineering, Faculty of Nuclear Sciences and Physical Engineering, Czech Technical University in Prague, Trojanova 13, 120 00 Prague, Czech Republic*

[2] *SVÚM a.s., Tovární 2053, 250 88 Čelákovice, Czech Republic*

[3] *RAPTECH s.r.o., U Vodárny 473, 330 08 Zruč-Senec, Czech Republic*

* *jiri.capek@fjfi.cvut.cz*

Abstract: The paper describes the influence of overlapping of tracks on the state of residual stresses and microstructural parameters of laser hardened railway axles made of EA1N steel. It has been found that the tensile residual stresses, large crystallite size and low microdeformation are located on the surface of overlapping tracks. This zone, accompanied by a decrease in hardness, can be a potential area for crack initialization.

Keywords: railway axles; laser hardening; X-ray diffraction; microstructure; hardness.

1 Introduction

This research is focused on the development of surface treatment technology of railway axle seats using laser hardening to increase resistance to fatigue failure. Railway non-powered axles made of standard EA1N steel were examined.

For single-track laser hardening, the compressive residual stresses (RS) were found out on the surface of the hardened area, inside the laser track. On the other hand, the bulk material (machined surface) could reach high tensile RS, see [1]. This zone with a significant change in RS is usually an area of potential surface crack initialization.

Using multi-track laser hardening with overlapping, the part of the hardened zone is re-hardened. Therefore, the different zones are generated in and near the overlapped zone, see Fig. 1. The re-hardened and tempered zones are usually accompanied by a decrease in hardness [2, 3]. With the presence of other undesirable values of surface integrity, especially tensile residual stresses and microstructural notches, these zones are typical areas of initialization of surface cracks [1].

Therefore, it is important to achieve optimal characteristics of surface integrity of the axle at critical zones, including fatigue resistance, which is positively affected by compressive RS [4].

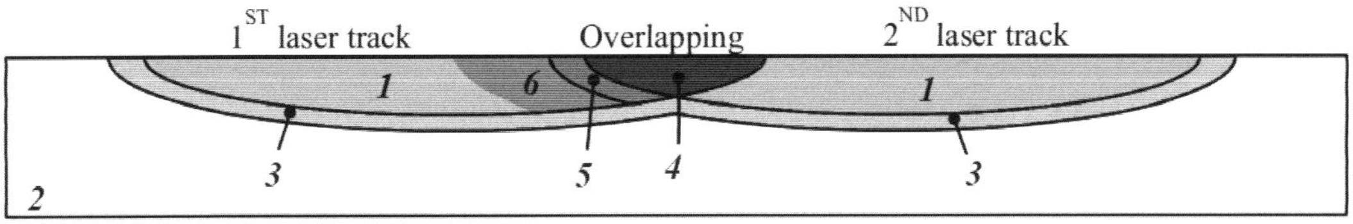

Fig. 1: Zones generated by multi-tracks laser hardening with overlapping: (1) hardened, (2) base material, (3) partially hardened, (4) re-hardened, (5) tempered partial re-hardened, and (6) tempered zones [2].

2 Experimental details

For experimental tests, a forged turned semi-finished product made of EA1N steel was used. The hardening was carried out circumferentially, following the hardening method, on axle seats with a controlled hardening temperature of 1150 °C, and a beam speed of 3 mm/s performed by the *Laserline LDM3000-60* in the company RAPTECH s.r.o. The width of the overlapping of tracks was 5 mm.

Metallographic specimens were hot embedded into conductive bakelite resin with the carbon filler and prepared by machine grinding and polishing. Structure was revealed by etching in 4% Nital (98 ml ethanol + 2 ml HNO_3). Microstructure was investigated by the optical microscope *Zeiss Axio Observer Z1M*.

Macroscopic residual stresses (RS) were determined using the *X'Pert Pro MPD* diffractometer by X-ray diffraction (XRD) and chromium radiation. The RS were analysed across the overlapped hardened tracks, see Figs. 2 and 3. The values of RS were calculated from lattice deformations determined based on experimental dependencies of $2\theta(\sin^2\psi)$ assuming a uniaxial state of residual stress, where θ is the diffraction angle, ψ the angle between the sample surface and the diffracting lattice planes. The diffraction angle was determined as the center of gravity of the Cr$K\alpha_1\alpha_2$ doublet diffracted by the lattice planes $\{211\}$ of the α-Fe phase. The X-ray elastic constants $s_1 = -1.25$ TPa^{-1}, $1/2 s_2 = 5.76$ TPa^{-1} were used for the stress calculation. The crystallite size (size of coherently diffracting domains) and microdeformation (inhomogeneous distribution of elasto-plastic deformation on a microscale; a grain size) were determined from the XRD patterns obtained using the cobalt radiation and the Rietveld refinement, performed in *MStruct* software. The irradiated volume was defined by experiment geometry, the effective penetration depth of the X-ray radiation (approx. 5 and 10 μm for chromium and cobalt radiation, respectively), and the pinhole size (4×0.5 mm^2).

Hardness was evaluated as Vickers hardness HV10 using the *Vickers Limited HTM*.

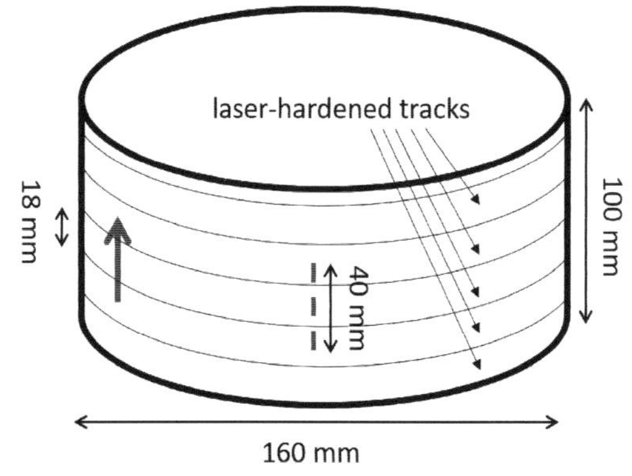

Fig. 2: Diagram of analysed sample, where arrow denotes the XRD measurement direction and dashed analysed line.

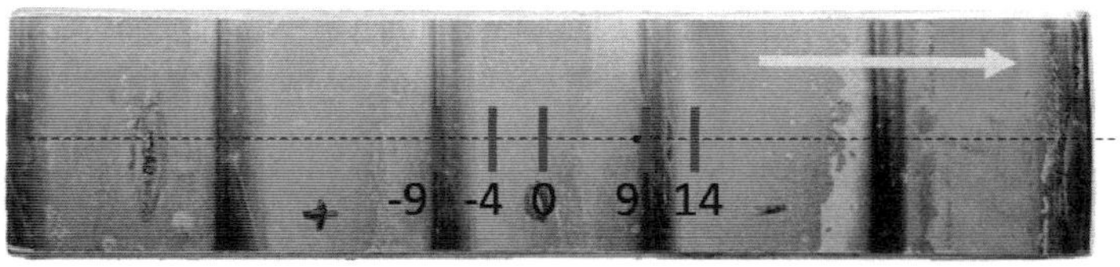

Fig. 3: Photo of multi-track hardening, the arrow represents the direction of overlapping tracks.

3 Results and discussions

Hardened depth evaluated from macrostructure was 1.5–2 mm, which is typical for laser hardening. The microstructure of the hardened tracks and the boundary between the tracks and base material was homogeneous in all cases. Microstructure of the hardened tracks corresponded to very fine-grained tempered martensite, in the direction of base material being continuously changed to the mixed martensitic and ferritic-pearlitic structure.

The microstructure of the base material was purely ferritic-pearlitic and corresponded to EA1N steel standard. Microstructure changes in the direction from hardened track to base material are documented in Fig. 4.

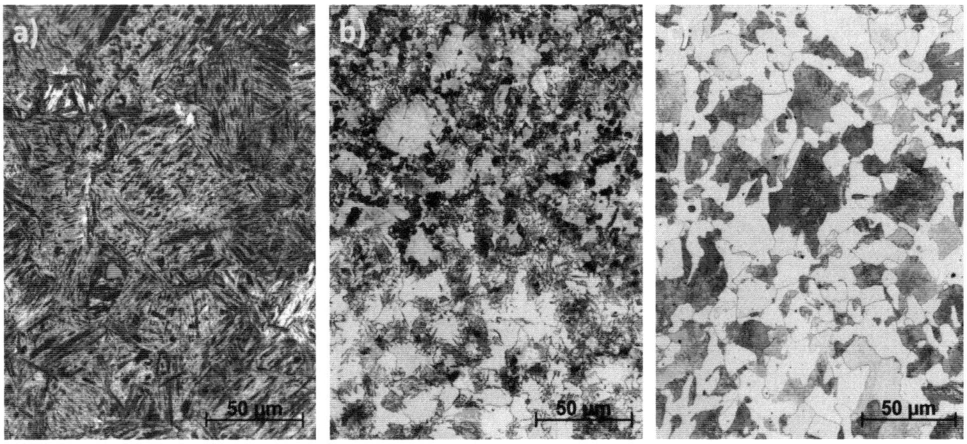

Fig. 4: a) Martensitic microstructure of the hardened track, b) mixed microstructure of partially hardened zone, and c) ferritic-pearlitic microstructure of base material.

Nevertheless, particular attention was paid to the overlapping of the laser tracks, as it is a zone where local deviations from both microstructure and hardness usually occur. Hardness analyses were performed across the overlapping and hardened zones, see Fig. 5, where the inhomogeneous course was found. This evolution confirmed the expectation of hardness decreasing in the tempered and re-hardened zones. Therefore, the hardness reduction in the overlapping zone is, generally, necessary to expect. What is, however, important is that in spite of some hardness decrease resulting from a partial tempering of the previously hardened material, the drop is not absolute in terms of the actual hardness value, which remains considerably higher than the hardness of the base material, which is approx. 170 HV.

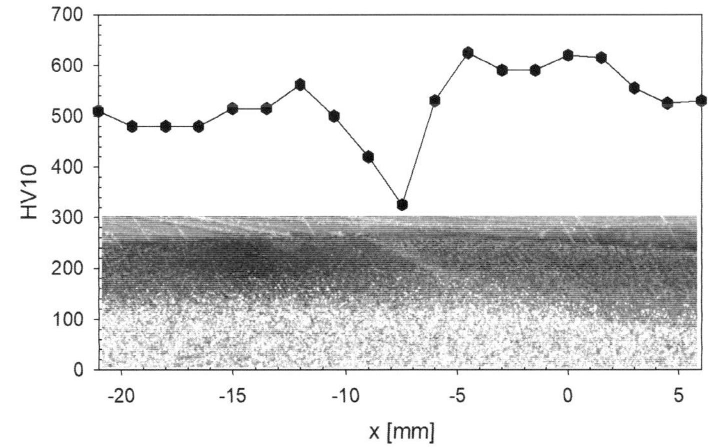

Fig. 5: Course of HV10 hardness in surroundings of laser tracks overlapping.

Detailed analyses of the surface macroscopic RS, microdeformation, and crystallite size of the laser-hardened sample with overlapped tracks are depicted in Fig. 6. The figure shows three overlapped tracks, hardened from left to right.

Stresses formed on the hardened surface because of rapid thermal heating and consequent cooling associated with phase transformation. Therefore, stresses occurred as a result of the superposition of thermal (tensile effect) and transformation (compressive effect) stresses and generally could reach high values.

Same as for single-track hardening [1], the compressive RS were found on the surface of the hardened track due to the dominant transformation effect. In the case of tracks' overlapping, the transition zones with different values of investigated parameters were observed. Fig. 6 shows the unfavourable surface tensile RS, large crystallite size, and low microdeformation localize at the boundary of the track (at a distance of 9 mm from the x-axis origin). At this distance (at the left boundary of the track), the temperature was not so high to phase transformation only the original microstructure was tempered. Therefore, the crystallite size was larger than in the track and heat-affected zone. The tensile character of RS was very likely caused by stress revealing from elevated temperature and subsequent cooling at this zone. For the same reason, a local minimum of microdeformation probably appeared. The lowest microdeformation were placed at the boundary of the track, where the thermal effect was dominant and, therefore, the grain distortion was reduced.

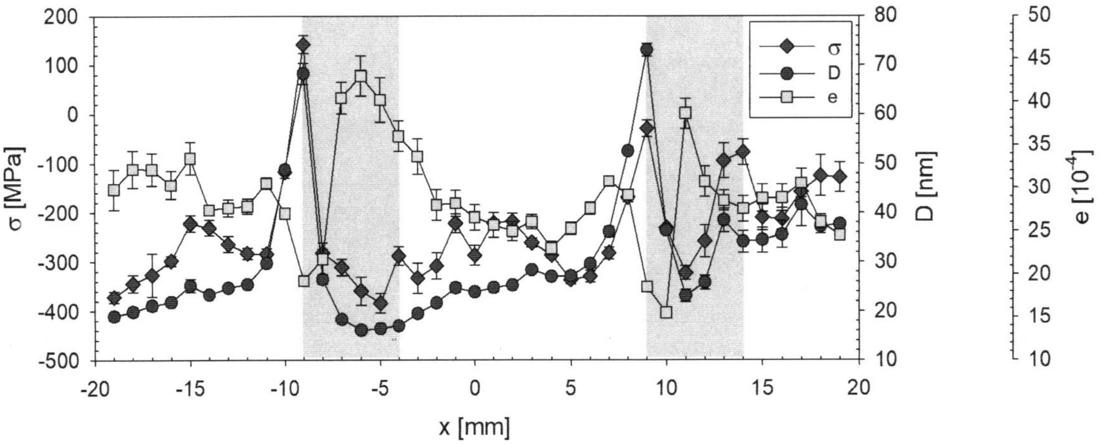

Fig. 6: Distribution of surface macroscopic residual stresses σ, crystallite size D, and microdeformation e across the hardened tracks in the direction perpendicular to the hardening. Grey areas denote the overlapping of tracks.

4 Conclusions

The overlapping (re-hardened) and surroundings tempered zones are accompanied by a decrease in hardness, microstructure changes and by extremes in values of RS, crystallite size and microdeformation, see Figs. 5 and 6. The distributions of investigated surface integrity parameters are in very good agreement with high-cycle fatigue damage, when the initiation of fatigue cracks occurred exactly at the boundary between two laser tracks, namely preceding track and subsequent — overlapping track.

Acknowledgement

Experiments were supported by the project TH04010475 of the Technology Agency of the Czech Republic.

References

[1] K. Trojan et al., Residual Stresses of Laser-Hardened Railway Axles, in proc.: Experimental Stress Analysis 2020 - Book of full Papers (2020), ed. M. Fusek et al., Technical University of Ostrava, online, 547–554.

[2] S. Martínez et al., Hardness simulation of over-tempered area during laser hardening treatment, Physics Procedia 83 (2016) 1357–1366, doi: 10.1016/j.phpro.2016.08.143.

[3] I. Černý et al., Advanced surface and thermal treatments for increasing the durability of railway components and tools (ASTRACOMP). Final EUREKA project report, SVÚM a.s. (2020), https://www.eurekanetwork.org/project-database/?search=Advanced+surface+and+thermal+treatments.

[4] V. Linhart, I. Černý, An Effect of Strength of Railway Axle Steels on Fatigue Resistance Under Press Fit, Engineering Fracture Mechanics 78 (2011) 731–741, doi: 10.1016/j.engfracmech.2010.11.023.

Experimental Support of Numerical Models for Verification of High-speed Impact Resistance of Composite Airframe

R. Doubrava[1,*], P. Bělský[1], M. Kadlec[1], B. Cabrnoch[1]

[1] *VZLÚ – Czech Aerospace Research Centre, Beranových 130, 199 05 Prague 9, Czech Republic*

* *doubrava@vzlu.cz*

Abstract: The composite structures are often subjected to impact damage during production, in-service operations, or maintenance. Testing all impact scenarios into a real structure is very expensive and time-consuming. The aim of this paper is to present an improved calibration and verification of numerical models by standard and non-standard experiments. The results were applied for analyses of bird strike simulation on a composite engine cowling of a newly developed high-speed helicopter.

Keywords: experiment; impact; high-speed; bird; composite.

1 Introduction

Safety is the ultimate aim in civil aviation. Therefore, the international civil aviation system evaluates with an acquired experience in the operation of aircraft, but of particular importance here are air accidents [1]. They are milestones for the new approaches to the operation of aircraft and demonstrate their resilience to the impact. The risk of foreign object impact that can decrease local strength property must be taken into account in the design of primary structures. Testing all impact scenarios onto real structures is expensive and impractical. Foreign object damage (FOD) such as bird, hail, debris etc. are important phenomena that must be taken into consideration when designing aircraft. The critical parts of planes or helicopters are windshield, nose, fuselage panels, wing and empennage leading edges, rotor blades, fan blades, engines cowlings or inlets [2].

However, the design of the aircraft structures usually involves many iterations of design-manufacturing-test and conducting bird impact experiments is not only time consuming but also costly.

The flat and simple curved test specimen used in the high-speed impact resistance verification tests was designed to provide confirmation of the performance of the selected composite material and to assist in the finite element modelling of the global structure [6]. All numerical models were calibrated based on experimental results gathered with an eye toward damage initiation and propagation behaviours. Fig. 1 shows the building block diagram (BBD) for the proposed approach.

2 Materials and methods

The test program aimed to obtain inputs for verification of simulation technique of complex composite airframe design from point of view of high-speed impact resistance. The test program was adapted to obtain inputs in terms of application of simplified models for explicit solver analysis (large size of FE elements for a decrease of critical calculation time) and also from point of view of easy production of the test specimen.

The standard and non-standard uniaxial tension and compression tests were used for input material properties onto numerical models on specimen level (see Fig. 1). The main focus was on inputs for simulation of honeycomb sandwich core and mechanical fasteners of composite parts of the airframe. The influence of high-speed load and inputs for the FE model of airframe and bird projectile was verified by flat and simple curved test specimens on an element level (see Fig. 1).

The results from calibration and simulation on specimen and element level were used for simulation and optimization of the real airframe on full-scale level of proposed BBD (Fig. 1).

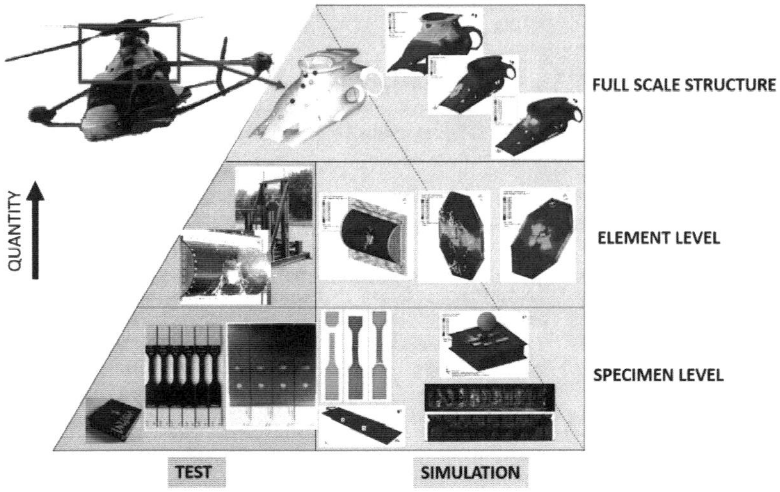

Fig. 1: Building block diagram for bird strike resistance analysis of composite cowlings.

2.1 Specimen level

The specimen level aimed to obtain mainly static material properties for input to the FE material models. In addition to standard tests of the properties of composite materials, attention was focused on the calibration of simplified replacement models of the honeycomb sandwich core and mechanical fasteners.

In the case of the design of the simplified model of a sandwich structure with honeycomb core for impact analysis in the explicit solver, attention was focused on the compression behaviour. The compression tests according to ASTM C365M standard [3] were performed in the VZLU test lab (Fig. 2), and the results of the tests were used for verification and calibration of the FE model (Fig. 3). The simplified FE model of honeycomb sandwich structure was proposed for simulation of complex airframe based on solid elements with ABAQUS crushable foam material model. The results from the tests and simplified FE model were also compared with the detailed FE model of the honeycomb core (Fig. 3).

Fig. 2: Compression test set-up with additional displacement measurement.

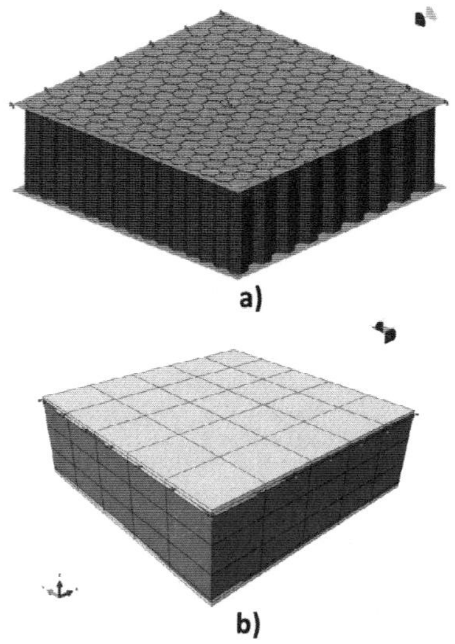

Fig. 3: FE models of test specimen – (a) detailed, (b) simplified.

Fig. 4 shows the schema of crushable foam hardening properties calibration for the simplified FE model based on known test results. The hardening parameters are changing in the three significant points:

1. Initiation of buckling.

2. Start of collapse phase.
3. Onset of densification.

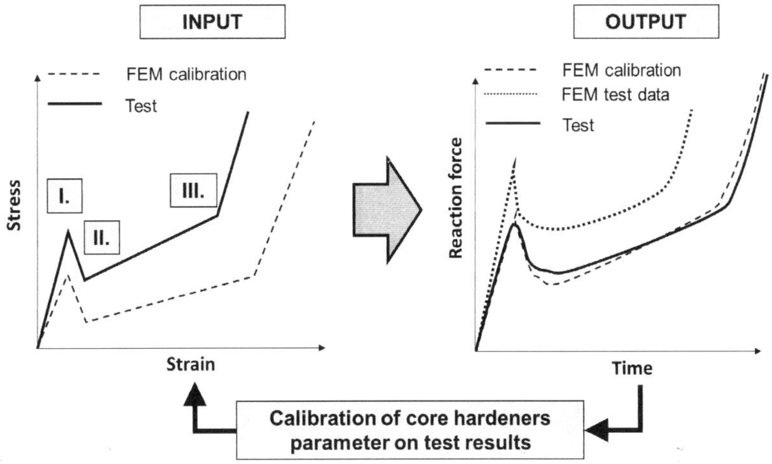

Fig. 4: Scheme of core material model parameter calibration for simplified FE model.

In terms of simplified modelling for the analysis of complex structures in the ABAQUS [4] explicit solver, attention was further focused on modelling and analysis of mechanical joints of composite parts. The results of the high-speed test of mechanical joints (Fig. 5) were used for verification and correlation of the simplified FE model of mechanical joints. Two types of 1D FE elements were compared from point of view of strain rate behaviour – BUSH and BEAM elements. FE model consists of shell elements (S4R) with mesh size 10 mm (size of global FE model), see Fig. 6. Mean unnormalized values at quasi-static load rate on the base of material equivalency test results [8] were used for high-load rate tests of mechanical joints.

Methodology for combined load fastener analysis is based on Bruhn [7].

Fig. 5: High-load rat test of mechanical joints.

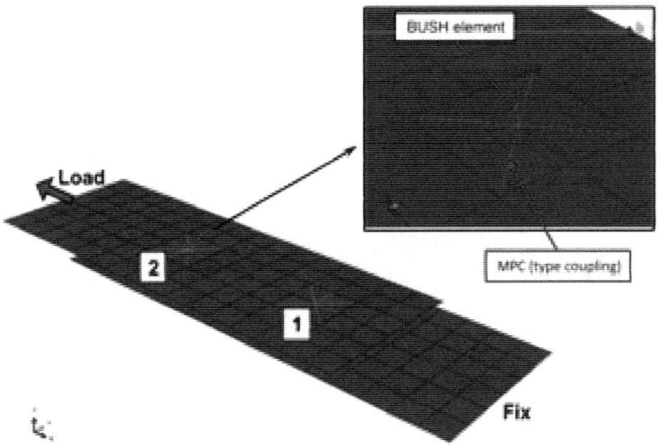

Fig. 6: Simplified FE model of mechanical join test.

2.1.1 Element level

Based on the input data from the specimen level and preliminary design of the structure, test specimens were designed to verify the impact resistance of the structure and numerical models. The experiments aimed to verify both numerical models of the ownership structure and the numerical model of the bird projectile. The flat and curved panels were defined in dependency on the preliminary design of the final part, bird mass and impact speed range. The flat test specimens represented monolithic and sandwich structure design. The simple curved test specimens represented sandwich structure design. The VZLU test rigs designed from point of view of numerical boundary conditions were used for attachment and instrumentation of the tests (Fig. 7).

For the flat test specimen, it was realised 6 perpendicular impacts and 6 declined impact 45° with a range of impact speed from 375 to 586 km/h. For curved test specimens, it was realised 3 bird strike tests with a range of impact speed from 326 to 450 km/h.

Fig. 7: VZLU test rigs to attached of flat and curved test specimen for bird strike tests.

The test results are compared to numerical simulations and the correlation of the models is performed. Altered parameters for material properties calibration on the base of test results were porosity of the projectile, fracture energy of composite material and hardness parameters of crushable foam material model of the sandwich core. The input parameters for calibration from the test results were maximal displacement during impact, reaction forces and results from non-destructive inspections (NDI). The results of the tests and numerical simulations were also qualitatively assessed on the basis of the high-speed camera pictures in terms of projectile behaviour and failure of the test specimens (Fig. 8).

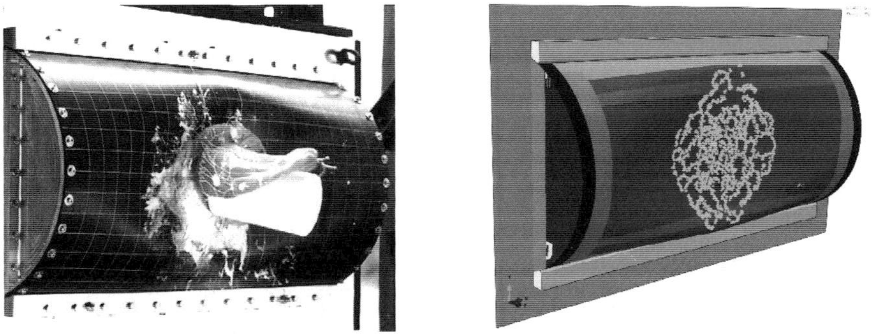

Fig. 8: Example of comparison high-speed camera measurement with numerical simulation during impact test on curved test panel.

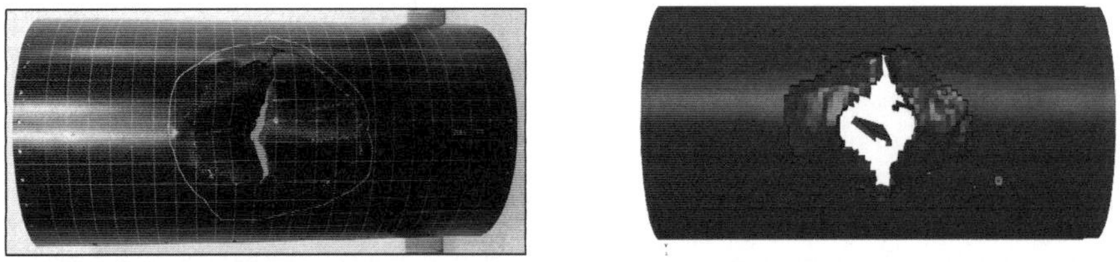

Fig. 9: Example of post-test damage and failure analysis of curved test panel and comparison with result of numerical simulation (failure of sandwich core).

2.1.2 Full scale structure

The analyses of full-scale composite cowlings were performed according to CS29.631 airworthiness requirements [5] for cruise speed and bird mass 1 kg. The design of composite cowlings and integrated parts of a structure such as frames, antennas, fasteners were analyzed for 11 bird trajectories and two lengths of the bird projectiles (Fig. 10).

The global FE model (Fig. 10) used the same parameters as used for verification and calibration of FE models of flat and curved panels such as size and quality of element, properties etc.

Parameters for evaluation of resistance of the structure to the bird impact from numerical models were:

- Maximal displacement – contact with inner structure or rotating parts (Fig. 11);
- Composite damage criterion – Hashin tensile fire criterion [6];
- Sandwich core damage;
- Loading of mechanical joints during the impact – analyzing of combined loading of mechanical fasteners by Bruhn methodology [7] (Fig. 12).

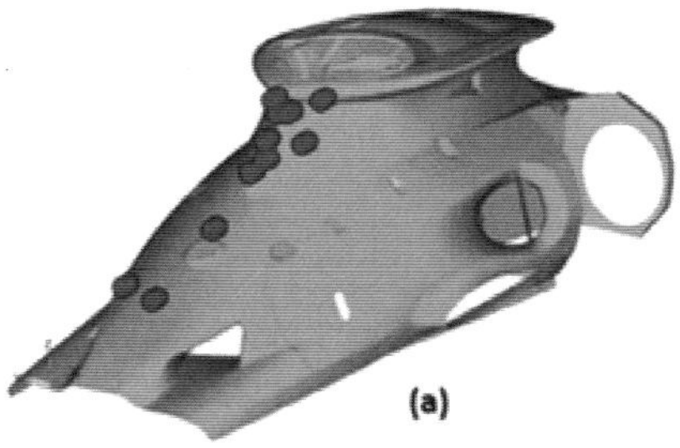

Fig. 10: Global FE model of cowlings with trajectory of bird projectile.

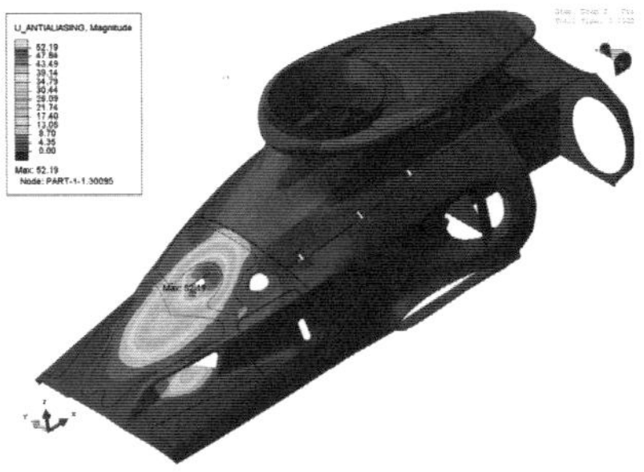

Fig. 11: Example of contour map of displacement.

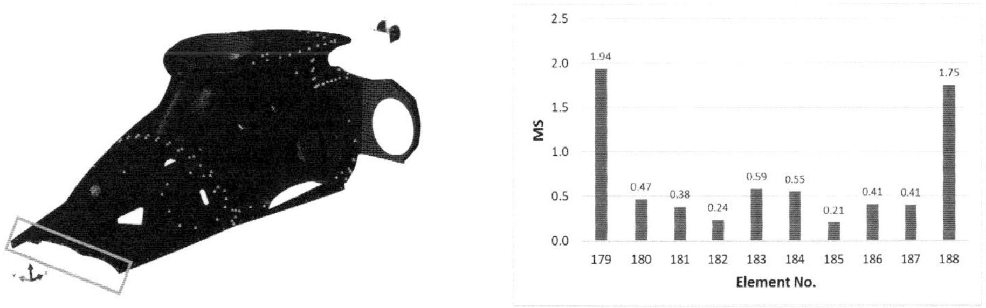

Fig. 12: Example of analysis of margin safety (MS) of the mechanical fasteners on the global FE model during the impact.

3 Conclusion

The proposed technique validation of FE models on the specimen and element levels enabled improvements and optimization in the design phase. The experiences obtained in the field of calibration of numerical models for high-speed impacts confirm different behaviour of isotropic and orthotropic material models. The application of verified numerical models in the field of emergency load cases using simple test specimens allows not only the optimization of the design in terms of compliance with airworthiness regulations, but also increasing the durability and safety of newly designed composite structures.

Acknowledgement

This project has received funding from the Clean Sky 2 Joint Undertaking under the European Union's Horizon 2020 research and innovation programme under grant agreement No 737955.

References

[1] R. Hedayati, M. Sadighi, Bird Strike – An Experimental, Theoretical, and Numerical Investigation, Woodhead Publishing, ISBN: 978-0-08-100093-9, Elsevier 2016F, 2002.

[2] R. Doubrava, M. Oberthor, P. Bělský, J. Raška, International Journal of Structural Integrity, Emerald Publishing Limited, 1757-9864 (2019), doi: 10.1108/IJSI-10-2018-0059.

[3] ASTM D7136M-20, Standard Test Method for Measuring the Damage Resistance of a Fibre-Reinforced Polymer Matrix Composite to a Drop-Weight Impact Event, ASTM International, West Conshohocken, PA, 2020.

[4] ABAQUS 6.14

[5] EASA, CS29.631

[6] Z. Hashin, ASME Journal of Applied Mechanics, Vol. 47, pp. 329–334 (1981).

[7] E.F. Bruhn, Analysis and Design of Flight Vehicle Structures, Jacobs Pub; ISBN-10: 0961523409, 1973 revised edition (June 1, 1973).

[8] B. Cabrnoch, High load rate tests, technical report DREAM-VZLU-RPT-0008, 2019.

Scissor Lift Structural Analysis

M. Dub[1,*], M. Havlíček[1], F. Lopot[1]

[1] *Department of Designing and Machine Components, Faculty of Mechanical Engineering, Czech Technical University in Prague, Technická 4, 160 00 Prague 6, Czech Republic*
* *martin.dub@fs.cvut.cz*

Abstract: The contribution deals with the structure analysis of the damaged scissor lift steel structure. Scoped scissor lift has two parallel triple-scissor mechanisms connected with several lateral beams. The motion is realised by means of two hydraulic cylinders. The FEM analysis is carried out in software RFEM 5 for several load cases, that represent the possible loadings during the lift operation. The structure is modelled predominantly by means of beam elements. Non-linear static analysis is carried out. Results reveal that the structure is insufficiently dimensioned and critical areas correspond to the damaged parts of real structure.

Keywords: scissor lift; structural analysis; FEM; structure failure.

1 Introduction

The contribution deals with the structural analysis of the scissor lift. This scissor lift was design according to customer requirements and it is not a serial solution. The structural analysis is carried out due to the frequent and large damages, which occurred on the structure during common operation of the machine. The main goal is to identify the causes of these damages.

Scissor lift structure failure is inadmissible state and it can also cause serious injury to the operator. This is the motivation for analysis of fall arrest forces on the scissor lift structure in [1]. Design and finite element analysis of the scissor lift structure is carried out also in [2], [3] and [4]. From these sources follows that safety design of scissor is an actual topic. Also the development of the detailed FE model is the important topic to support and verify theoretical computations. FE models can be also used for the structure optimization and comparison of various design solutions, [6]. These computations can be supported by means of measurement by means of strain gauges. This measurement will provide the information about the deformations of the steel structure during various load cases.

Scissor lift is the lifting device and the structural computations should be in accordance with the appropriate standards. The assessment of this design is carried out with respect to the standard ČSN EN 1493+A1 Vehicle Lifts [5]. The loading of the scissor lift is realised in several load cases and also with respect to the dynamic loads during departure and breaking.

2 Problem description

The scissor lift consists of two parallel triple-scissor mechanisms connected with several lateral beams. The motion is realised by means of two hydraulic cylinders driven by means of 5.5 kW electric motor. Those are connected to the scissor structure by means of welded consoles with a considerable eccentricity. The spatial beam model of the structure appears in Fig. 1.

The scissor lift has the stroke 1800 mm and load capacity 1500 kg. The load area dimension is 1300 x 1200 mm (length x width). The structure of the scissors is realised by means of standard cross-sections 100x50x5, lateral beams are made of 100x100x5 profile, both made of steel S355. The beams of scissors are coupled by the pins with diameter 30 mm.

Damage occurred during common operation in quite short time and repeatedly after several reparations. The critical areas revealed in the pin joint surrounding (Fig. 3) and near the welds of the console and the lateral beam (Fig. 2). As the scissor lift was not overloaded, there was a suspicion of insufficient dimensioning of the steel structure. In further paragraphs the detail structural analysis is described.

Fig. 1: The 3D beam model of scissor lift, hydraulic cylinders are modelled by means of spring members.

Fig. 2: Damage of the console for the hydraulic cylinders.

Fig. 3: Damage of the pin joint.

3 Structural analysis

3.1 Model description

Structural analysis is carried out by means of finite element analysis in RFEM 5, [7]. The structure is modelled by means of member elements. Linear elastic material model is used for construction steel S355 (E = 210 000 MPa, $\nu = 0.298$, σ_y = 350 MPa). The structure is modelled by means of 190 beam elements with higher order definition. The pin joints of the beams are modelled by means of hinges with appropriate degree of freedom.

Hydraulic cylinders are modelled by means of non-linear springs to reach arbitrary position of the scissor lift in whole range. By means of these special members it is also possible to simulate asymmetric shift of the cylinders. This load case may occur due to the asymmetric placement of the load and also due to the design of the hydraulic circuit. Due to the usage of these members, the non-linear analysis has to be carried out.

The pin joint is then modelled more detailed by means of shell elements to get stress field around the hole for the pin. Internal forces obtained from the member elements are used for loading of the detailed shell model of the pin joint. In case of shell model of the pin joint, the basic elastic-plastic material model for steel S355 is used. Only a part of the profile is modelled. Total 11289 shell elements are used in RFEM 5. Global element size is set to 5 mm. Mesh refinement to 1 mm is used around the hole for the pin. Both beam and shell model simulations are computed on basic laptop with negligible computational time.

3.2 Load cases

The top platform is modelled as rigid and it is used for loading. The structure is loaded by means of self-weight and in addition by means concentrated force which represents the maximum weight of the load 15000 N. Within three load cases, there are simulated three different load positions of the force, Fig. 4. These load cases are based on the possible positions of the palette. Load case 1 (LC1) represents the ideal middle position of the palette. LC2 simulates the limit position of the palette, which is positioned to the edge of the top platform. LC3 simulates the limit position of the rotated palette. Furthermore, the special load cases are designed to simulate asymmetric shift of the cylinders. These load cases simulate alternative to the LC1 - LC3. This state is simulated by means of non-linear spring members. The shift difference is set to 10 mm. Together six load case variants are simulated. The worst assumed load case supposed to be LC3 with asymmetric shift of cylinders.

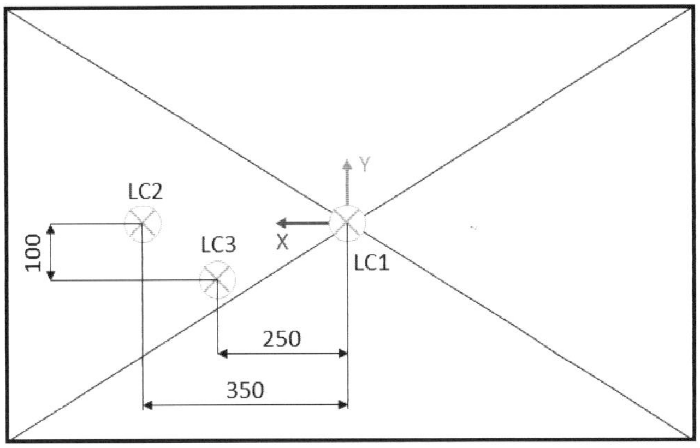

Fig. 4: Schema of the external loads on the top platform – three load cases.

4 Results

4.1 FEA results

As was mentioned above, results are calculated for three different load cases and of two configurations of hydraulic cylinders, total six simulated situations. The von Mises stress is evaluated both in member and shell model. The internal forces from member models are used for loading of the corresponding shell models.

In Fig. 5 and 6 there is the von Mises stress field on the whole structure (member model) and corresponding stress field around the hole for the pin (shell model) for load case LC1 (loading in the center of gravity) with asymmetric shift of the cylinders. During this load case the yield stress of the material is reached although no effect of dynamics is considered. The maximum stress occurs near the welded joint of the scissor beam and lateral beam. In the results from shell model, there can be seen very good agreement of the area in plastic state with the damaged area around the pin joint.

4.2 Evaluation

In case of the scissor lift structure, the limits given in standard [5] should be taken as a recommendation. This document can be used to determine the allowable stress σ_A. For the steel S355 the allowable stress is $\sigma_A = 237$ MPa for the basic material. This stress should be decreased due to the dynamic effects by means dynamic coefficient ϕ.

$$\phi = 1.1 + 0.34v, \tag{1}$$

where v is the speed of motion, which is for this scissor lift 85.7 mm/s. The final allowable stress σ_{Af} is then

$$\sigma_{Af} = \frac{\sigma_A}{\phi} = 210 \text{ MPa}. \tag{2}$$

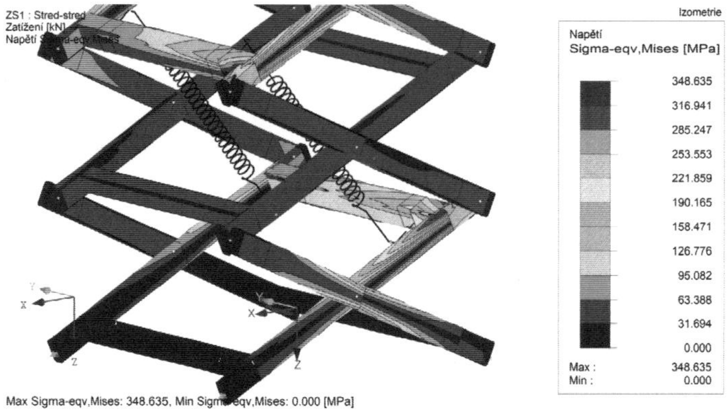

Fig. 5: Von Mises stress of the beam structure.

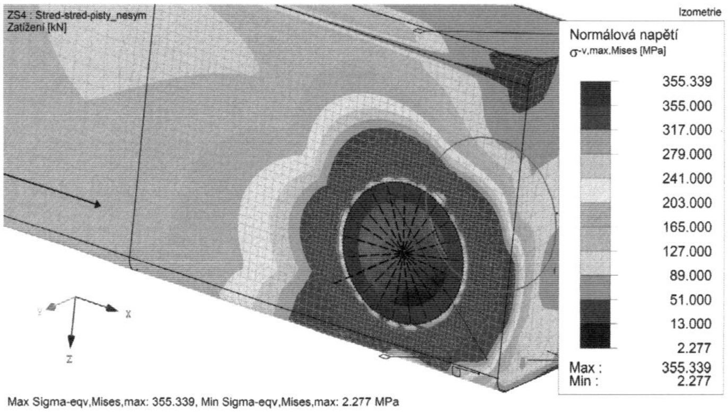

Fig. 6: Von Mises stress around the hole for the pin.

The maximum von Mises stress obtained from member model summarizes the Tab. 1.

Tab. 1: Maximum von Mises stress in the beam structure for all the load cases.

von Mises stress [MPa]	Symmetric shift of cylinders	Asymmetric shift of cylinders
LC 1	252	348
LC 2	262	357
LC 3	278	359

If the values of von Mises stress calculated in member model are compared with the value of the allowable stress, it is evident that at the point of connection of the lateral reinforcement carrying hydraulic motors does not meet the allowable stress condition of the structure in either load case, even in the case of symmetrical shift of both pistons. In the case of asymmetric shift of the hydraulic motors by 10 mm, the stress value oscillates around the yield strength of material S355.

The maximum von Mises stress obtained from shell model of the pin joint summarizes the Tab. 2.

Tab. 2: Maximum von Mises stress in the shell model of the pin joint for all the load cases.

von Mises stress [MPa]	Symmetric shift of cylinders	Asymmetric shift of cylinders
LC 1	194	355
LC 2	205	355
LC 3	224	355

The shell model results then shows the exceeding of allowable stress in load case LC3 (for the same shift of hydraulic cylinders). In case of asymmetric loading the plasticization of large areas around the hole for the pin occurs. In this case the material anisotropy of the welded joints is not taken into account. This will cause further decrease of the allowable stress.

Reached results proved the assumption that the worse load case is LC 3. But it can be stated that the structure does not fulfil the requirements for the scissor lift in the none of simulated load cases. Simulation of asymmetric cylinders shift leads to the yield stress reach.

5 Conclusions

Within the structural analysis three load cases and two configurations of the hydraulic cylinders were modelled and tested. The scissor lift was modelled at first by means of member elements then more detailed analysis of the pin joint was carried out using shell model. Performed structural analysis proved the underestimation of the scissor lift structure. The areas with maximum stress correspond well to the places of damage. The values of the von Mises stress were compared with the maximum allowable stress according standard for vehicle lifts [5]. Detailed analysis of the design calculations of the producer showed excessive simplification of the computational model. This simplification led to the wrong results and their misinterpretation. The simplification of the mechanical model is possible but it has to be taken into account during determination of the safety factor respectively the maximum allowable stress.

References

[1] J. Harris, et al., Fall arrest characteristics of a scissor lift, Journal of Safety Research, 41, (2010), 213-220, doi: 10.1016/j.jsr.2010.01.004.

[2] S.A. Manoharrao, R. S. Jamgekar, Design and Analysis of Hydraulic Scissor lift by FEA, International Research Journal of Engineering and Technology, 3.10 (2016).

[3] R. Helmi, et al., Design Review of Scissors Lifts Structure for Commercial Aircraft Ground Support Equipment using Finite Element Analysis, Procedia Engineering, 41, (2012), 1696-1701, doi: 10.1016/j.proeng.2012.07.370.

[4] C. Kiran, J. Sruthi, Design and Structural Analysis of Scissor Jack using ANSYS Workbench, CVR Journal of Science and Technology, 15 (2018), 101-105.

[5] ČSN EN 1493 + A1. Vehicle Lifts. Prague: Czech Office for Standards, 2010.

[6] F. Starý, et al., Comparison of Different Construction Hoist Masts by FEM Analysis, in proc.: Proceedings of 58th International Conference of Machine Design Departments – ICMD 2017 (2017), Česká zemědělská univerzita v Praze. Technická fakulta, Praha-Horoměřice, 344-347.

[7] Dlubal Software GmbH, RFEM 5.26.02 (2020)

Experimental Investigation of Pull-out Load Bearing Capacity of Threaded End Pieces for Composite Pull Rods

M. Dvořák[1,*], D. Blaha[1], N. Schmidová[1], K. Doubrava[1], J. Chromý[2], O. Uher[2]

[1] *Department of Mechanics, Biomechanics and Mechatronics, Faculty of Mechanical Engineering, Czech Technical University in Prague, Technická 4, 160 00 Prague 6, Czech Republic*

[2] *Compo Tech PLUS, spol. s r.o., Nová 1316, 342 01 Sušice, Czech Republic*

* *milan.dvorak@fs.cvut.cz*

Abstract: The topic of this paper is an experimental research of pull-out load bearing capacity of a threaded end pieces for composite pull rods, which are made by a filament winding technology from a carbon/epoxy composite. The original pull rod design utilizes glued end pieces. The new design solution is based on an embeddable end pieces, which are integrated into the pull rod body directly during its production. A total of four variants of end pieces were tested. The optimal variant from the point of view of a maximal load bearing capacity and the optimal variant from the point of view of a composite production technology were found.

Keywords: pull rod; carbon composite; filament winding; glued joint; pull-out force.

1 Introduction

The paper is focused on an experimental development of end pieces for composite pull rods. For many industrial applications, it is still very effective to use a conventional mechanical force transmission. The advantages are simplicity, reliability, independence from a failure of a transmission path source (electricity, hydraulics, etc.), and a low implementation price. Power transmission by means of pull rods and levers is commonly applied in parts of aircraft structures (aircraft control systems or structural reinforcing rods) [1], in the automotive industry (suspension and steering mechanisms etc.), in systems of production machines (textile machines, manipulators, shifting and connecting mechanical drives, etc.). Pull rods are usually mounted on bearings and must primarily transmit tensile or compressive loads. The limiting factor is usually the buckling behavior under a compressive loading, or a natural frequency for dynamically excited systems. In a case of a traditional design of the composite pull rod, which consists of a composite tube and a glued end pieces, the weak point is the glued joint itself. The load bearing capacity of the glued joints is limited by the value of the achievable shear strength, which is approx. 20-30 MPa for the epoxy-based structural adhesive systems. However, the quality of the glued joint is strongly tied to the quality of the manufacturing technology [2]. In an addition, the subsequent inspection of the finished joint quality is difficult and often impossible. The solution is to develop an end piece, which is an integral part of the composite rod body itself. This would make it possible to achieve a composite pull rod design combining the advantage of a simpler production technology and a higher operational reliability.

2 Experimental configuration

In this first phase of the experimental work, the load bearing capacity of the end pieces of four different types under the axial pulling force loading was evaluated. This first experimental test set contained a total of 14 specimens.

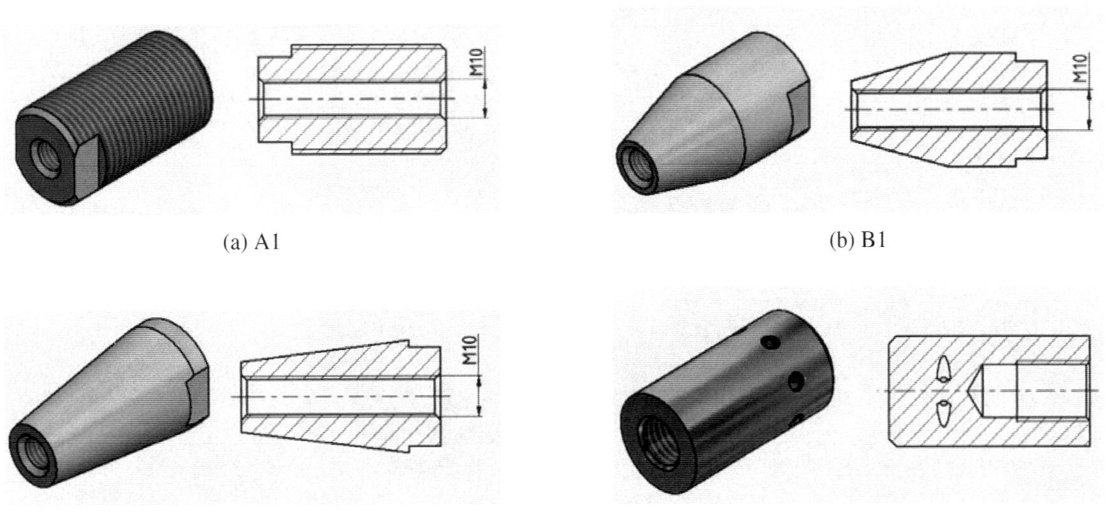

(a) A1 (b) B1

(c) B2 (d) C1

Fig. 1: Configuration of threaded end pieces.

2.1 Specimen description

Specimen bodies of a tubular shape were made of carbon/epoxy composite using the filament winding technology. One end of the specimen was fitted with an end piece and the other end was reinforced with a wound fiberglass reinforcement to prevent damage from the clamping grips of the testing machine. The end pieces were made of steel and provided with an internal thread. Four different types of end pieces were made (marked A1, B1, B2, C1), differing in their external shape, as can be seen in the Fig. 1. They were integrated into the composite structure of the pull rod during the winding process.

The A1, B1 and B2 types rely on a simple shape connection between the end piece and the composite rod body, without additional fixing elements. The C1 type is equipped with pins inserted into radially oriented holes. This is advantageous in terms of the load bearing capacity, because the carbon composite tows can be wrapped around these pins. Of course, this is a demanding solution from the point of view of assembly and the following operational life, because each of the pins needs to be reliably secured against movement.

2.2 Testing method

Heckert UFP 400 hydraulic testing machine with a maximum loading force of 400 kN was used for an experimental evaluation of a pull-out force in an axial direction. Specimens were loaded quasi-statically in a displacement loading mode until the damage occurs. Specimen loading rate was 1 mm/min. The specimen and loading configurations are shown in the Fig. 2.

3 Results and discussion

The results for each type of end piece are plotted in the graphs in the Fig. 3. It can be seen that C1 specimens have sligthly lower stiffness response, compared to A1, B1 and B2 type specimens. This is due to the smaller amount of a load-bearing carbon fibres in the end piece area. The measured pull-out forces were compared on a relative scale, where 100 % is the highest determined force of all specimens. In this experimental set, maximum measured force was indicated for the C1 type end piece specimen C1-1-1. The average load bearing capacity of the A1 type reaches about 69.3 % of the maximal determined force. In the case of the end piece types B1, B2, and C1 it is 78.4 %, 85.9 % and 86.4 % respectively. Although the tested set is still relatively small, it is possible to evaluate the B2 end piece type as a good compromise between the pull-out load bearing capacity and the requirements for the simplest possible production technology. If we try to compare the current concept of glued end piece (cylinder glued inside the tube) with the integrated end piece solution, we come to the following.

Fig. 2: Composite pull rod specimen configuration; experimental configuration (right).

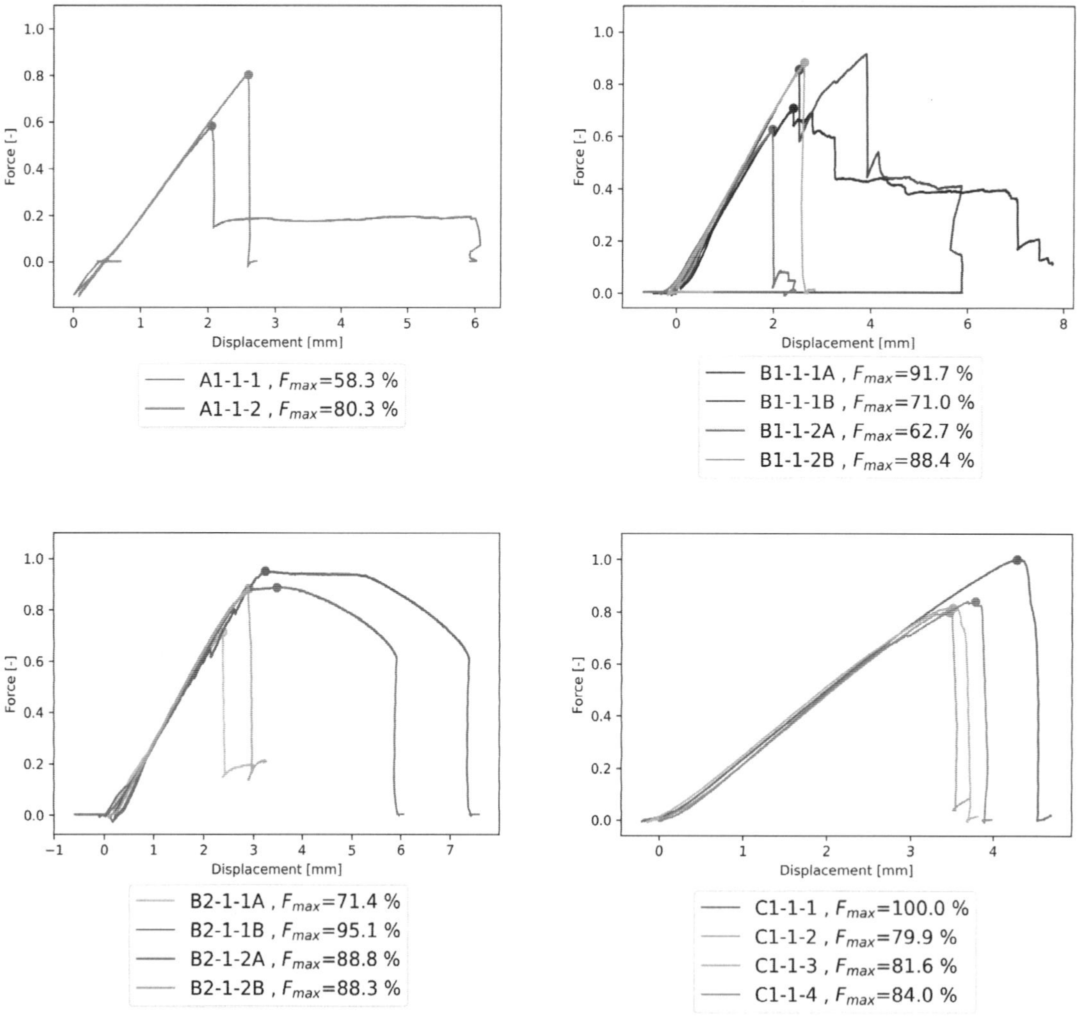

Fig. 3: Summary of experimentally determined pull-out forces.

While maintaining the size of the glued surface (or a functional length and a diameter of the end piece), if we needed to transfer the pull-out load corresponding to the individual types of integrated end pieces, the shear stress in the glued joint would reach 19, 21, 23 and 24 MPa for the average maximum pull-out force of the A1, B1, B2 and C1 end piece type respectively. Due to the quasi-static nature of the load, these values are quite high and would not provide sufficient safety margin for a normal operation of the pull rod, where the end pieces are also loaded dynamically.

4 Conclusion

Specimens with four different types of integrated end pieces were tested. The aim was to identify the shape of the end piece with the greatest pull-out load bearing capacity. Due to the small number of specimens so far, the results are preliminary. However, it can be seen that from the point of view of the application oriented to the maximum load bearing capacity, the C1 type is the best. From the point of view of the effectivity of the manufacturing technology in combination with sufficient load bearing capacity, the B2 type is the best choice. At the same time, it was confirmed that the integrated types of the end pieces will safely transmit a greater pull-out force than the glued end piece type with the similar dimensions, for which the upper limit of shear strength of the adhesives used would be reached.

Acknowledgement

The authors would like to thank the Technology Agency of the Czech Republic for supporting this research with project no. FW01010450.

References

[1] I. Hendarko, et al., Design of Flight Control System Mechanism for National Glider GL-1. IOP Conference Series: Materials Science and Engineering 2019, 645, 012016, doi: 10.1088/1757-899X/645/1/012016.

[2] M.A. Wahab, Joining Composites with Adhesives: Theory and Applications; DEStech Publications, Inc, 2015, ISBN 978-1-60595-093-8.

Optical Experimental Methods and their Implementation and Comparison in Conditions of Experimental Stress Analysis

P. Frankovský[1,*], P. Sivák[1], I. Delyová[1], J. Kostka[1]

[1] *Department of Applied Mechanics and Mechanical Engineering, Faculty of Mechanical Engineering, Technical University of Košice, Letná 1/9, 042 00 Košice, Slovakia*

* *peter.frankovsky@tuke.sk*

Abstract: This paper deals with a brief overview of the application and mutual comparison of optical experimental methods for the analysis of deformations and mechanical stresses. In addition to the now traditional PhotoStress optical method, such modern methods as the digital image correlation, holographic interferometry, digital speckle interferometry, digital speckle shearography, including their various special modifications, etc., also come to the fore. These methods find their specific application as sophisticated non-contact methods mainly in the development of new methods for determining deformations, stresses and material characteristics, mainly in the field of fracture mechanics, residual stresses, analysis of elements in the elastic-plastic area of deformations, etc. The paper also presents related specific application examples, respecting the technical and software equipment of the experimental workplace of the Department of Applied Mechanics and Mechanical Engineering, Faculty of Mechanical Engineering, Technical University in Košice. These examples formed part of practical tasks or research programs for the development of new experimental mechanical methods.

Keywords: optical experimental method; PhotoStress; digital image correlation; holographic interferometry; digital speckle interferometry.

1 Introduction

Optical experimental methods (OEM) are an integral part of the analysis of deformations and stresses of structural elements and entire structures. They exist in addition to today's classical analytical, modern numerical methods and, of course, in addition to other experimental methods, such as tensometry, the method of brittle varnishes, etc. The above 3 fundamental approaches to experimental analysis of stresses (EAN, ESA) and deformations coexist in symbiosis on the basis of mutual verification, comparison and complementation. At the experimental workplace of the Department of Applied Mechanics and Mechanical Engineering, Faculty of Mechanical Engineering, Technical University in Košice, they found their application not only in teaching, but also in the field of scientific research and also in conditions of engineering practice. In addition to the clearly classic or already relatively widespread OEMs, such as the Moiré method of mechanical-optical interference, photoelasticity or PhotoStress, sophisticated optical methods such as digital image correlation, holographic interferometry, digital speckle interferometry, digital speckle shearography, including their various special modifications, etc., also come to the fore [1–7].

2 General characteristics and application of OEM

The essence of OEMs for the analysis of mechanical stresses and deformations, the applied physical principle and the area of operation was and is determined by gradual development of engineering and technology. They are based on different specific principles, are at different levels of development, have their specific focus and application. Their common feature is that they are either partially contact or completely contactless and are based on different properties of electromagnetic waves and their interference. The application of OEMs in experimental mechanics is advantageous mainly by contactless nature, variability and a high degree of processing of measured, currently practically completely digitized data. They typically have a high measurement sensitivity and often allow direct measurement, which does not require further analytical or numerical processing of the results. OEM finds application in fracture mechanics, research of composite and other modern materials, analysis of residual stresses, determination of local elastic-plastic deformations under cyclic stress, biomechanics, high-speed applications, vibration analysis, etc.

Of course, as with any experiment, OEMs are relatively time-consuming and technically demanding compared to other types of modeling and simulations. On the other hand, they can take actual physical conditions of the analyzed object into account.

Another general feature and, at the same time, a disadvantage of practically all OEMs is that their application, due to the principle and technical means used, is limited to relatively sterile laboratory conditions in practical implementations. They are generally unsuitable for difficult real conditions of structures in the field with significantly negative disturbing influences and for the analysis of dynamic events in general with a stochastic character.

3 Relevant OEMs today and examples of their use

3.1 Moiré mechanical-optical interference method

Due to the comprehensive overview and completeness, the Moiré mechanical-optical interference method can also be mentioned in part within the OEM. Moiré light effect with visible stripes is created by overlapping two optical gratings or rasters (Fig. 1a with Moiré stripe lines m_1 to m_7). By moving the original raster and the new deformed raster relative to each other (Fig. 1b) it is possible to approximately detect deformations on the surface of a loaded and deformed object. The application and extension of the Moiré method is currently very rare due to its limited possibilities.

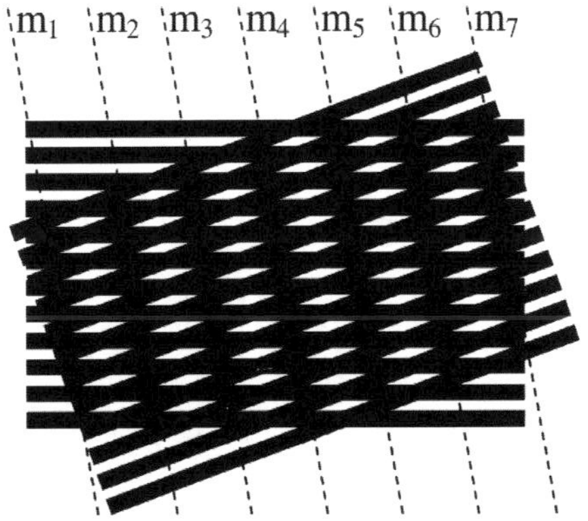

(a) Overlap of two grids (rasters) with visible Moiré stripes m_1 to m_7.

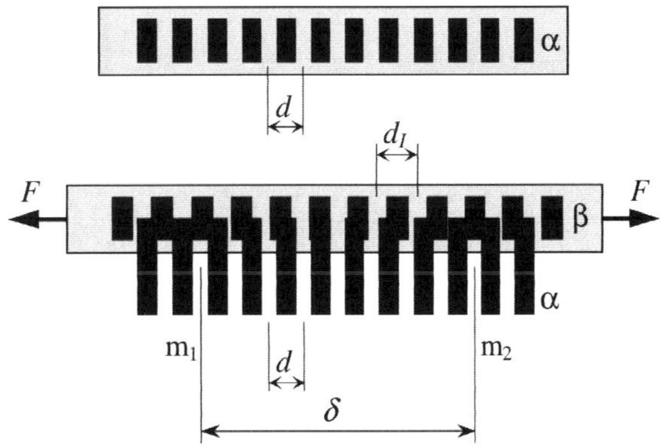

(b) Overlap of two Moiré rasters: Original (undeformed) α and new (deformed) β.

Fig. 1: Grids in the Moiré mechanical-optical interference method.

3.2 Method of transmission and reflection photoelasticity

One of the historically first and, at the same time, now classic OEMs is the transmission or reflection method (PhotoStress) or planar or spatial photoelasticity. It is based on the use of the properties of polarized light and the capabilities of some transparent, optically sensitive materials. At present, these are materials based on epoxy resins. These materials change from optically isotropic to temporary optically anisotropic under load and thus simultaneous deformation by a change in the refractive index due to a change in the speed of light propagation in the respective direction. The result is the emergence of observable, registrable and measurable interfering optical structures, so called isolines (optical interpretation), related either to the magnitude or direction of the respective deformations and mechanical stresses (interpretation of mechanics). Thus, by methods of photoelasticity, it is generally possible to determine magnitudes and directions of main relative elongations and main normal stresses, as well as the bevels and shear stresses at discrete points or on a certain surface. It is possible to identify immediately and relatively easily, in particular, the maximum circumferential stresses at the unloaded edges of the component. The method is therefore suitable for distinguishing the magnitude of nominal stress, concentrations and gradients of stresses and strains, including the identification of entire areas with high or low stress levels. It detects and identifies critical points in the structure and allows us to optimize stress distribution in the structure. Stress can also be examined in the event of plastic deformations, or even until a crack occurs.

Photoelasticity can be divided according to various criteria. In terms of how the beam passes through the optically sensitive layer, it is possible to distinguish between transmission photoleasticity for transmitted light and reflective photoelasticity, usually identified as PhotoStress, either for vertically or obliquely reflected light. According to the criterion of locating the measuring point, photoelasticity can be planar for measurement at any point of the planar model or in the coating on the surface of the spatial model, and spatial for measurement at any point of the spatial model.

Unlike transmission photoelasticity, in which the experimental object can only be a model, completely made of optically sensitive material, in case of PhotoStress, the experimental object does not have to be only a model, but also a real object or a structural element with applied optically sensitive and reflective layer (coating). The PhotoStress method also makes it possible to identify and measure surface and residual stresses. The methods of photoelasticity require appropriately selected and consistently implemented means, methods and processes for coating, calibration, compensation, correction, separation, etc.

OEMs based on photoelasticity have already experienced significant development in former Czechoslovakia. This is also evidenced by the transmission polariscope FMB 53 (Fig. 2), a product of former Meopta plants, Bratislava, Elektropodnik Praha and the Research Institute TOZ Čelákovice from the beginning of the 1950s. At that time, it was a unique technical experimental facility, which won a gold medal at the EXPO World Exhibition 1958 in Brussels, Belgium.

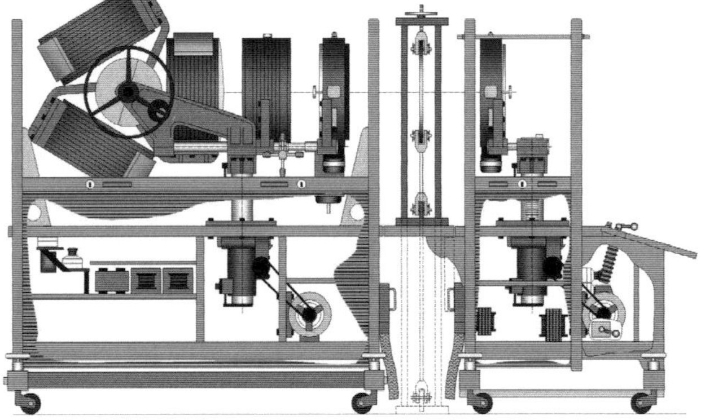

(a) Polariscope FMB 53 when viewed from the side and in partial section.

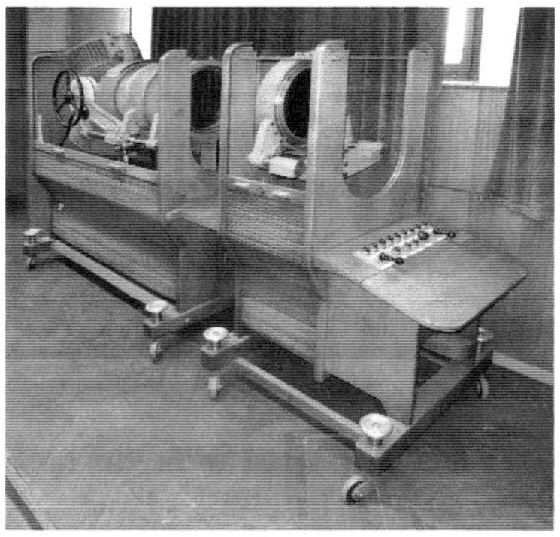

(b) Overlap of two Moiré rasters: Original (undeformed) α and new (deformed) β.

Fig. 2: Polariscope FMB 53 in the TU laboratory in Košice.

Fig. 3 shows isochromatic fields created by the illumination of an object with white light by a reflective polariscope with linear polarization. The images were created during the development and verification of the methodology and the creation of an experimental measuring chain for the determination of simulated residual stresses by PhotoStress. At the same time, relevant procedures were applied in accordance with the methodology according to ASTM E 837-13a, both by blind cylindrical drilling and the Ring-Core groove method.

With a special arrangement or configuration, it is also possible to perform measurements to a limited extent during movement or dynamic measurements, e.g. within the analysis of cyclic stress of rotating parts. In such cases, it is necessary to place a sensor on the measured component, providing impulses for switching on the stroboscopic lamp, which serves as a light source. Using the stroboscopic effect, it is then possible to observe a time-stable image. Special methodologies for the analysis of various non-periodic dynamic effects in the form of various loading pulses, shocks, etc. are also developed [8].

Fig. 4 and Fig. 5 show isochromatic fields created by the illumination of an object with white light by a reflective polariscope with linear polarization. The images were created during the development and verifi-

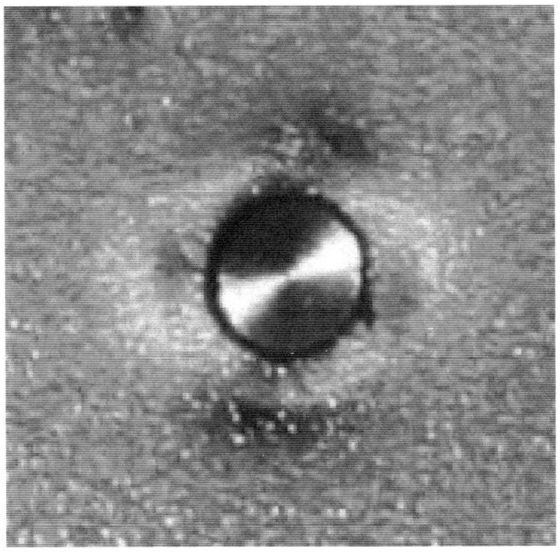

(a) Field of isochromatics around a blind cylindrical hole with a depth of 0.3 mm.

(b) Field of isochromatics around a groove with a depth of 2.0 mm for the Ring-Core method.

Fig. 3: The field of isochromatics around the blind hole or around the groove when illuminated with white light by a reflective polariscope with linear polarization in the analysis of residual stresses by PhotoStress.

cation of the methodology and the creation of an experimental measuring chain for the dynamic analysis by PhotoStress. Two samples of different shapes and a constant thickness were used, which were subjected to rotational analyzes at different angular velocities. The first test specimen had the shape of a modified S with two axes of antisymmetry (Fig. 4). The second test specimen had the shape of a disk with holes with four axes of symmetry (Fig. 5).

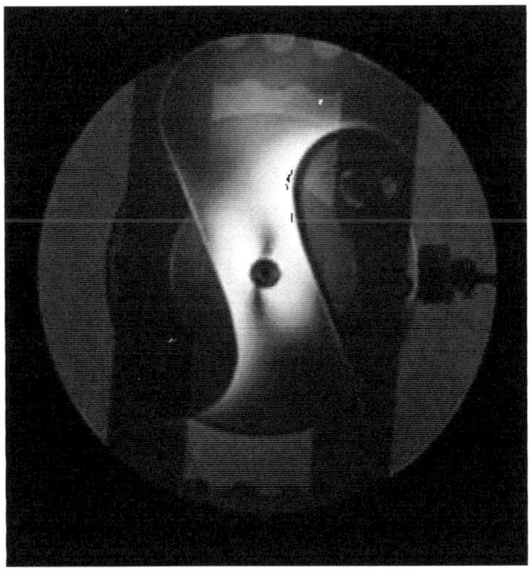

(a) Field of isochromatics at angular velocity $\omega = 125.7$ rad·s^{-1}.

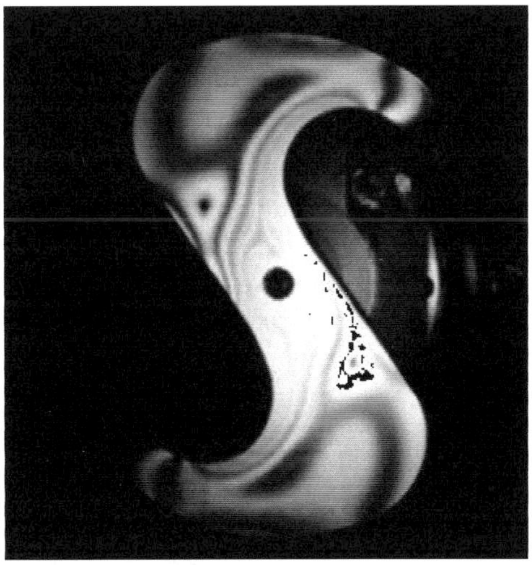

(b) Field of isochromatics at angular velocity $\omega = 712.1$ rad·s^{-1}.

Fig. 4: Field of isochromatics when illuminated with white light by a reflective polariscope with linear polarization on a sample of S-shape of constant thickness as it rotates at a certain constant angular velocity.

The main technical and software tools used were reflective polariscope Vishay LF/Z-2, and in the case of dynamic analyzes, the stroboscopic white light source Strobex Model 135M-11, digital tachometer Laser Tacho, electric motor HSM 60, signal generator and digital camera. LF/Z-2 polariscope also included a digital compensator model 832 and software PSCalc^{TM}, preferably designed to obtain values of principal strains and principal normal stresses. The experimentally measured objects were made of a photoelastic film Vishay PS-1A with a reflective layer, with mechanical characteristics h = 3.05 mm, E = 2.5 GPa, μ = 0.38 and with an optical characteristic – sensitivity K = 0.15.

(a) Field of isochromatics at angular velocity ω = 230.4 rad·s^{-1}.

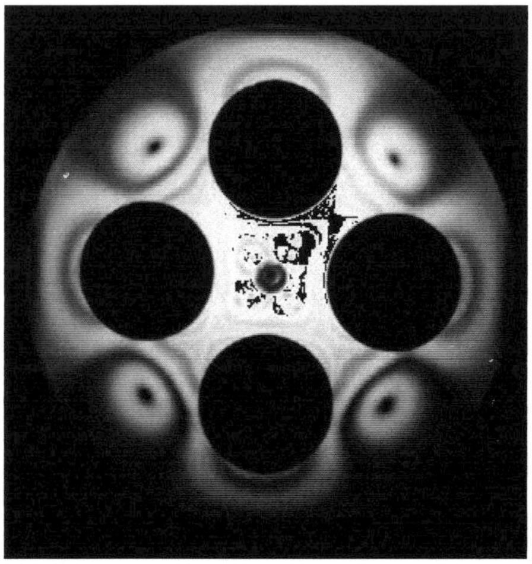

(b) Field of isochromatics at angular velocity ω = 816.8 rad·s^{-1}.

Fig. 5: Field of isochromatics when illuminated with white light by a reflective polariscope with linear polarization on a sample of a disc of constant thickness with holes as it rotates at a certain constant angular velocity.

3.3 Digital image correlation method

Current modern so-called "basic" OEMs which are preferably dedicated for displacement analysis include the digital image correlation (DIC) method. DIC correlates images of a stochastic pattern created on the surface of the examined object in the process of its loading and thus also stressing. The total deformation of the object's surface is given by a set of deformations of individual image elements – facets. Therefore, the accuracy of the DIC is based on the displacement analysis algorithm used. The comparative process consists in finding the extremes of the correlation functions of the corresponding facets. Basic applied correlation algorithms include the translational tracking algorithm and the method of included surfaces. The first of the algorithms does not consider the deformation of the facets but only their displacement. The second requires knowledge of the correlation coefficients of the pixels inside the 3×3 facets. The Newton-Raphson algorithm, in turn, works with a minimum of correlation function. The algorithm based on optical flow is used to analyze displacements with subpixel accuracy. At present, the correlation algorithm using pseudo-affinity transformation is relatively widespread. Correlation errors of a statistical or systematic nature can affect the accuracy of measurements. For 2D observation and measurement, one camera is used, placed perpendicular to the measured sample or the object. For observation and measurement in the 3D image correlation mode, two cameras are used, which capture the observed area from different directions [9].

An example of the application of the DIC method is the analysis of deformations within a contact task. It was modeled on a bracket, tightly inserted into a rectangular groove on one side and stressed by bending on one side. Fig. 6a shows an experimental object with a stochastic pattern applied to the surface. Fig. 6b shows the obtained field of bevel sizes (shear deformation) γ. Fig. 7 then depicts obtained field of vectors – magnitudes and directions of principal strains ε_1 and ε_2.

The main technical and software tools used were Dantec Dynamics Q-400 optical system, as one of a series of Q-400, Q-400 Micro, Q-450 and Q-480 systems, and the Istra4D software. The Q-400 system typically consists of two cameras with CCD sensors, and up to eight cameras can be used in the extended mode. The illumination is usually realized by a halogen reflector with white light or a HILIS LED illuminator with homogeneous cold light. Istra4D software performs all processes related to measuring, processing and evaluating results. Since the analysis by the DIC method was preceded by the analysis by the PhotoStress method, the experimentally measured objects on which the stochastic sample was sprayed were also preserved. The material was therefore a photoelastic film PS-1A from Vishay with mechanical characteristics h = 3.05 mm, E = 2.5 GPa, μ = 0.38.

(a) Stochastic sample.

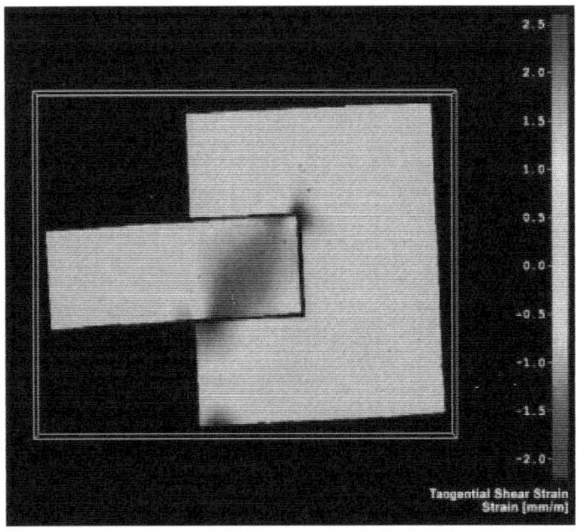

(b) Bevel size filed γ $[10^{-3}]$.

Fig. 6: Stochastic sample and field of bevel sizes (shear deformation) γ in the analysis of deformations within a contact task by the DIC method.

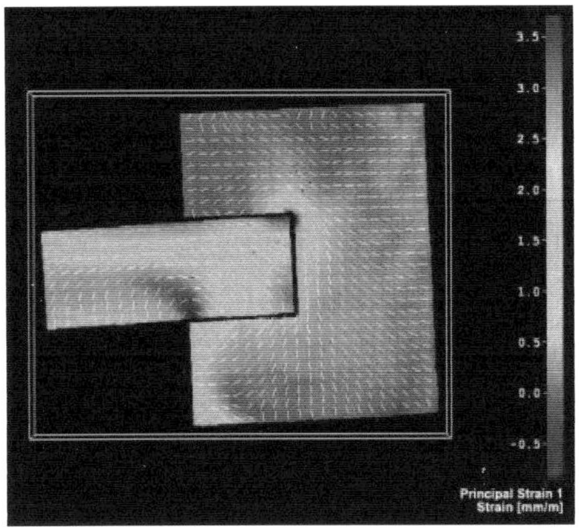

(a) Stochastic sample.

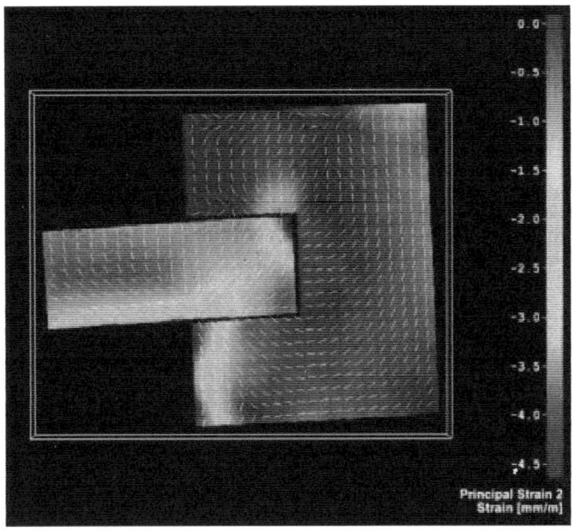

(b) Bevel size filed γ $[10^{-3}]$.

Fig. 7: Field of vectors – magnitudes and directions of principal strains ε_1 and ε_2 during the analysis of deformations within the contact task by the DIC method.

3.4 Holographic interferometry method

Other modern OEMs are methods based on the holographic principle. They are based on the creation of three-dimensional images - holograms, e.g. Fresnel, image, Frauhofer or Fourier. Their creation is based on capturing an object using coherent light and waves reflected from the object. It includes e.g. holographic interferometry (HI). Here, during reconstruction of the hologram, the information wave or the wave in question is reconstructed. If the reconstruction wave is an exact copy of the reference wave, the information wave with the same amplitude and phase is restored. If the examined object is removed during the reconstruction, it is possible to see its image in exactly the same place and in the same shape as it took when creating the hologram. If the object under investigation is not removed during the reconstruction, then there are two waves behind the hologram. The first comes from the reconstructed hologram, the second is spreading from the object. These are coherent waves that interfere with each other. The created interference pattern, the so-called holographic interferogram, will capture those changes in the object that occurred between the creation of the hologram and its reconstruction. If the position, dimensions or shape of the object change during the reconstruction, the interference pattern also changes. This is HI in real time. In the second variant HI, two or more subject waves corresponding to different states of the same object are gradually registered in one recording environment. During the reconstruction of the hologram, the recovered waves interfere with each other and form interference patterns. This is therefore a method of two or more exposures [9, 10].

HI is a typical two-step process in which information in the form of an observable macroscopic image can be obtained by reconstructing a record in which information characterizing at least one physical state of the object under investigation is encoded.

The following is an example of applying the HI method to a newly developed methodology for the analysis of deformations and stresses around the crack front (root, tip). At the same time, it is a matter of obtaining data for experimental determination of the coefficient or stress intensity factor K_I, or K_{II} or K_{III} for Modes I or II or III of crack opening and material separation. Fig. 8a shows a Mode I model of material separation and Fig. 8b shows a preparation for describing stress around the crack front (root, tip) in both Cartesian and polar coordinate systems.

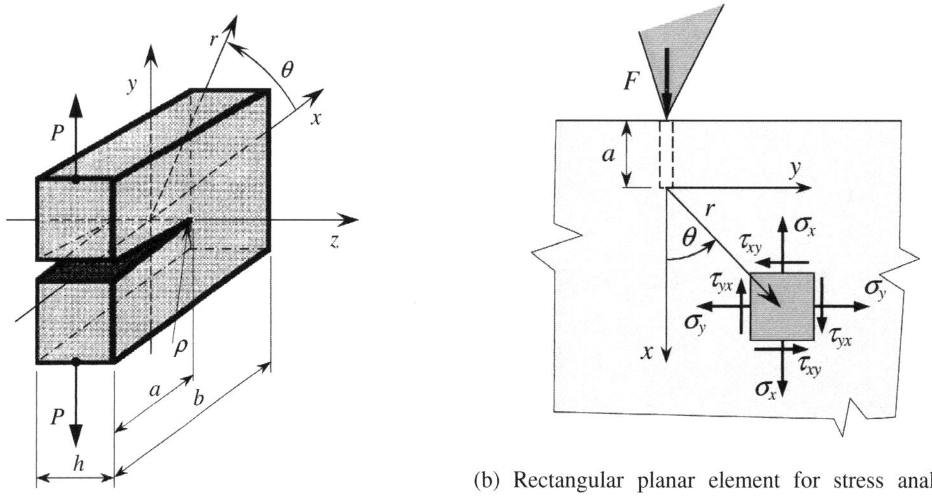

(a) Crack opening and material separation for Mode I.

(b) Rectangular planar element for stress analysis around the crack front (root, tip) in both Cartesian and polar coordinate systems..

Fig. 8: Crack opening model and rectangular planar element around the crack front (root, tip) for stress and strain analysis by HI method.

Fig. 9a shows an image of a holographic interferogram from the vicinity of mutual force interaction of two objects according to Fig. 8b. For a better identification of individual isopachs, the real image in Fig. 9a has a slightly adjusted color density. Experimentally investigated were various situations of force applied to an object without a crack and with a defined crack or notch. Fig. 9b depicts a field visualization of discrete isopachs 1 to 8 obtained from the holographic interferogram in Fig. 9a in the angular range 0° to 180° from the vicinity of the examined point for a certain parameter (polar coordinate) r.

Fig. 10 shows angular dependences of the normal stress σ_y and the shear stress τ_{xy} around the crack face (root, tip) for a certain parameter r, obtained by the HI method. With the help of these data and the use of appropriate analytical relations of fracture mechanics for stress analyzes from the crack environment, it is then possible to express the coefficient or stress intensity factor K_M for the respective Mode M and for the respective stress or load condition of a certain material.

The main technical means used was a coherent radiation source with a helium-neon laser with a radiation wavelength of 632.8 nm and a CCD camera. Other elements of the optical system used were continuous optical laser beam splitters, lenses and a holographic plate. The experimental measured object was made of natural styrene-butadiene rubber with mechanical characteristics E = 4.24 MPa, R_e = 0.77 MPa and R_m = 3.01 MPa.

3.5 Digital speckle interferometry

The approach characterized in the previous paragraph makes it possible to organically incorporate the method of interference of laser spots, i.e. Speckle interferometry (SPI), into the uniform classification of interference techniques. The correlation SPI method is based on the coherent superposition of speckle fields formed by a light wave reflected from the optically rough surface of the examined object and a reference wave. The presence of a reference wave during the recording of a speckle image induces an intensity distribution in the imaginary plane of the optical system depending on the relative phase shift of the interfering speckle fields. Deformation of the object's surface will cause a change in the intensity distribution in the resulting speckle

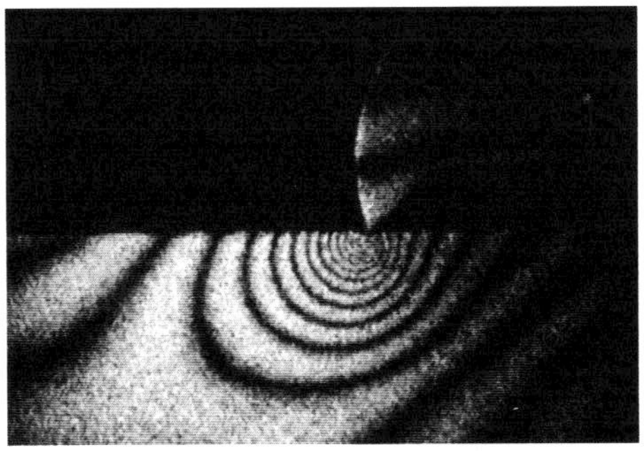

(a) Real image of holographic interferogram from the environment of mutual force interaction of two objects.

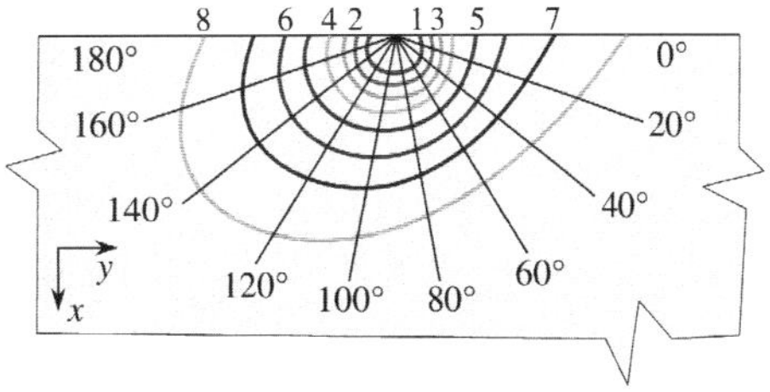

(b) Visualization of the field of discrete isopachs 1 to 8 of the holographic interferogram in the range of 0° to 180° from the vicinity of the examined point.

Fig. 9: Holographic interferogram and corresponding isopachs obtained for stress and strain analysis around the crack front (root, tip) by the HI method.

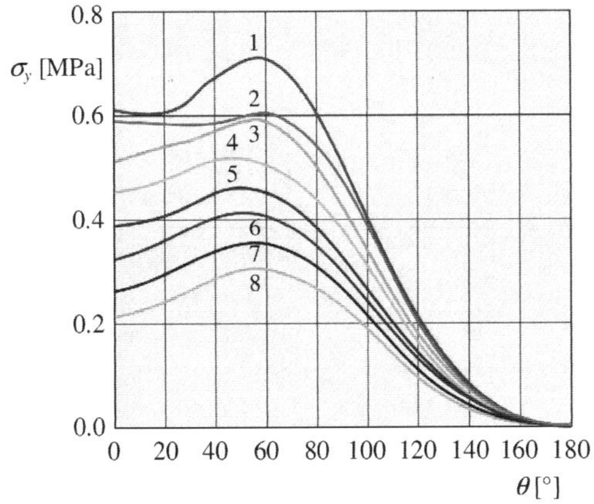

(a) Stress angular dependence σ_y.

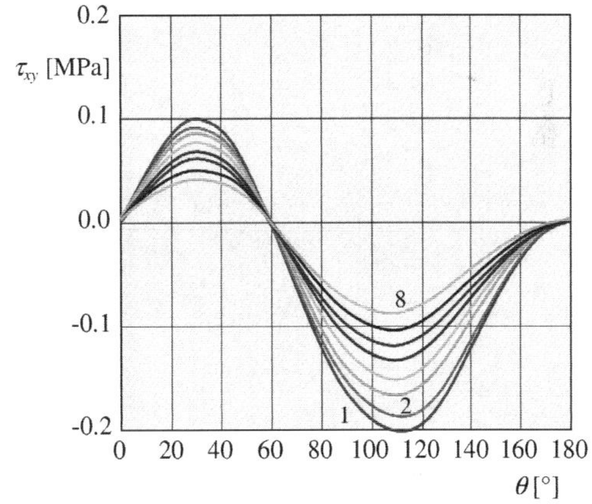

(b) Stress angular dependence τ_{xy}.

Fig. 10: Stress angular dependence σ_y and τ_{xy} around the crack front (root, tip) in distance r gained by the HI method.

image in this case. The phase-shifting SPI method, in turn, consists of double-recording two images of an object, one of which is slightly shifted relative to the other. An improved method of SPI is the Electronic Speckle Patten Interferometry (ESPI), also known as digital SPI (DSPI) or TV holography, using a camera with a CCD or CMOS sensor as a recording device. The speckle image recorded by the digital camera can be considered a hologram, but instead of reconstructing it, the speckle images are correlated. Paradoxically, the speckle effect is considered an undesirable phenomenon in traditional HI that needs to be eliminated or at least appropriately suppressed. ESPI measurements can be performed either in-plane or out-of-plane. The superposition of the reference light and the light reflected from the surface of the object creates a new speckle image – the interferogram [2, 11]. Overall, the possibilities of using SPI and ESPI are essentially identical to the areas of application of HI, while ESPI is used mainly to measure very small displacements and relative deformations.

The following is an example of the application of the ESPI method to a newly developed methodology for the analysis of deformations and stresses in place of stress concentrators on structural elements. Fig. 11 shows a design of the shape and the method of specimen loading by shear stress for the purposes of stress and strain analysis by the ESPI method. Fig. 12 then shows the resulting stress distribution field σ_3 and τ_{xy} in the test specimen under its shear stress obtained by the ESPI method.

The main technical and software tools used were the white spray Diffu-Therm, optical system Q-100 from Dantec Dynamics as one of a series of Q-100, Q-300, Q-500 and Q-600 systems, and the Istra software. The Q-100 system has a miniaturized 3D ESPI sensor that uses four different illuminating directions and one central CCD camera, so it has up to four sensitivity vectors. The Istra software then ensures all processes of measuring, processing and evaluating the results. Stainless steel sheet with mechanical characteristics $R_{p0.2}$ = 300 MPa, R_m = 658 MPa and A_{80} = 63.3% was used as the material of the test experimental sample.

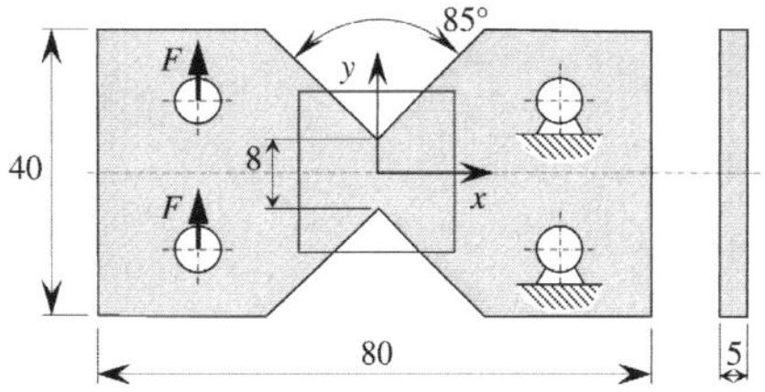

(a) Test sample with a sharp V-notch with an opening angle of 85 ° – the so-called butterfly shape.

(b) Load and clamping jig (frame).

Fig. 11: Design of the shape and the method of specimen loading by shear stress for the purposes of stress and strain analysis by the ESPI method.

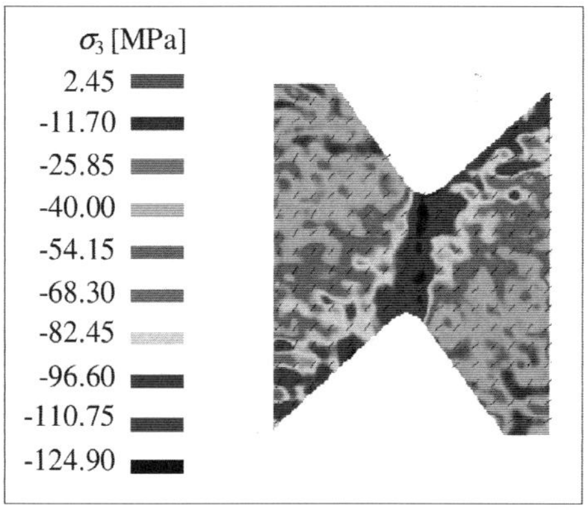

(a) Field of distribution of principal normal stresses σ_3 [MPa].

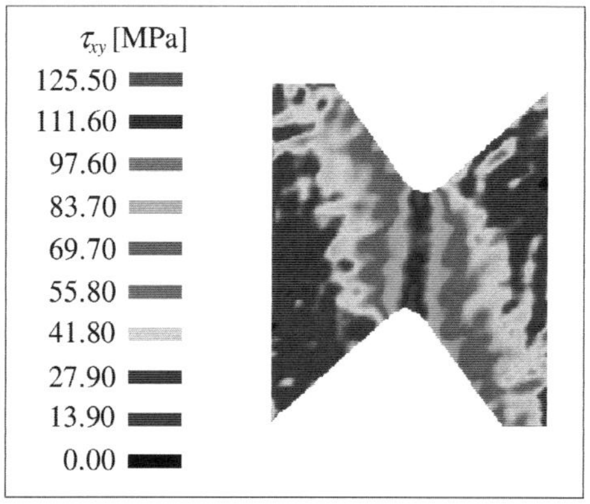

(b) Field of distribution of shear stresses τ_{xy} [MPa].

Fig. 12: Field of distribution of stresses σ_3 and τ_{xy} inside the test specimen during shear loading, obtained by the ESPI method.

3.6 Digital speckle shearography

Another relatively new and specific OEM method is Digital Speckle Shearography (DSSG). It uses a coherent laser light and speckle effect to measure deformation, but unlike holographic methods, it is only able to record a deformation gradient. Compared to SPI, however, it is more resistant to disturbing environmental influences, e.g. shocks, but, at the same time, its sensitivity is extremely high. The essence of the measurement lies in the comparison of the reference state of the surface of the object with the state of the surface when it is deformed by mechanical or thermal load. Then the interferograms captured before and after deformation are superimposed. The result is a stripe pattern, the so-called shearogram, corresponding to the deformation gradient on the surface. The method is applicable to a wide range of materials from steel, ceramics to plastics and composites, in which it relatively successfully identifies various defects, discontinuities, anomalies, delaminations, cracks, etc. A typical representative of measuring devices for the DSSG method is the optical system Q-800 of the Q-800 and Q-810 series from Dantec Dynamics with the appropriate software [7]. This system is based on a shearographic sensor with an integrated high-resolution CCD camera.

4 Origin of presented application examples

All of the mentioned OEM application examples or their key stages were realized using technical and software means, i.e. HS and SW equipment exclusively of the experimental workplace of the Department of

Applied Mechanics and Mechanical Engineering, Faculty of Mechanical Engineering, Technical University in Košice. They formed part of practical tasks or research programs for the development of new experimental methods for the analysis of material properties and the analysis of the response of structural elements or entire structures to loading, as well as the assessment of limit states [12–14].

In cases where the nature of the problem allowed, the results obtained by OEM methods were verified by other experimental methods, most often by tensometry or non-experimental, i.e. analytical or numerical methods, always with comparable results.

5 Conclusion

In this paper, individual OEMs were briefly presented in approximately chronological order of their origin, development and application. Although their development is rapid, their current state, technical and software resources, or even the principle does not yet allow their deployment for instance in the area of measurement of dynamic events, especially the measurement of stochastic character and, at the same time, under difficult conditions of negative influences, e.g. as a replacement for dynamic strain gauges. Their use is therefore limited mainly to laboratory conditions.

However, OEMs find their specific application as sophisticated non-contact methods mainly in the development of new methods for determining deformations, stresses and material characteristics, mainly in the field of fracture mechanics, residual stresses, elemental analysis in the field of elastic-plastic deformations, limit state analysis, etc.

Acknowledgement

This work was supported by the Ministry of Education of Slovakia Foundation under Grant project VEGA No. 1/0290/18, VEGA No. 1/0500/20, KEGA 018TUKE-4/2020 and KEGA 027TUKE-4/2020.

References

[1] J.S. Pérez-Huerta, et al., Digital Holographic Interferometry Applied to the Investigation of Ignition Process, Optics express 25 (2017) 13190–13198.

[2] U. Schnars, et al., Digital Holography and Wavefront Sensing. Principles, Techniques and Applications. Second Edition, Springer 2015.

[3] F. Trebuňa, F. Šimčák, The Handbook of Experimental Mechanics (in Slovak), TypoPress Košice 2007, ISBN 970-80-8073-816-7.

[4] F. Trebuňa, et al., The Use of Optical Methods in Experimental Mechanics 1 (in Slovak), TypoPress Košice 2014, ISBN 978-80-553-1863-9.

[5] F. Trebuňa, et al., The Use of Optical Methods in Experimental Mechanics 2 (in Slovak), TypoPress Košice 2015, ISBN 978-80-553-2273-5.

[6] O. Ostertag, P. Sivák, Degradation Processes and Lifetime Prediction (in Slovak), TypoPress Košice 2010, ISBN 970-80-553-0486-1.

[7] Retrieved from: http://www.dantecdynamics.com.

[8] P. Frankovský, et al., Experimental Analysis of Stress Fields of Rotating Structural Elements by Means of Reflection Photoelasticity, Applied Optics Vol. 56, Issue 11 (2017) 3064–3070.

[9] P. Frankovský, et al., The Use of Digital Image Correlation in a Strain Analysis, Applied Mechanics and Engineering, Vol. 18, no. 4 (2013) 1283–1292. ISSN 1425-1655.

[10] P. Frankovský, et al., Application of Holographic Interferometry in the Analysis of Stress States in a Crack Root Area, Applied Optics Vol. 59, Issue 13 (2020) 170–178, doi: 10.1364/AO.377952.

[11] M. Kalina, F. Šimčák, Strain and Stress Analysis Around the Notches Using ESPI Method, Applied Mechanics and Materials, Volume 611, Zurich TTP (2014) 484–49, ISSN 1662-7482, doi: 10.4028/www.scientific.net/AMM.611.484.

[12] F. Bárnikm, et al., Measurement and comparison study of deformation using extensometer and 2D DIC technology. In IOP Conference Series: Materials Science and Engineering, Vol. 776, No. 1, p. 012065.

[13] J. Živčák, R. Hudák, M. Schnitzer, T. Kula, Numerical simulation and experimental testing of topologically optimized PLA cervical implants made by additive manufacturing methodics. acta mechanica et automatica, 12(2), (2018) 141–144.

[14] J. Živčák, et al., Development and Application of Advanced Technological Solutions within Construction of Experimental Vehicle. Applied Sciences, 10 (15), (2020) 5348.

Analysis of the Effect of Changing the Position of a Unique Drilling Device for Residual Stress Analysis using Digital Image Correlation

M. Hagara[1,*], M. Pástor[1], R. Huňady[1]

[1] *Technical University of Košice, Faculty of Mechanical Engineering, Department of Applied Mechanics and Mechanical Engineering, Letná 1/9, 042 00 Košice, Slovakia*

* *martin.hagara@tuke.sk*

Abstract: The quantification of residual stresses by the hole-drilling method is one of the most widely used techniques. Nowadays, there is an effort to replace strain gauge sensors with full-field optical systems, which requires intervention in the design of the drilling machine or the development of its own. The paper reports on a unique drilling/measuring device that is being developed at the authors' facility. In the case of the analysis of thick samples, it is necessary to mill a blind hole in the material in several steps. The principle of the digital image correlation (DIC) method, i.e. recording and correlating images at each measurement step, requires a mutual change of position between the milling cutter and the camera system. Therefore, this paper focuses on the analysis of the influence of the change of the position of the stereo camera correlation system on the quality of the strain/stress analysis results around the milled hole. The conclusion from the analyses performed is that there is no significant influence of the obtained fields and thus the results obtained by the DIC method can be accepted.

Keywords: residual stress; hole drilling; drilling device; positioning; digital image correlation.

1 Introduction

Residual stresses arise in the material as a result of various technological processes, sudden temperature changes, or surface structure changes such as oxidation or corrosion. As these stresses are easily overlooked, they can superimpose with operational stresses, and thus their determination is an important part of the experimental mechanics' work.

Methods for quantifying residual stresses can be classified into three groups:

- non-destructive,
- semi-destructive,
- destructive.

A frequently used semi-destructive method is the drilling method, where a blind or through-hole is made in the analyzed structure [1–3]. In the presence of residual stresses, the surroundings of the milled hole are deformed, which is recorded by special strain gauge rosettes, and based on the methodology described in ASTM E837-13a [4], the levels of residual stresses in thin and thick samples, respectively, are determined.

In order not only to speed up the measurement itself but also to be able to evaluate the deformations released at more than one point, the combination of the drilling method with non-contact full-field methods is now becoming more common. In particular, optical measuring systems based on electronic speckle interferometry or digital image correlation are used. Since commercially produced drilling devices are designed to work with strain gauge apparatus, their construction does not allow simple application of the above-mentioned optical systems, which leads to the design and implementation of custom portable drilling devices [5, 6].

The authors of the paper proposed a unique drilling device allowing the application of a single or stereo camera correlation system. Unlike almost all custom drilling devices, this device is unique in such a way that it allows horizontal and vertical movement of not only the milling cutter (MC) but also the camera system

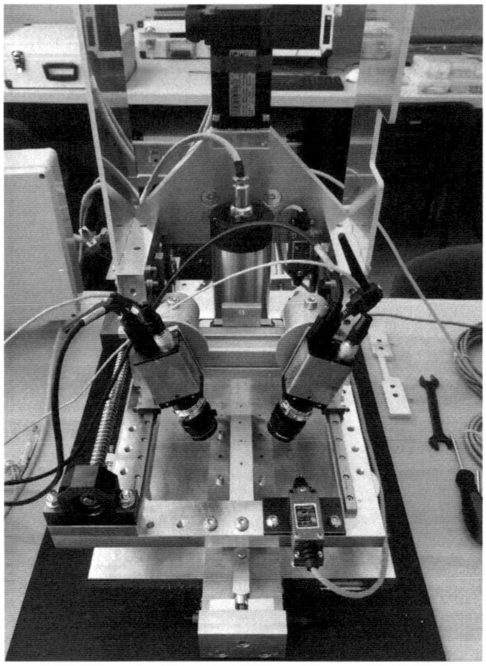

Fig. 1: Drilling/measuring device for quantification of residual stresses using 3D DIC.

(Fig. 1). As a consequence, images can be acquired with the required quality from a suitable distance for various specimen shapes and thicknesses [7].

As in the process of measuring residual stresses by the hole-drilling method, it is mainly necessary to create a blind hole in a stepwise (sequential) manner, the accuracy of the results obtained by the correlation system is influenced by the accuracy with which the positioning device can return the camera system to its initial position. It has been proven that even a very small rotation of the camera system can cause a large correlation error, especially when a single camera system is used [8]. However, in the case of very small displacement deviations (level on the order of a few μm), it can be assumed that the influence of correlation errors will not be so serious that the values of the quantified residual stresses will be significantly affected.

2 Experimental analysis

Full-field analysis of displacements and stresses was performed on a specimen made of Vishay PS-1D material with a thickness of $t = 0.5$ mm, which is used as a coating in the PhotoStress method. The mechanical properties of the analysed material are Young's modulus of elasticity $E = 2{,}500$ MPa and Poisson's ratio = 0.38. The specimen, whose shape and dimensions are shown in Fig. 2a, was subjected to uniaxial tensile loading. The magnitude of the tensile force was recorded by an HBM RSCC-50kg sensor and displayed on the display of Vishay P3 strain indicator and recorder (Fig. 2b).

The full-field non-contact measurement was performed using a Q-400 Dantec Dynamics stereo camera correlation system consisting of low-speed cameras with an image resolution of 5 Mpx. Digital image correlation is a modern optical method that allows the analysis of displacement or strain fields based on the comparison (correlation) of digital images captured during the loading process of the analysed object. The image correlation is carried out over small image elements, which have mostly squared shape, reach the size of several pixels and are called facets. It should be noted that the quality of the results of the analyses performed by Dantec Dynamics correlation systems depends on many aspects, described e.g. in [8–10]. Among the main factors influencing the results of an experimental analysis performed with correlation systems, the following can be included:

- obtaining images with the highest possible quality (resolution, uniform and sufficient illumination, etc.),
- the accuracy of the calibration of the camera system,
- correct adjustment of correlation parameters such as facet size or smoothing level.

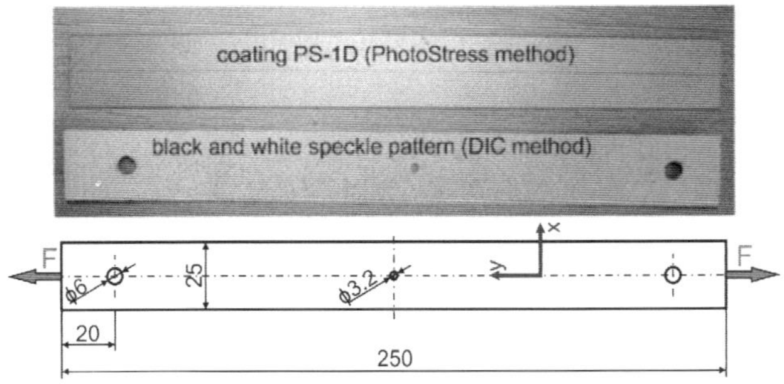

(a)

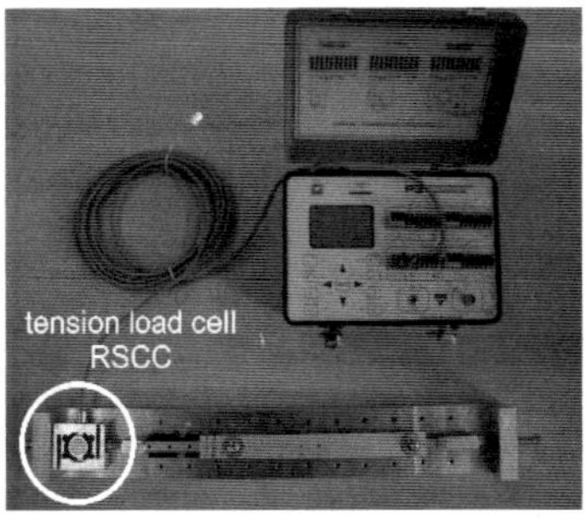

(b)

Fig. 2: (a) Dimensions and shape of the specimen to be analysed, (b) measuring string recording the magnitude of the tensile force.

As the results of the analyses performed by the correlation devices also depend on the quality of the acquired images (which need to be acquired with the highest possible image resolution), the correlation system was attached to the proposed drilling device in a position allowing the recording of the surroundings of the milled hole, as shown in Fig. 3. According to the manufacturers of the Dantec Dynamics digital image correlation systems, the largest correlation error occurs at the edges of the analysed objects and therefore the evaluated area around the hole was reduced to an area of approximately 19 × 19 mm^2 shown in Fig. 3 in yellow. The above-mentioned images, recorded with an image resolution of 2452 × 2056 px (pixel density of approximately 83 px/mm), are reference images of the analysed specimen (i.e. the images of the specimen captured before its loading).

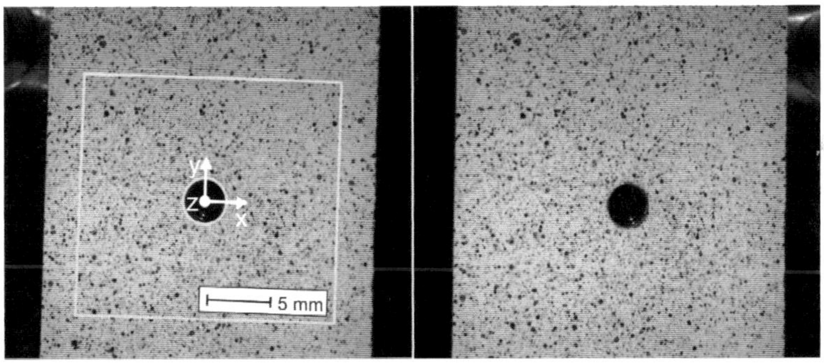

Fig. 3: Reference images captured by a Dantec Dynamics Q-400 stereo camera correlation system with a marked area of evaluation.

The analysed specimen was loaded in 10 steps (1 step corresponded to a force increase of 12 N) with a tensile force that gradually increased from a value of 0 N to a final value of 120 N. It has been shown that the results of strain-stress analyses performed by digital image correlation systems depend significantly on the size of the facet, which has to ensure that each correlated part is unique (i.e. each facet has to contain a unique part of black-and-white speckle pattern created at the surface of the specimen) and on a corresponding level of smoothing that can be set in the Istra4D software supplied with the Dantec Dynamics correlation systems [10]. In Fig. 4, the displacement fields and the equivalent von Mises stress field obtained by the MOSTRAN v.1.0 software developed at the authors' department can be seen for a chosen facet size of 28 × 28 px. The 6 px overlapping of the facets ensured that the data, determined at the centre of each facet, was acquired with a higher resolution.

By the above-mentioned procedure, a stress state characterizing the simulated (residual) stress was induced in the specimen material. In the next phase of the measurement, such a procedure was implemented to simulate the process used to quantify the residual stresses in thick specimens, i.e. drilling a blind hole and sequentially recording the released strains in 20 steps until a drilled hole depth of 2 mm was reached. This approach thus leads to 20 mutual position changes between the milling cutter (MC) and the correlation system (Fig. 5).

Under ideal conditions, i.e. if the correlation system still returned exactly to its initial position (or did not

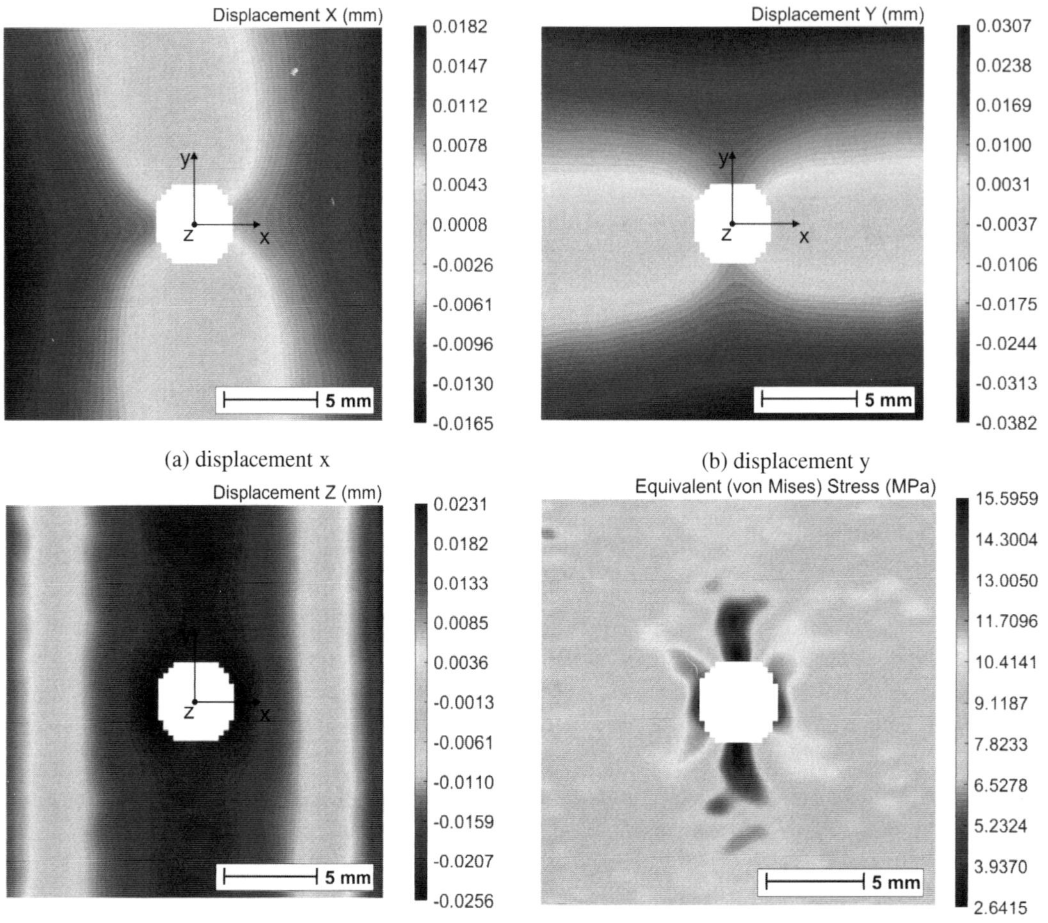

(a) displacement x

(b) displacement y

(c) displacement z

(d) equivalent von Mises stress

Fig. 4: Displacement and stress fields obtained after maximum loading of the specimen.

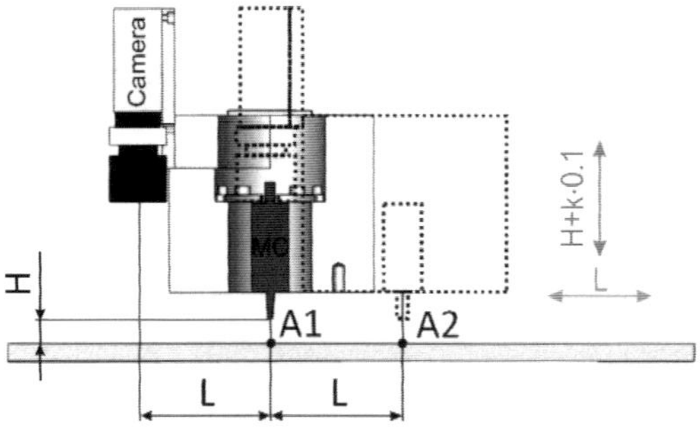

Fig. 5: The mutual position between the milling cutter (MC) and camera system and their necessary movements in the horizontal and vertical direction.

move from it), the digital image correlation system should evaluate approximately the same displacement and strain fields as in the analysis described above (the mean relative deviation found by the authors is approximately 1% and 3% for the displacement and strain fields, respectively, when capturing 30 images under the same loading and measurement conditions). The entire drilling/measuring system was moved in two directions – in the horizontal one (forward and backward) by a distance L = 150 mm, which characterizes the distance between the milling cutter axis and the focus of the camera lenses, and in the vertical one (down and up) by a distance $H + k \cdot 0.1$ mm, where H = 10 mm is the distance of the milling cutter from the specimen surface, k is the measurement step number, and 0.1 mm is the increment of material being removed at each measurement step (Fig. 5). After each return of the camera system to the starting position, images of the analyzed specimen were captured by the digital image correlation system.

As the measurement continued under the same conditions as the above-mentioned displacement/stress analysis, information on the accuracy with which the camera system returned to its original position was provided by the correlation system itself, based on the comparison of the 3D coordinates of the evaluated surface points of the specimen. This process has already been used and verified in the initial testing phase of the drilling device when the results obtained by the correlation system were compared with the deviations recorded by the inductive displacement sensor and have achieved a high level of similarity [11]. Fig. 6 shows the mean deviations of the specimen surface points' 3D coordinates compared against the condition at the specimen maximum loading. Based on the presented plot, it can be concluded that the camera system returned to its original position with an evaluated maximum deviation of approximately 10 μm, observed in the direction of the largest drilling/measuring device shift, which corresponds to the positioning accuracy already established [11].

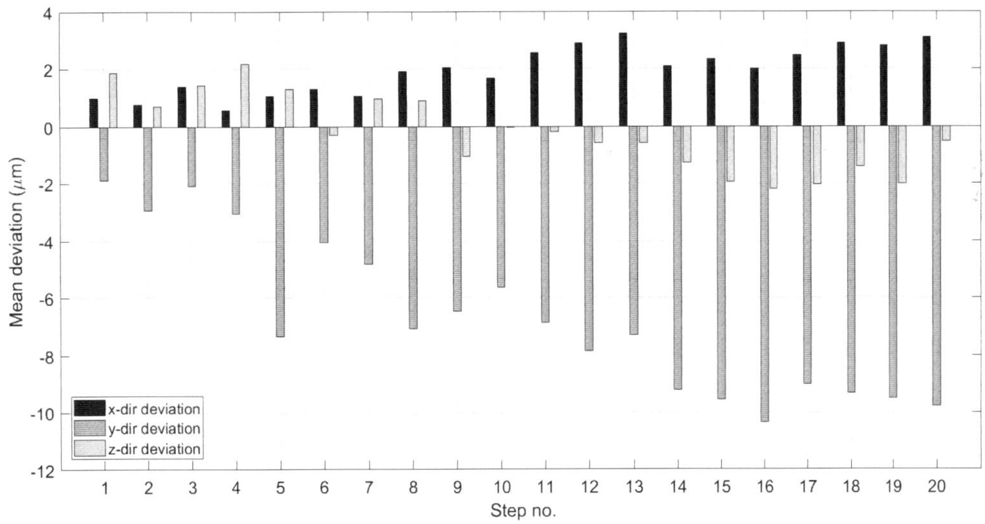

Fig. 6: The mean deviations from the initial position of the correlation system obtained at each measurement step after the necessary change of position between the drilling and measuring device has been carried out.

The obtained 20 images were correlated under the same above-mentioned conditions and the acquired displacement or equivalent von Mises stress fields were compared with the results obtained at the maximum loading of the specimen. It should be noted that rigid motions were filtered out from the fields of the individual components of displacement and thus the compared fields should not be affected by the motion of the stereo camera system, but should only correspond to the deformation of the specimen itself. The relative deviations of the displacement components and the equivalent von Mises stress obtained as mean values from the whole evaluated specimen area are shown in Fig. 7 and Fig. 8. While the maximum deviations in the measured displacement components reach higher levels (ca. 10%) for displacement x and z, respectively, the smallest deviation is observed for displacements obtained in the loading direction of the specimen, i.e. in the y-direction. The mean deviations observed in the equivalent von Mises stress reach levels between 4–6%, which can be considered acceptable when expecting the value of about 3%.

To find the optimal region around the hole, which could be further used for residual stress quantification through the proposed system, the obtained deviations were subjected to an analysis that documents how the investigated quantities change as a function of the distance from the center of the drilled hole. The relative deviations of the compared quantities obtained in 20 measurement steps were averaged over 15 concentric areas A_i with a width of 2 facets, i.e. approximately 0.5 mm (see Fig. 9d). Based on the obtained results, it can be concluded that the mean deviation acquired in the displacement x (Fig. 9a), displacement z (Fig. 9c) and equivalent von Mises stresses (Fig. 9d), respectively, increases with an increasing distance from the center of the hole. When analyzing the deviations in the displacement y (Fig. 9b), the same conclusion cannot be established. However, in this case, the level of deviations is significantly lower than in the other ones.

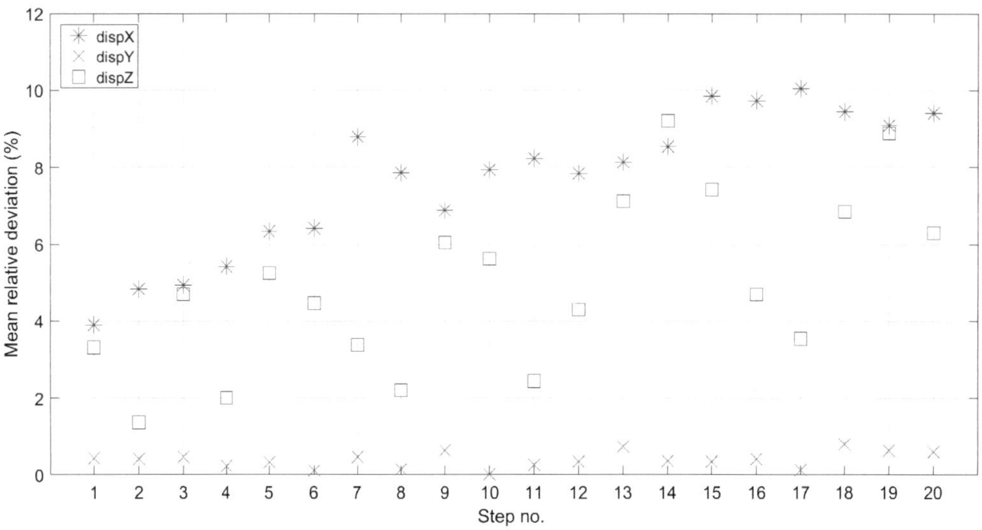

Fig. 7: The mean relative deviations of the displacement components obtained on the evaluated specimen area at each measurement step after the necessary change of position between the drilling and measuring device has been carried out.

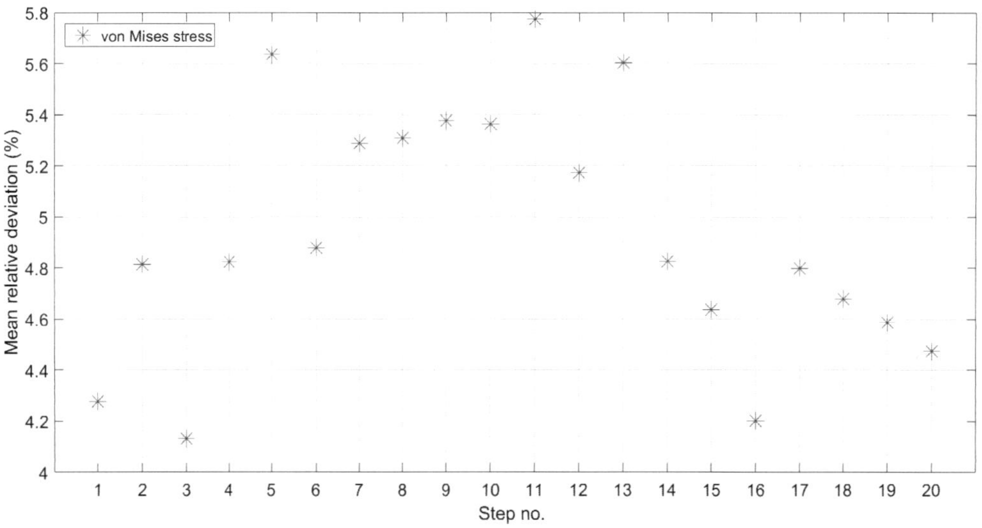

Fig. 8: The mean relative deviations of the equivalent von Mises stress obtained on the evaluated specimen surface at each measurement step after the necessary change of position between the drilling and measuring equipment has been carried out.

3 Conclusion

The paper describes the possibilities of performing analysis of the released full-field deformations or stresses in the vicinity of the hole using a custom drilling device combined with a stereo camera digital image correlation system Q-400 Dantec Dynamics on thick specimens when a blind hole is milled in the analysed material in a stepwise manner. In previous analyses, it has been shown by the authors that even a small change in the position (rotation) of the correlation systems can result in significant correlation errors that negatively affect the results of the analyses. For this reason, it was necessary to design the drilling/measuring device so that when the position between the milling cutter and the correlation system is changed relative to each other (to ensure milling the hole and capturing the images in each of twenty measurement steps), the measuring system returns to the initial position as accurately as possible. The results obtained show that even if the correlation system returns to its original position with a deviation of the order of a few μm, the evaluated displacement,

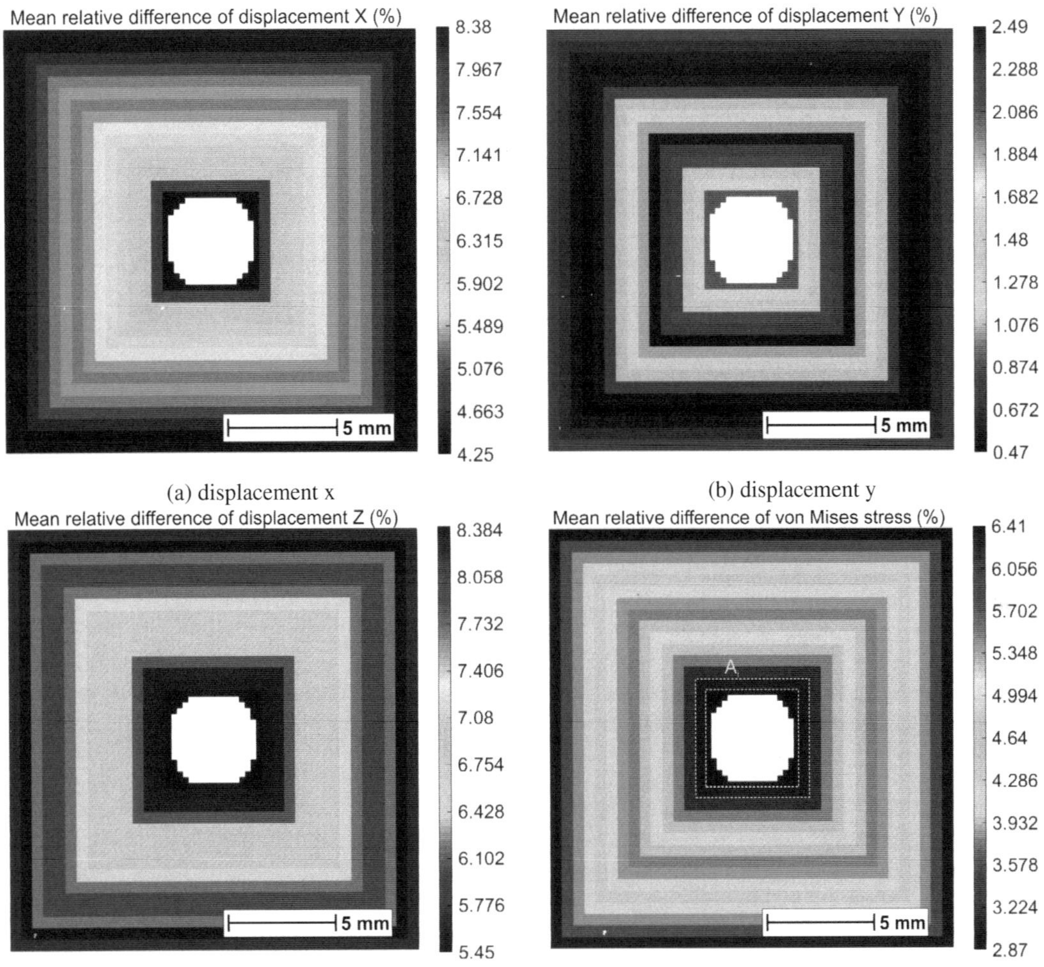

(a) displacement x

(b) displacement y

(c) displacement z

(d) equivalent von Mises stress

Fig. 9: Planar distribution of mean relative deviations as a function of distance from the centre of the drilled hole.

strain or stress fields will be affected. However, in the case of a stereo camera system, this deviation is not so significant that the moving camera approach (which allows the recording of images with the required quality for any shape and thickness of the specimens) cannot be used for the analysis of residual stresses.

As the use of a stereo camera digital image correlation system always introduces some image distortion into the measurements, the authors plan to perform the same analysis using a single camera system. It will also be necessary to perform similar analyses for other types of stress states, e.g. biaxial stress state. Once the above-mentioned analyses have been performed, it will be possible to decide more accurately which type of digital image correlation system is preferable to use for image capturing, how large the area evaluated in the vicinity of the drilled hole is appropriate to use for calculating residual stresses, and from what quantities (i.e. displacements or strains) the residual stress values will be determined.

Acknowledgement

This research was funded by Slovak Grant Agency VEGA 1/0355/18.

References

[1] N.J. Rendler, I. Vigness, Hole-drilling strain-gage method of measuring residual stresses, Experimental Mechanics 6 (1966) 577–586, doi: 10.1007/BF02326825.

[2] F. Trebuňa, P. Šarga, Using of Program MEZVYNA for Measurement of Residual Stresses to Solving Practical Problems, Acta Mechanica Slovaca 15 (2011) 14–21, doi: 10.21496/ams.2011.013.

[3] F. Menda, F. Trebuňa, P. Šarga, New Method of Residual Stress Evaluation and its Advantages in Comparison with More Common Hole-drilling Method, Acta Mechanica Slovaca 17 (2013) 64–70, doi: 10.21496/ams.2013.048.

[4] ASTM E837-13a: Standard Test Method for Determining Residual Stresses by the Hole-Drilling Strain-Gage Method. ASMT international, West Conshohocken, PA (2013).

[5] M.R. Viotti, et al., A portable digital speckle pattern interferometry device to measure residual stresses using the hole drilling technique, Optics and Lasers in Engineering 44 (2006) 1052–1066, doi: 10.1016/j.optlaseng.2005.09.004.

[6] T. Brynk, B. Romelczyk- Baishyaa, Residual stress estimation based on 3D DIC displacement filed measurement around drilled holes, Procedia Structural Integrity 13 (2018) 1267–1272, doi: 10.1016/j.prostr.2018.12.259.

[7] M. Pástor, P. Čarák, M. Hagara, Development of the Device with a High Positioning Accuracy Serving for Residual Stress Quantification using Optical Methods, Acta Mechanica Slovaca 23 (2019) 24–29, doi: 10.21496/ams.2020.005.

[8] M. Hagara et al., Analysis of the aspects of residual stresses quantification performed by 3D DIC combined with standardized hole-drilling method, Measurement 137 (2019) 238–256, doi: 10.1016/j.measurement.2019.01.028.

[9] R. Huňady, M. Hagara, M. Schrötter, Impact Assessment of Calibration Parameters on Accuracy Method of Digital Image Correlation, Acta Mechanica Slovaca 16 (2012) 6–12, doi: 10.21496/ams.2012.012.

[10] M. Hagara, J. Bocko, The Aspects of Stress Analysis Performed by Digital Image Correlation Method Related with Smoothing and its Influence on the Results, Acta Mechatronica – International Scientific Journal about Mechatronics 8 (2018) 15–21, doi: 10.22306/am.v3i4.45.

[11] M. Pástor, et al., Design of a Unique Device for Residual Stresses Quantification by the Drilling Method Combining the PhotoStress and Digital Image Correlation, Materials 14 (2021) 1–27, doi: 10.3390/ma14020314.

Functionality Validation of New Three-point Hitch Dynamometer with Close Six Measuring Rods to Measure the Draft Requirement of Mounted Implements

M. Havlíček[1,*], R. Pražan[2], F. Lopot[1], L. Jílek[2]

[1] *CTU in Prague, Faculty of Mechanical Engineering, Department of Designing and Machine Components, Technická 4, 166 07 Prague 6, Czech Republic*

[2] *Research Institute of Agriculture Engineering, p.r.i., Drnovská 507, 161 01 Prague 6, Czech Republic*

[*] *martin.havlicek@fs.cvut.cz*

Abstract: This paper is focused on a new and special method used in agriculture, for observation of emerging forces between tractor and attachment connected through three-point hitch. These emerging forces are hard to capture by ordinary measuring device, so a special frame was designed for this measurement. This special frame consists of two triangular frames and these frames are connected together with six measuring rods. Connection of frames with rods is designed as fixed, so no motion is allowed between them. All measuring rods are made from steel with circular cross section and all rods are equipped with strain gauges to capture the tension and compression. These values are than recalculated and acting forces between tractor and attachment are obtained. FEM analysis of this frame was made additionally to find out adequate cross section of rods and for comparison if data from measurement correlate with data from analysis.

Keywords: dynamometer; three-point hitch; frame; tractor; rod; force, FEM analysis.

1 Introduction

Agriculture is one of the most important industry in the world. In agriculture and forestry, work machines integrated into the energy means are used by means of a three-point hitch. Knowledge of detailed force and moment effects acting on the machine or energy device during work is necessary for the creation of analyzes, knowledge know-how, verification of work or design of innovative new generation machines. Research and development in the field of agriculture and forestry focuses on the development of machines with higher working widths, the grouping of operations into one, variable tillage or new tillage technologies. The proposed measuring equipment should meet the above requirements for the development and verification of the work of new machines and technologies.

Main goal of this special measurement is to validate acting forces in rods during testing, compare these results with FEM analysis results and calculate calibration constant . In future, this dynamometer can help to optimize selection of tractor or the attachment for farmers. This could help them to enhance fuel consumption and also enhance a distribution of pressure from wheels acting on a ground. These forces acting between tractor and attachment are measured by a special frame which is designed only for this purpose. Tensile and compressive forces are measured by strain gauges placed on rods which connect two triangular frames.

2 Instrumentation

For this experiment is designed a special frame (see Fig. 1) which can be connected to a tractor with 200 horsepower and work machines by three-point hitch. This special frame is more complex than so far designed frames [1–3]. This frame consists of one welded front frame which connects to a tractor (see Fig. 1–1), one welded rear frame which connects to work machines (see Fig. 1–2) and six measuring rods (see Fig. 1–3). These frames are coupled together with measuring rods with spherical hinge at the end to limit bending

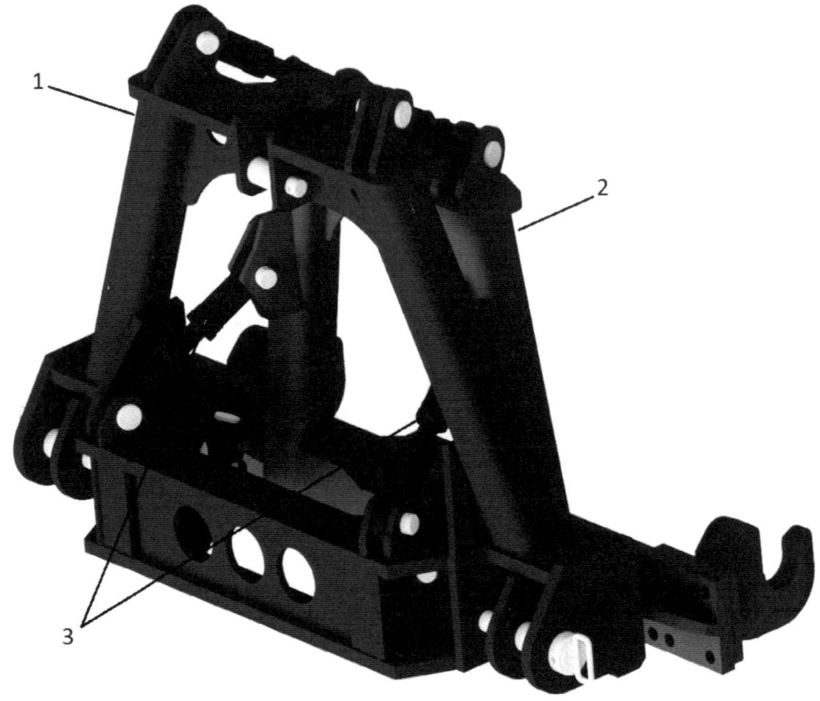

Fig. 1: Measuring frame (1 front frame, 2 rear frame, 3 measuring rods (6 pcs)).

moment. These measuring rods are similar as are used in tram [4]. Due to the rods, frames are fully fix with each other.

Measuring rod is an assembly which connects two welded frames. These rods consist of a rod with circular cross section, two spherical hinges, two hexagonal nuts, two strain gauges and a cover. For this measurement are used strain gauges with one measuring grid [5]. Two strain gauges are installed on a smooth surface of a rod with circular cross section with a diameter of 36 millimeters. Thanks to the strain gauges, the deformation of rod is measured and data are recalculated to forces. Because the dynamometer is designed as space rod system which can capture forces acting in more direction, it is made a FEM analysis to find out acting forces in rods to choose adequate diameter of the rod. Thanks to this is elected suitable strain gauges which are connected in quarter bridge and cables transfer data to measuring device. To protect these strain gauges from dust, protective cover is applied on them.

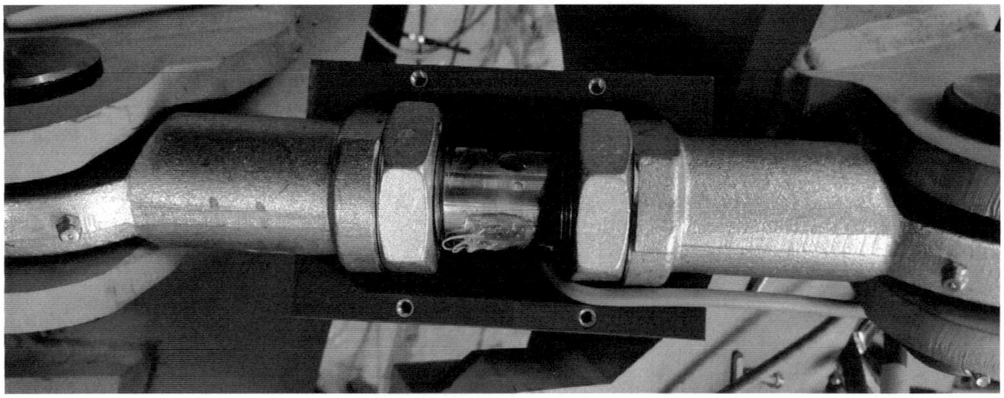

Fig. 2: Measuring rod.

3 Calibration of measuring rods

Fig. 3: Calibration of rods.

Before the measurement started, the frame is disassembled because it is necessary to calibrate all the measuring rods. This step is done to enhance accuracy of data obtained from the measurement. All rods are calibrated separately.

Firstly, one spherical hinge is connected to a forklift truck by rope to fix a rod in one position. On the second spherical hinge, a weight of 802 kilograms is attached to cause a deformation of a rod. Thanks to this deformation, the resistance of strain gauges differs from catalog data of strain gauges. All data are captured by measuring device.

All data are saved and then postprocessed to find out actual value of tension acting in a rod in comparison with real value from the weight. During the comparison is found out calibration constants which is used in next measurements to make measurement more accurate.

4 Validation of welded frame by FEM analysis

When calibration of measuring rods is done, the complementation of dynamometer starts. Thanks to the 3D model is determined an exact position of both frames with each other. To fix frames in one position with each other, measuring rods are used and are jointed by pins to both frames. First test of this special frame took place in laboratory of Research Institute of Agricultural Engineering, p.r.i.. The frame is fully fixed to a forklift truck and to simulate acting force on this frame in z axis, the weights are fastened to the middle part of the frame(see Fig. 4). Acting force is caused by weight of 160 and 450 kilograms. To make sure that this weight is correct, HBM measuring device force transducer U10M is attached between weight and measuring frame (see Fig. 4) [6]. In case that the force transducer U10M is connected between frame and weight, the connection differs from the situation without a force transducer U10M (see Fig. 4). Due to this weight acting in z axis are captured data of tension or compression forces in measuring rods. To validate, if these forces are appropriate or if the frame is welded and connected precisely, FEM analysis of the frame is made again. FEM analysis is made in RFEM software [7].

To validate forces emerging in rods, virtual model for FEM analysis is made (see Fig. 5). Model is supported on one side of a frame which is connected to tractor by three-point hitch and load is placed in same position as the weight of 160 and 450 kilograms is placed. Thanks to this model, acting forces in measuring rods are obtained.

Tab. 1: Validation of dynamometer.

	Rod	1	2	3	4	5	6
Load 160 [kg]	Real force [N]	855	436	-871	224	227	-1 293
	RFEM force [N]	726	726	-758	47	47	-758
Load 450 [kg]	Real force [N]	811	3 759	-2 771	-128	-47	-1 522
	RFEM force [N]	3 142	3 142	-2 900	-188	-188	-2 908

Values in Tab. 1 represent values in rods obtained during the first test where the frame is loaded by weight of 160 or 450 kilograms and values obtained by FEM analysis. As it can be seen, for both results, the tension and compression forces are in same rods, but values differ. FEM analysis shows that forces are symmetric in

(a) 160 kg (b) 450 kg

Fig. 4: Testing of frame.

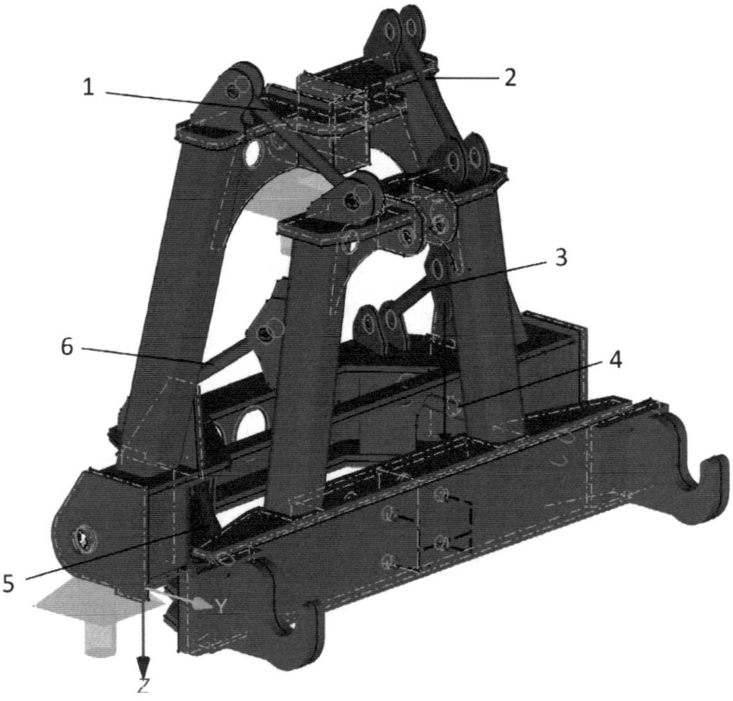

Fig. 5: Model in RFEM.

YZ plane. On the other hand, forces emerging in measuring rods are not symmetric. These results represent that there are some imperfections in the frame.

5 Conclusion

Thanks to this special frame connected between a tractor and agriculture attachment, acting forces in axis x, y and z to a tractor can be determined. This special frame consists of two welded frames which are connected by six measuring rods. These rods measure deformation of a rod by strain gauges and all data are recalculated by equations to a tension or compression force acting in a rod. Rods were firstly calibrated by weight to ensure that other measurement will be more accurate. After first measurement, the results were compared with FEM analysis to find out if this special dynamometer measure precisely. As the comparison was made, it was found out that results obtained from the measurement don't correlate with data from virtual model and it was impossible to determine calibration constant between real value and RFEM values. Thanks to this was made a conclusion that there is some imperfection on the frame. First imperfection can be caused by wrong positioning of two welded frames that are connected with the rods. Second imperfection can be caused by wrong welding of the frames themselves. Although there were found some differences between testing and FEM analysis, it can be said that this device can be tested in fields connected to a tractor after small inspection because dynamometer acts as it was expected.

Acknowledgement

This paper was created with grant the support project of long time development of Research Institute of Agricultural Engineering p.r.i. no. RO0621.

References

[1] P.Procházka, et al., Evaluation of measuring frame for soil tillage machines draught force measurement. Agronomy Research, (2015), 13.1: 186-191.

[2] Y.S.Kim, et al., Development of a real-time tillage depth measurement system for agricultural tractors: application to the effect analysis of tillage depth on draft force during plow tillage. Sensors, (2020), 20.3: 912.

[3] H.F. Aljalil, et al., Design and performance of an adjustable three point hitch dynamometer, Soil & Tillage Research, (2001), 62: 153–156.

[4] M. Dub, et al., Experimental Analysis of Stress State and Motion of Tram Gearbox Hinge, The 52nd International Conference on Experimental Stress Analysis, 2014.

[5] Retrieved from: https://www.hbm.cz/wp-content/uploads/S01265.pdf.

[6] Retrieved from: https://www.hbm.cz/wp-content/uploads/B01444.pdf.

[7] Dlubal Software GmbH, RFEM 5.26.02 (2020)

Heat Transfer Analysis of MRE Under Compressive Cyclic Loading

M. Hdaib[1,*], I. Petríková[1], B. Marvalová[1], T. H. Nam[1]

[1] *Faculty of Mechanical Engineering, Technical University of Liberec, Studentská 1402/2, 461 17 Liberec 1, Czech Republic*

* *Mohammad.yousef.hdaib@tul.cz*

Abstract: The self-heating phenomenon is one of the essential parameters in the design and analysis of composites, as the accumulated heat can dominate the fatigue process under particular circumstances. This paper presents an uncoupled finite element (FE) simulation of the heat accumulation inside an isotropic magnetorheological elastomer (MRE) under compressive cyclic loading. The FE model was solved solely as a heat transfer problem based on dissipated energy data from an experiment. The theoretical analysis, FE modeling, and results were presented and discussed. The simulation results showed good agreement with experimental data.

Keywords: magnetorheological elastomer; finite element method; self-heating; energy dissipation.

1 Introduction

Magnetorheological elastomers (MREs) are smart materials whose mechanical properties can be modified under the presence of a magnetic field. MREs are usually fabricated by mixing micron-sized iron particles with a rubber-like matrix before the matrix is fully cured. MREs can be either isotropic or anisotropic according to applying a magnetic field during the curing process [1]. Under the presence of a magnetic field, the particles will be regularly aligned and form an anisotropic MRE. In contrast, the particles will be distributed randomly without a magnetic field, creating an isotropic MRE [2]. MREs are used in applications where a variable stiffness is required, such as tunable vibration absorbers, sensors, vehicles seats, and electronic devices development [3, 4].

Many studies have investigated the temperature effect on the deterioration of mechanical properties and referred is as thermal fatigue. Ratner et al. [5] illustrated the difference between thermal and mechanical fatigue by presenting a criterion for critical failure stress. This criterion compares the failure stress with the critical one to decide whether the dominant failure mode is thermal or mechanical. So that understanding the heat transfer mechanisms, portions of dissipated energy, and the temperature distribution are crucially essential for fatigue and design purposes.

During compressive cyclic load, the temperature rises steeply at the first cycles then gets stabilized when the dissipated energy is fully transferred to heat [6]. In literature, the ratio of dissipated heat to the total dissipated energy is referred to as the conversion ratio. Pichon et al. [6] demonstrated that a portion of dissipated energy is dissipated at the first cycles as a change of the material microstructure. Then, it is totally dissipated to heat when the temperature stabilizes. Rittel et al. [7] illustrated the dependence of conversion ratio on the strain and frequency.

There are several studies in the literature that used FE modeling to investigate the self-heating phenomenon. Generally, there are two types of FE Models: the first one is the fully coupled approach. The heat transfer problem is solved simultaneously with the mechanical one, and the mechanical properties deterioration with temperature rise is evenly updated. While in the second approach, the problem is solved solely as a heat transfer one with information from the experiment regarding the hysteresis loops areas or sample temperature distribution. Bazkiaei et al. [8] illustrated the different approaches in more detail and proposed a FE model with different heat source distributions.

In this study, an uncoupled finite element model was presented and compared with experimental data from Nam et al. [4]. Firstly, the experimental setup of Nam et al. had been illustrated, and the used mechanical and thermal analysis were described, then the Numerical FE model was explained in detail. Finally, the results, discussion, and conclusion were presented.

2 Theory

In this study, the Experimental setup proposed by Nam et al. [4] was simulated solely as a heat transfer problem. Nam et al. [4] investigated the self-heating phenomenon of isotropic MRE filled with carbonyl iron particles (CIPs) under compressive cyclic loading. The CIPs were added with a volumetric fraction of 0.27 and subjected to compressive cyclic loading for two hours, at variable strain amplitude (1%–5%), variable pre-strain (5%–20%), and variable frequency (10 Hz–50 Hz).

Assuming an incompressible rubber and normalized stress response ($\sigma(t)$) upon the strain amplitude ($\Delta\varepsilon$) during cyclic loading, then the stress response can be estimated as [9]:

$$\sigma(t) = \frac{F(t)}{A} = \sigma_0 + \Delta\varepsilon \left[E' \sin(2\pi f t) + E'' \cos(2\pi f t)\right], \tag{1}$$

where E' and E'' are the storage and loss moduli, which can be evaluated using the lag angle (δ) between the applied stress and the resulted strain, as shown in Eqs. 2 and 3.

$$E'(\varepsilon_0, f, \Delta\varepsilon) = \frac{\Delta\sigma}{\Delta\varepsilon} \cos(\sigma), \tag{2}$$

$$E''(\varepsilon_0, f, \Delta\varepsilon) = \frac{\Delta\sigma}{\Delta\varepsilon} \sin(\sigma). \tag{3}$$

The storage modulus represents the strain energy stored within the sample, while the loss modulus represents the dissipated energy. The volumetric dissipated energy (q_{diss}) can be estimated from the stress-strain loops areas [4], as shown in Eq. 4:

$$q_{diss} = \int_0^T \sigma(t)\dot{\varepsilon}(t)dt = \pi \Delta\varepsilon^2 E''. \tag{4}$$

Then the conversion ratios (η) can be estimated from the temperature distribution and by applying the First Law of thermodynamic (Conservation of energy principle) on the sample boundaries, as shown in Eq. 5 [10]:

$$\dot{q}_g = \frac{\rho C_p \Delta T}{t} = \dot{q}_{net} - \dot{w}_{net}, \tag{5}$$

where $\dot{q}_g$ represents the variation of the total energy stored in the system. In this study, it is considered the accumulated heat during the experiment. While ρ, C_p and ΔT are sample density, specific heat, and the temperature difference. The net heat done on the system ($\dot{q}_{net}$) is considered the summation of conduction, convection, and radiation heat rates. In contrast, the net work done by the system ($\dot{w}_{net}$) is regarded as the dissipated energy (q_{diss}) multiplied by the conversion ratio (η) divided by the time interval (t). So that, Eq. 5 can be rewritten as:

$$\dot{q}_g = \eta \dot{q}_{diss} - \dot{q}_c - \dot{q}_{con} - \dot{q}_r. \tag{6}$$

Using this one-dimensional equation with information about temperature distribution, one can predict the conversion ratio of dissipated power to heat (η). A similar approximation was used by Balandraud et al. [11], Samaca et al. [12] and Loukil et al. [13].

3 Finite Element Model

In this study, a transient 2d finite element model was built using Comsol Multiphysics [14] to simulate the experimental data of Nam et al. [4]. The geometric structure in a 2D axisymmetric configuration and the thermal boundary conditions were described, as shown in Fig. 1. The used MRE sample is cylindrical with 13 mm height and 28 mm diameter. FE meshing has been applied to the MRE sample, Teflon platens, and steel grips. The following boundary conditions were used:

- Heat conduction between MRE sample and Teflon platens, and between Teflon platens and steel grips.

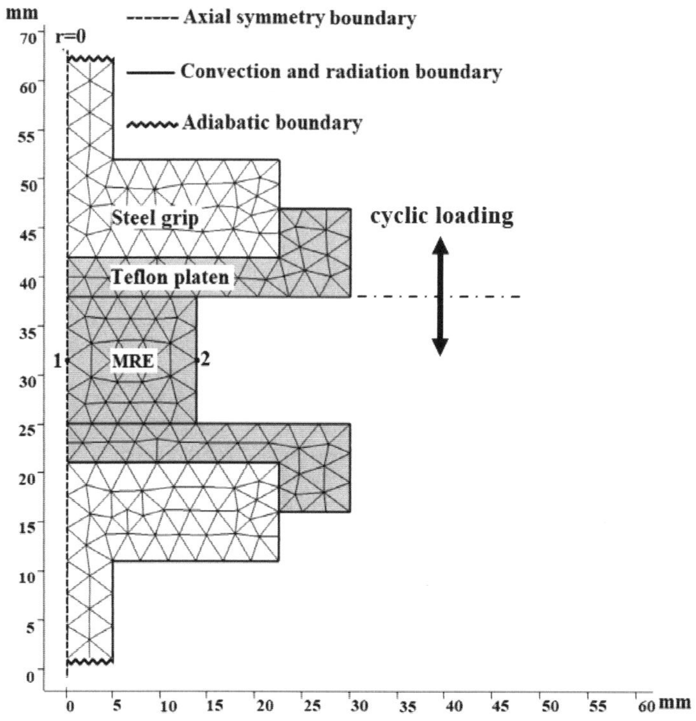

Fig. 1: FE Model as described by Nam et al. [4].

- Convection and radiation on the exposed surfaces.
- Adiabatic at the end of steel grips, where the temperature is close to the ambient one.
- The initial temperature of all components was considered as the ambient temperature.
- Axial symmetry.

The thermal contact resistances were ignored as the Teflon platens have a soft surface and are in pressure contact with the steel and MRE sample [10]. For horizontal exposed surfaces, the convection coefficient (h) was calculated as [10]:

$$h_u = 5.15 \left(\frac{T_s - T_\infty}{L T_\infty} \right)^{1/4}, \tag{7}$$

$$h_l = 1.5 \left(\frac{T_s - T_\infty}{L^2 T_\infty} \right)^{1/5}. \tag{8}$$

The subscripts (u) and (l) represent upper and lower according to the surface direction. In contrast, for vertical surfaces, the convection coefficient is estimated as [10]:

$$h_v = \frac{0.025}{0.01295} \left[0.825 + 17.04 \left(L^3 \frac{(T_s - T_\infty)}{T_\infty} \right)^{1/6} \right]^2, \tag{9}$$

where (L) is the characteristic length which can be estimated by dividing area over the perimeter or volume over the area. Tab. 1 illustrates each component's conductivity, emissivity, and specific heat [4].

4 Results and discussion

The experimental data of stress-strain loops areas were used as a transient variable for heat generation inside the MRE sample multiplied by the conversion ratio (η), and the generated heat was assumed to be evenly distributed through the sample [6–8]. The internal temperature and surface temperature were compared with

Tab. 1: Thermal properties of MRE sample, Teflon platens, and steel grips.

Material	Emissivity	Conductivity	Specific heat
Emissivity [-]	0.95	0.22	0.85
Conductivity [W/mK]	0.775	44.5	0.24
Specific heat [J/KgK]	585.08	475	1,050

experimental ones, as shown in Figs. 2, 3, and 4. The comparison was performed at point (1) and point (2) mentioned in Fig. 1.

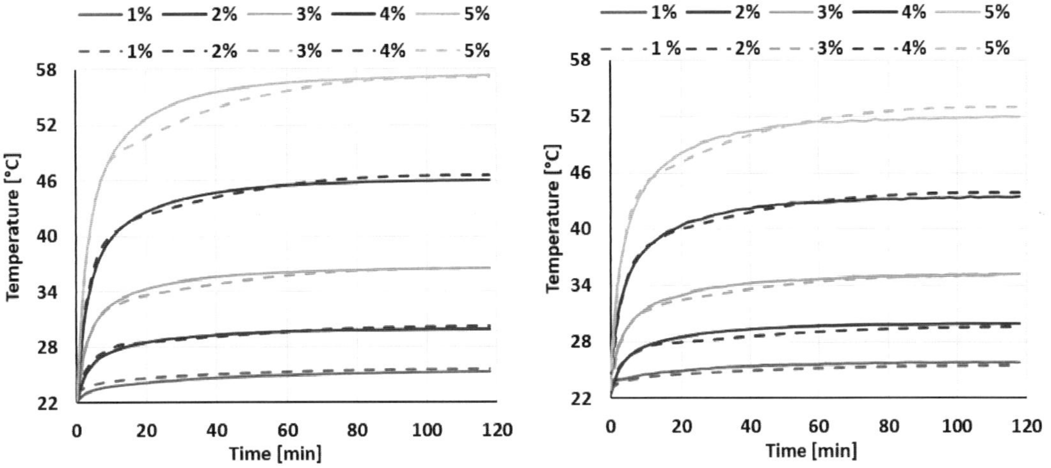

(a) Internal temperature (b) Surface temperature

Fig. 2: Internal and surface temperature at different strain amplitudes (1% to 5%), pre-strain of 20%, and frequency of 40 Hz. Solid lines (experimental) and dashed lines (simulation).

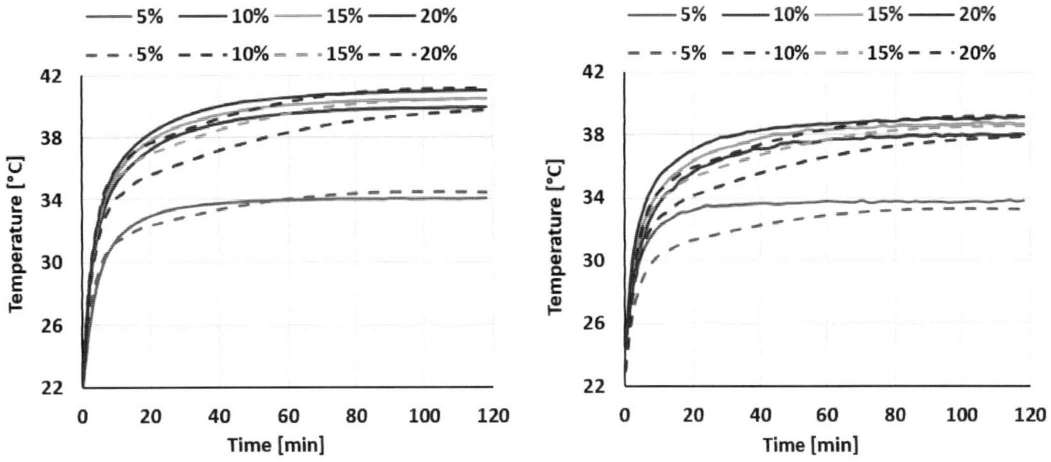

(a) Internal temperature (b) Surface temperature

Fig. 3: Internal and surface temperature at different pre-strains (5% to 20%), strain amplitude of 3%, and frequency of 50 Hz. Solid lines (experimental) and dashed lines (simulation).

The results showed good agreement with the experiment and literature. The average relative error was less than 2% for all cases, and the maximum absolute error was observed at the pre-strain effect testing group with a value of 8%. Most of the uncoupled FE studies mentioned in the literature were able to predict the final internal and surface temperatures accurately and underestimate transient temperatures. Because these studies either ignored the radiation effect, horizontal plane of symmetry, or used a constant conversion ratio, such approximations could result in a significant error as the test time interval is relatively long. While in this study, the transient self-heating was successfully predicted by solving the problem solely as a heat transfer one. Moreover, the temperature contour lines shown in Fig. 5 showed good agreement with the infrared camera data mentioned by Nam et al. [4].

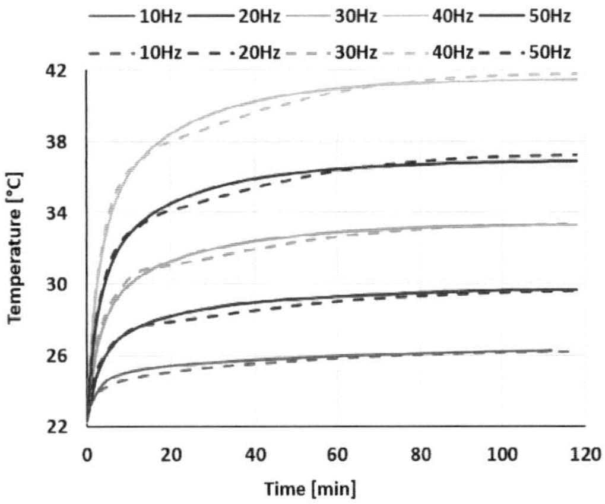

(a) Internal temperature

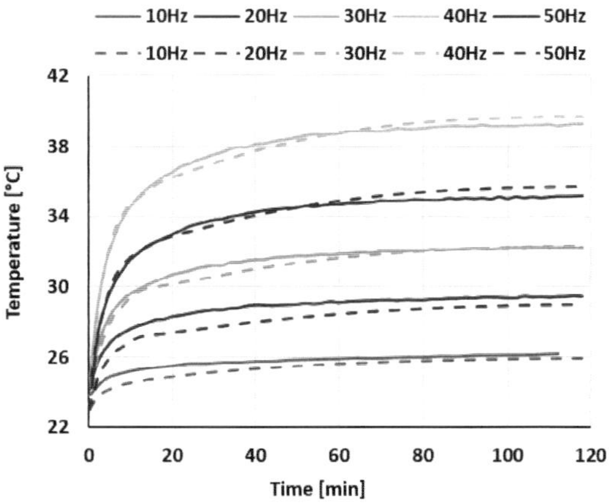

(b) Surface temperature

Fig. 4: Internal and surface temperature at different Frequencies (10 Hz to 50 Hz), pre-strain of 20%, and strain amplitude of 3%. Solid lines (experimental) and dashed lines (simulation).

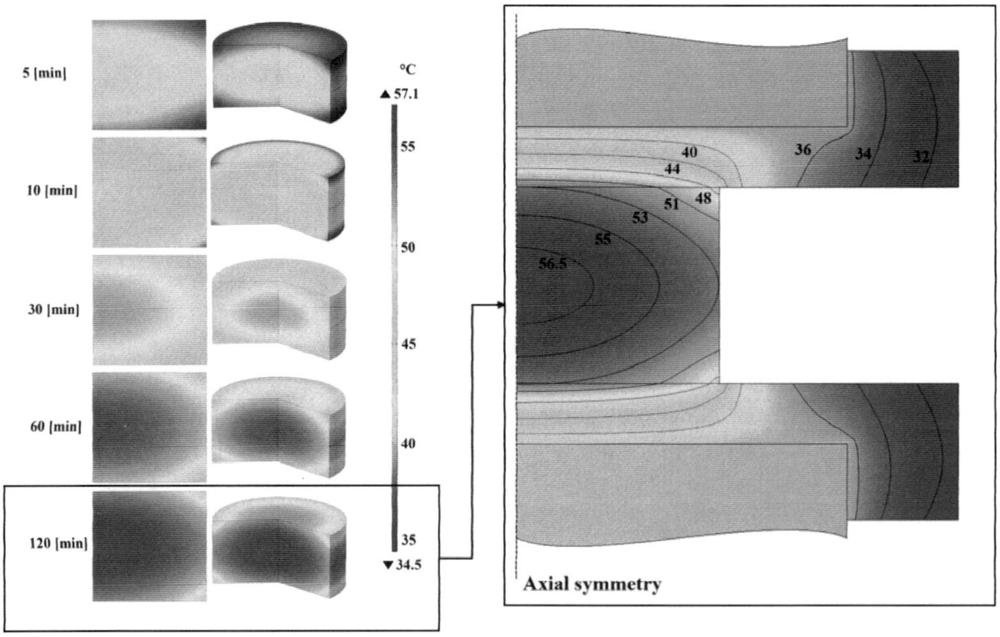

Fig. 5: Temperatur evolution with time and contour lines at a frequency of 40 Hz, pre-strain 20%, and 5% strain amplitude.

5 Conclusion

In this study, an uncoupled 2d FE simulation was performed to investigate the self-heating phenomena of isotropic MRE under compressive dynamic loading. A 1d-model was proposed using the First Law of thermodynamics to calculate the conversion ratios based on the experimental temperature distribution. The stress-strain loop areas were inserted simultaneously multiplied by the conversion ratios into the heat transfer problem to simulate the transient effect accurately.

Moreover, the conversion ratios significantly depend on the frequency, which agrees with Rittel et al. [7] study. Also, this study shows that the transient effect was predicted with a small error by considering precise boundary conditions and temperature-dependent thermal coefficients.

Acknowledgment

This work was supported by the Ministry of Education, Youth and Sports of the Czech Republic and the European Union European Structural and Investment Funds in the frames of Operational Program Research, Development and Education project Hybrid Materials for Hierarchical Structures (HyHi, Reg. No. CZ.02.1.01/0.0/0.0/16_019/0000843). This work was partly supported by the Student Grant Competition of the Technical University of Liberec under project No. SGS-2019-5072.

References

[1] T. Liu and Y. Xu, Magnetorheological Elastomers: Materials and Applications, Smart Funct. Soft Mater., 2019, doi: 10.5772/intechopen.85083.

[2] S.S. Kang, K. Choi, J. Do Nam, and H.J. Choi, Magnetorheological elastomers: Fabrication, characteristics, and applications, Materials (Basel)., vol. 13, no. 20, pp. 1–24, 2020, doi: 10.3390/ma13204597.

[3] R. Ahamed, S.B. Choi, and M.M. Ferdaus, A state of art on magneto-rheological materials and their potential applications, J. Intell. Mater. Syst. Struct., vol. 29, no. 10, pp. 2051–2095, 2018, doi: 10.1177/1045389X18754350.

[4] T.H. Nam, I. Petríková, B. Marvalová, and M.Y. Hdaib, Self-heating and dynamic mechanical behavior of silicone rubber composite filled with carbonyl iron particles under cyclic compressive loading, J. Compos. Mater., p. 002199832110370, 2021, doi: 10.1177/00219983211037055.

[5] S.B. Ratner, V.I. Korobov, and S.G. Agamalyan, Mechanical and thermal fracture of plastics under cyclic strains, Sov. Mater. Sci., vol. 5, no. 1, pp. 66–70, 1972, doi: 10.1007/BF00721313.

[6] P.G. Pichon, M. Boutaous, F. Méchin, and H. Sautereau, Measurement and numerical simulation of the self heating of cross-linked segmented polyurethanes under cyclic loading, Eur. Polym. J., vol. 48, no. 4, pp. 684–695, 2012, doi: 10.1016/j.eurpolymj.2012.01.005.

[7] D. Rittel, On the conversion of plastic work to heat during high strain rate deformation of glassy polymers, Mech. Mater., vol. 31, no. 2, pp. 131–139, 1999, doi: 10.1016/S0167-6636(98)00063-5.

[8] A.K. Bazkiaei, K.H. Shirazi, and M. Shishesaz, Thermo-hyper-viscoelastic analysis of a rubber cylinder under cyclic deformation, J. Rubber Res., 2021, doi: 10.1007/s42464-020-00068-2.

[9] B. Marvalová, I. Petríková, and T.H. Nam, Fatigue of filled rubber due to hysteretic heating, Const. Model. Rubber XI - Proc. 11th Eur. Conf. Const. Model. Rubber, 2019, pp. 374–378, 2019, doi: 10.1201/9780429324710-65.

[10] F.P. Incropera, Fundamentals of Heat and Mass Transfer. Hoboken, NJ, USA: John Wiley & Sons, Inc., 2006.

[11] X. Balandraud and J.B. Le Cam, Some specific features and consequences of the thermal response of rubber under cyclic mechanical loading, Arch. Appl. Mech., vol. 84, no. 6, pp. 773–788, 2014, doi: 10.1007/s00419-014-0832-3.

[12] J.R. Samaca Martinez, J.B. Le Cam, X. Balandraud, E. Toussaint, and J. Caillard, Mechanisms of deformation in crystallizable natural rubber. Part 2: Quantitative calorimetric analysis, Polymer (Guildf)., vol. 54, no. 11, pp. 2727–2736, 2013, doi: 10.1016/j.polymer.2013.03.012.

[13] M.T. Loukil, G. Corvec, E. Robin, M. Miroir, J.B. Le Cam, and P. Garnier, Stored energy accompanying cyclic deformation of filled rubber, Eur. Polym. J., vol. 98, no. August 2017, pp. 448–455, 2018, doi: 10.1016/j.eurpolymj.2017.11.035.

[14] C. Multiphysics, Introduction to COMSOL multiphysics®, COMSOL Multiphysics, Burlington, MA, accessed Feb, vol. 9, p. 2018, 1998.

Crank-based Cycling Powermeter – Construction and Validation

J. Heczko[1,*], J. Krystek[1], T. Kroupa[1]

[1] *NTIS - New Technologies for the Information Society, Faculty of Applied Sciences, University of West Bohemia, Technicka 8, 301 00 Pilsen, Czech Republic*

* *jheczko@ntis.zcu.cz*

Abstract: We describe a single-sided crank-based cycling powermeter intended for mountainbike (MTB) use. The device is based on a printed circuit board (PCB) supplied by its manufacturer, but its assembly differs from the official instructions. The main difference is that the strain gauges are applied on the inside of the crank spindle. This was achieved by using a dedicated clamp that ensures easy application and precise positioning of both of the strain gauges relative to each other and to the other parts of the assembly as well. Other aspects of our design include mainly space considerations and waterproofing. The validation step was performed by comparing power readings to a consumer-grade indoor trainer under various riding conditions. The most important outcomes are, therefore, the design of the clamp used for gluing the strain gauges and the potential in teaching mechanics using a complex project.

Keywords: bicycle powermeter; design; validation.

1 Introduction

Powermeters are nowadays ubiquitous in almost any form of competitive cycling, their use ranging from track, triathlon and road racing to BMX, MTB and ultra-endurance events. The use of powermeters has been proven to bring many benefits, some being evident, such as objective effort quantification, better pacing or improving pedaling dynamics, while some being not so obvious, such as optimizing diet or aerodynamics [1, 2].

The working principle of different products may differ, but the most common one is to measure torque or force applied to a structural member of the drivetrain (such as pedal, crank arm, crank spindle, chainring spider or rear hub) and the corresponding angular velocity. The torque/force is usually measured by means of strain gauges. The design of a powermeter might provide measurement of total power output or the power output of each leg separately or the power output of one leg only (so-called single-sided powermeter), in which case it is assumed that both legs deliver the same amount of power.

Our choice of the working principle and design was driven by the availability of a ready-made electronic module provided by its manufacturer and its low price. It is a single-sided powermeter built into the inside of a hollow crank spindle. Our paper therefore describes the installation of the supplied module into a suitable crankset, its calibration and validation of power readings.

2 Design, manufacture, and assembly

We obtained the SG53 powermeter module and strain gauges, see Fig. 1, from their manufacturer, Sensitivus Gauge, within their project "DIY power meter". The website of the project [3] provides basic instructions for the build and calibration of the powermeter. The suggested procedure, however, includes placing the strain gauges to the outside of the crank spindle and routing the wires through a 2 mm hole drilled in the spindle. We rejected this design to avoid premature failure of the crankset in the harsh conditions of MTB racing (both in terms of loading and environment). Instead, we decided to place the strain gauges on the inside surface of the spindle, which results in all the components being sealed inside and no additional weakening of the structure. This approach, however, brings the following problems:

- to bond the strain gauges reliably from inside the crank spindle (inner diameter 20 mm);
- to position the two strain gauges relative to each other with sufficient precision.

Fig. 1: The parts supplied by the electronics manufacturer: strain gauges (top), lithium-ion battery (left), SG53 powermeter module (right). Scale shown for size comparison.

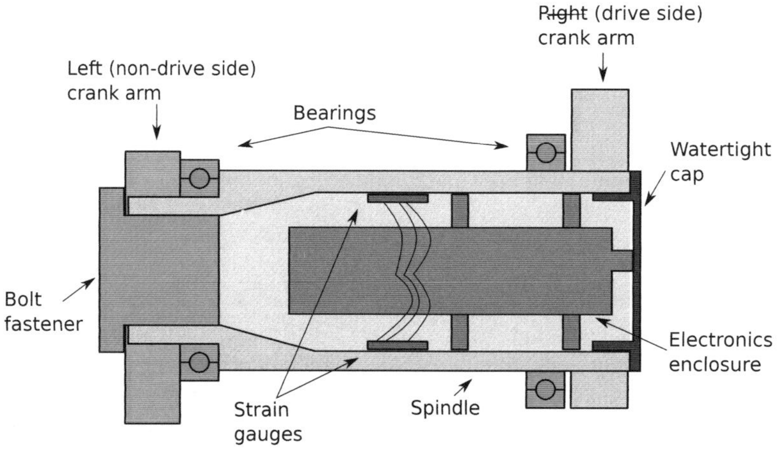

Fig. 2: Schematic depiction of the Sram GXP bottom bracket with the powermeter assembly.

Other challenges are common with the official design:

- to design the parts in a way that they can be assembled together;
- waterproofing;
- placing the antenna so that the wireless signal is not blocked.

The powermeter was intended to be built into a basic Sram S1000 crankset that uses the Sram GXP bottom bracket. The interface between bottom bracket bearings and the spindle is of 24 mm diameter on the drive side and 22 mm on the non-drive side (see Fig. 2). Both the radial and the axial loads are transmitted to the non-drive side bearing whereas only radial loads are transferred to the drive-side bearing. The inner diameter of the hollow spindle is 20 mm on the drive-side opening and decreases toward the threaded non-drive side end.

Fig. 2 provides a schematic depiction of the assembly. The main parts are the strain gauges, wiring, an enclosure for the electronics, and a watertight cap. The exact shapes of the enclosure and the cap were developed using 3D printed prototypes and a dummy copy of the spindle. The dummy copy was used for dry-fitting the parts together and testing the assembly process. See Fig. 3 for an overview of the prototypes.

The only resource that the authors could find regarding placement of strain gauges inside a hole uses an inflatable membrane [4]. We propose an alternative approach, in which proper placement of the strain gauges is

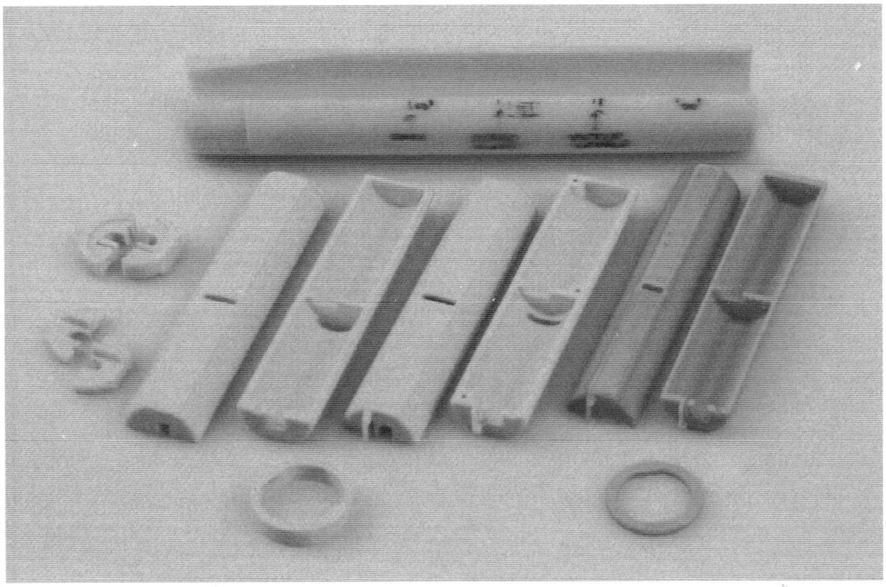

Fig. 3: Prototypes of the enclosure, end cap, and the dummy copy of the spindle.

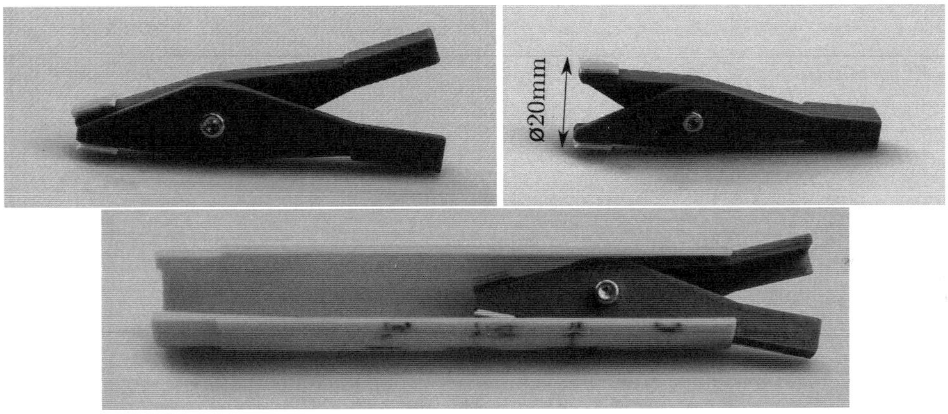

Fig. 4: Clamp for gluing the strain gauges.

achieved by using the custom-made clamp shown in Fig. 4. Our clamp is, again, 3D printed with a small amount of soft padding at the ends that ensure uniform pressure distribution onto the strain gauges during the gluing process. The strain gauges were bonded to the padding by a small amount of double sided tape, which released easily after the strain gauges bonded to the spindle so that the clamp might have been removed. Moreover, the surface of the padding was covered by a plastic foil that did not stick to the strain gauges glue, thereby reducing the chances of damaging the strain gauges while removing the clamp. The shape of the clamp enables to fix the outer end until the glue cures and also fits the spindle in a way that ensures that the strain gauges are placed in the exact desired distance from the outer end of the spindle.

3 Calibration and validation

The build of the powermeter is finished by its calibration, i.e. setting the internal parameters of its firmware in a way that the power readings are accurate. Next is the validation by comparing its power readings to another calibrated and trusted powermeter. These steps are described in detail in the following two paragraphs.

3.1 Calibration

Calibration was performed according to the official recommendations, i.e.:

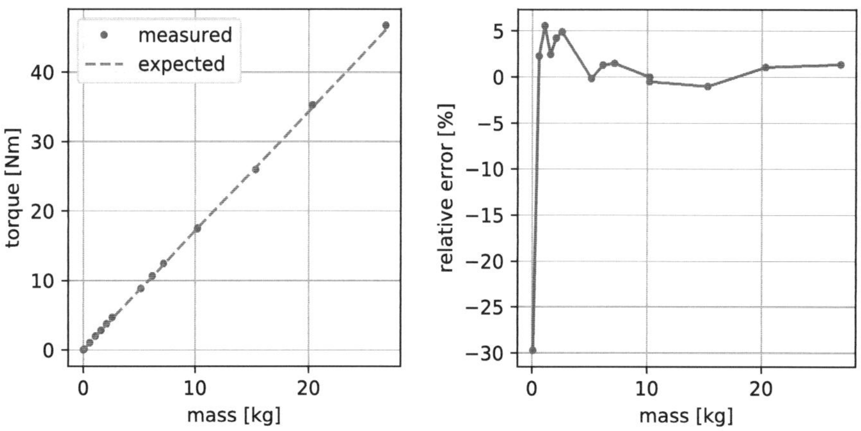

Fig. 5: Comparison of measured and expected torque values with different calibration weights.

1. Zero the powermeter with no load on the pedals using the developers mobile app.
2. Put the cranks into horizontal position.
3. Hang a known weight, $m \in \langle 10, 20 \rangle$ kg, to the left crank arm.
4. Read the value of torque, M_{measured}, in the mobile app.
5. Compare that value to the expected value of torque

$$M_{\text{expected}} = m\, g\, l\,, \tag{1}$$

   where $g = 9.81\,\text{N/kg}$ is the gravity of Earth and $l = 0.175\,\text{m}$ is the length of the crank arm.
6. Set the calibration parameter in the app to $M_{\text{expected}}/M_{\text{measured}}$.

We calibrated the powermeter using $m = 10.12\,\text{kg}$ and then compared the torque readings at different weights, see Fig. 5. The results show linear behavior and acceptable relative error of less than $2\,\%$ for $m > 5\,\text{kg}$. With heavier weights ($m > 10\,\text{kg}$), however, it became increasingly difficult to perform the measurement due to balancing, positioning issues and noticeable deformation of the bicycle frame. For this reason, it is good practice to use a dedicated test stand for powermeter calibration.

3.2 Validation

The Wahoo Kickr Core indoor trainer was used for comparison of the power readings. It is mounted in place of the rear wheel to the bicycle (see Fig. 6), uses a $5.44\,\text{kg}$ flywheel, an electromagnetic break, and estimates pedaling cadence from slight changes in power output. Its claimed accuracy is $\pm 2\,\%$ and maximum measurable power $1800\,\text{W}$.

The obvious consequences of this testing setup are:

- Pedaling asymmetry results in errors in power readings due to the tested powermeter being single-sided.
- Friction losses in the drivetrain may cause a slight difference in power readings.

The first issue is an inherent property of the design of the tested powermeter and cannot be eliminated. Friction losses were reduced by cleaning the drivetrain and lubricating the chain (using a paraffin wax bath) prior to testing.

Since the head-unit (cycling computer) can only be paired with one powermeter at a time, a mobile app was used to record the data from the indoor trainer over the Bluetooth Low Energy protocol. Each recording was then saved in a `.fit` file and read using the `fitparse` python library [5].

Fig. 7 shows the recorded quantities, namely power and pedaling cadence. The ride starts with a 10-minute warm-up, which includes easy riding around $100\,\text{W}$, variations in pedaling cadence ($50 - 110\,\text{RPM}$), single legged pedaling, and a short sprint ($> 700\,\text{W}$). The single legged efforts are clearly recognizable by: (1) the

Fig. 6: The bicycle and indoor trainer as used for validation (top). Close-up on the installed powermeter with the watertight end cap removed (bottom).

pedaling cadence estimated by the indoor trainer being dropped to about one half and (2) the power readings of the tested powermeter being dropped to zero during the right leg effort and being increased twice during the left leg effort. There is also a discrepancy at high cadence, say above 100 RPM. This might have to do with altered pedaling dynamics or asymmetry and will be investigated further.

After a short pause, the test continues by several 30 s-intervals at intended constant power output from 50 W to 350 W, spaced by 50 W. Data from this part of the test are shown in Fig. 8. The power readings show good agreement in general. One noticeable aspect is that it becomes more difficult to control the power output at harder efforts. This is mostly caused by insufficient training of the rider, but increasing fatigue and changing gears also play a role.

The final stage of the test consists of easy riding at 100 W with a short maximum effort sprint in the middle. Data from the sprint are shown in Fig. 9 and display very good agreement considering the achieved power output of more than 800 W. The slight disagreement is most probably caused by altered pedaling dynamics due to the effort, by different data collection on the two recording devices, and by the cadence calculation together with power averaging.

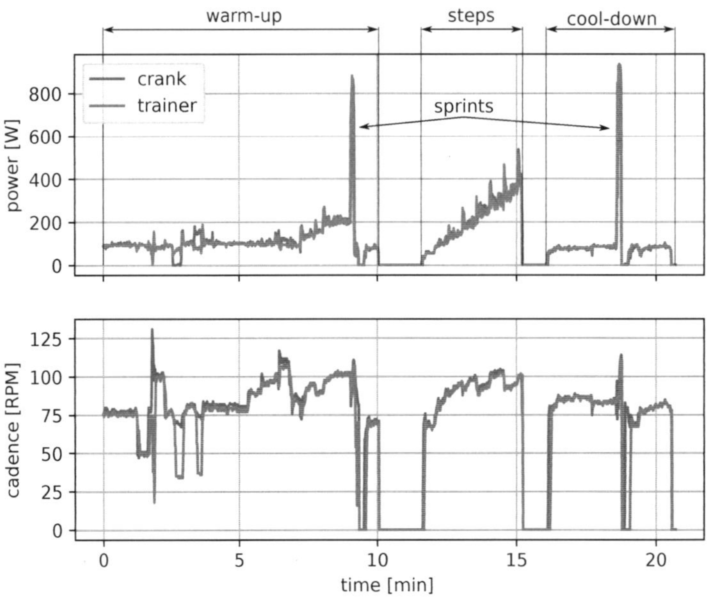

Fig. 7: Data from the validation ride.

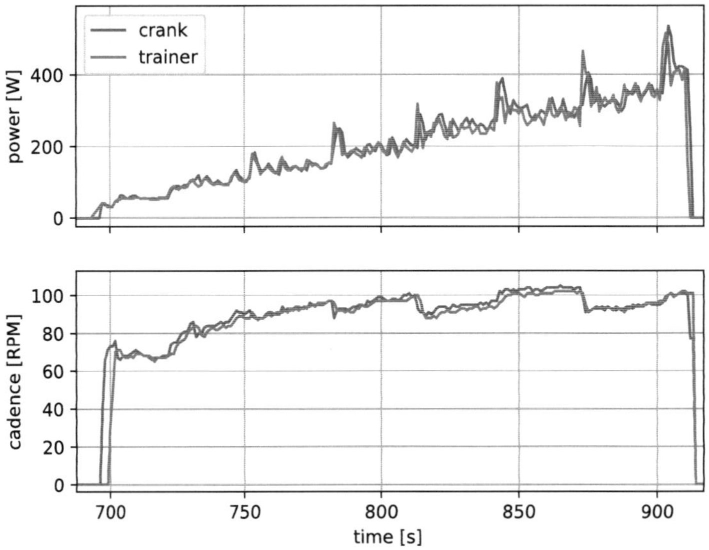

Fig. 8: Data from the validation ride – steps at constant power.

4 Summary

The cycling powermeter build described in this work resulted into several valuable outcomes. The first one is the development and testing of the described method of placing a set of strain gauges inside a deep narrow tube using a dedicated clamp. Another valuable result is the design of a waterproof end cap that was 3D printed from a flexible material. This approach proved superior to the use of o-rings in the context of spatial constraints of the assembly and mechanical durability. Our solution is, however, not fully tested yet as long-term degradation of the 3D printed material in outdoor conditions might prove crucial.

Another point worth emphasizing is the benefit of similar projects for one's technical skills and understand-

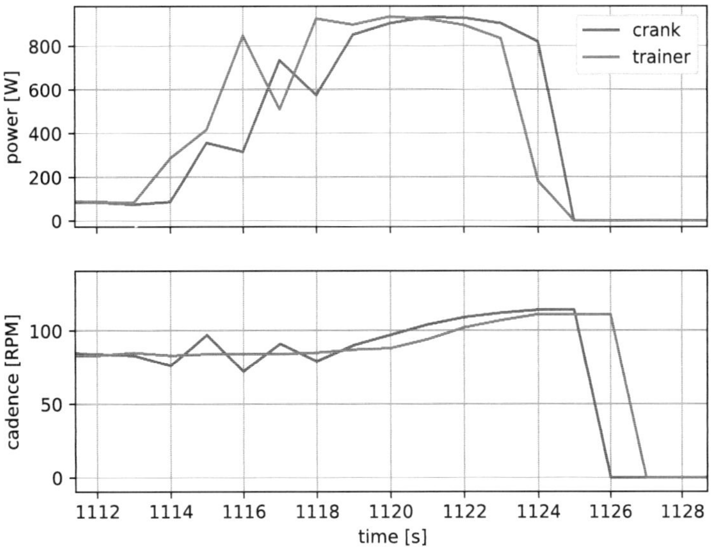

Fig. 9: Data from the validation ride – maximum effort sprint.

ing of mechanics. The key property is the complexity of the project, which means that the student needs to combine knowledge and skills from different areas – e.g. 3D drawing, industrial design, electronics, theory – and even gain some completely new ones. The difficulty of the project may be tailored in order to match the skills of the students and expected gains, e.g. by using (or not using) ready-made components (such as the SG53 module in our case) or by specific guidance (or the lack of it) regarding design or manufacture of the parts. Validation is an important step of the project in order to evaluate the design choices and other aspects of the prototype.

Future development and use of these results may include:

- use of similar clamps for strain gauges placement in production;
- changes in firmware in order to fully analyze the collected data, including accelerometer, pedaling smoothness, and torque effectiveness data;
- use of similar projects in teaching experimental mechanics and industrial design;
- development of own PCB module, possibly extending it for independent measurement of power output of both legs;
- use of power measurement in biomechanics for testing the hypotheses of muscle activation or the mechanisms of muscle fatigue;
- use of power measurement in sports education and medicine.

References

[1] H. Allen, A. Coggan, S. McGregor, Training and Racing with a Power Meter, 3rd ed. Boulder: VeloPress, 2018, ISBN 9781937715939.

[2] S. Jobson, D. Irvine, Ultra-Distance Cycling: An Expert Guide to Endurance Cycling, Bloomsbury Sport, 2017, ISBN 9781472919878.

[3] Sensitivus, DIY Power Meter, 2020, https://sensitivus.com/products/diy-power-meter/.

[4] Vishay Precision Group, VMM-6 How to Install a Strain Gage in a Small, Deep Hole, 2010, https://micro-measurements.com/knowledge-base/application-notes.

[5] D. Cooper, Python-fitparse 1.2.0, 2021, https://dtcooper.github.io/python-fitparse/.

Validation of Methodology for Numerical Modeling of 3D Printed Structures Reinforced by Long Carbon Fibers

L. Horňáček[1], R. Halama[1,*], D. Krzikalla[1,2], J. Měsíček[1], J. Pauldoss[1], N. James[1]

[1] *VSB-Technical University of Ostrava, FME, 17. listopadu 15, 708 00 Ostrava, Czech Republic*

[2] *Tatra Trucks, Areál Tatry 1, 742 21 Kopřivnice, Czech Republic*

[*] *radim.halama@vsb.cz*

Abstract: This paper describes the main experimental and numerical results gained for 3D printed composites prepared by FDM technology. The matrix is considered from onyx – short carbon fiber reinforced nylon. In a dual nozzle system, long carbon fibers are prepared in the required orientation. Realized tensile tests were used to identify the elastic properties of the matrix and long carbon fibers. Gained experimental stiffness of the shear specimen under three loading modes was used to verify the proposed numerical strategy, making possible the introduction of contour lines of carbon fibers in the computational structural analysis.

Keywords: 3D print; carbon; electronic speckle pattern interferometry; digital image correlation; FEM.

1 Introduction

Additive Manufacting (AM), the main technology behind manufacturing of these structures is the method of linking materials to result in a 3D model. This technology makes it easier to construct larger ranges of prototypes or components with much complex geometrical structures or those structures which are quiet complex to be manufactured by conventional methods. There are various techniques used behind Additive Manufacturing, but in this case, the Fused Deposition Modelling (FDM) technique is used to fabricate the parts due to its low wastage and ability for material change post-fabrication [1]. There are many industrial applications of 3D printed composites due to their excellent strength-to-weight and stiffness-to-weight ratios.

The 3D printed composite materials bring the required properties. Mark forged X7 printer offers a unique possibility of printing composites, in which a plastic matrix, applied using FDM technology and a reinforcing fiber is continuously inserted into it through a dual nozzle. The laser unit controls the quality and height of the layer during printing to ensure the highest possible dimensional accuracy. 3D printer offers combining a matrix material in the form of nylon, nylon reinforced with short carbon fibers (Onyx), or nylon with additives against burning (Onyx-FR) or electrostatic discharge (Onyx-EDS) with carbon, glass, or aramid long fibers. Parts can be printed as solid or filled with a honeycomb structure. Structural parts with high demands on accuracy and durability can be produced. However, the elastic properties and numerical modeling approaches of 3D printed composites are still open questions. Some useful results obtained for onyx were published in [2–4]. Mechanical properties of 3D printed composite materials with carbon fibers were tested in tension [5] as well as flexural and indentation conditions [6].

This contribution presents the current state of mechanical testing realized on specimens made of onyx reinforced by long carbon fibers (CF) at VSB-Technical University of Ostrava. Tensile tests were realized considering different fiber volume fractions to estimate the elastic properties of the matrix and fibers. Then, a shear-type specimen was used in the measurements via optical methods to gain a basis of validation data for developing a suitable numerical modeling approach.

2 Analytical approach for tensile load

Analytical approach is an easy and quick way of Young's modulus calculation for various volume fractions. Estimation of Young's modulus for longitudinal and lateral loading of a unidirectional fiber composite are depicted below.

2.1 Expression for longitudinal loading

Let us consider the tensile load of F_c acting along the vertical direction in the longitudinal fiber direction as shown in the Fig. 1. The Fig. 1 shows the blue region which is the carbon fiber orientation direction, and the white region is the Onyx region. Because of the application of the tensile load, the composite tends to elongate with the distance of δL as shown in the Fig. 1. Crucial idea is that both the fibers and the matrix are experiencing the same elongation along the loading direction, which leads to establish the relation as follows.

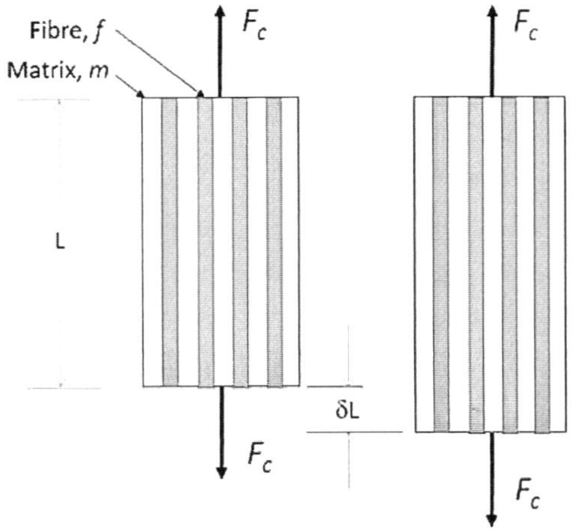

Fig. 1: Unidirectional fiber in Longitudinal direction.

Equilibrium of the forces is described by the equation

$$F_C = F_F + F_M, \tag{1}$$

where F_C, F_F and F_M are the total force in composite, the force in carbon fibers, and the force in the matrix, respectively. After substituting the relationship between an axial stress and corresponding cross-sectional area

$$\sigma_C A_C = \sigma_F A_F + \sigma_M A_M, \tag{2}$$

where $\sigma_C, \sigma_F, \sigma_M$ are stresses acting in matrix, fiber and composite respectively, and A_C, A_F, A_M are cross-sectional areas of composite, long fibers and matrix, respectively. Due to the geometrical constraints total strain in the composite ε_C is the same as in fibers ε_F and matrix ε_M, i.e.

$$\varepsilon_C = \varepsilon_F = \varepsilon_M. \tag{3}$$

The stress-strain relationship is given by

$$\begin{aligned} \sigma_c &= E_C \varepsilon_C \\ \sigma_F &= E_F \varepsilon_F \\ \sigma_M &= E_M \varepsilon_M \end{aligned} \tag{4}$$

where E_M, E_F, E_C is Young's modulus obtained in matrix, fiber and composite respectively, By combining the Eqs. 4 with the Eq. 2, we get

$$E_C \varepsilon_C A_C = E_F \varepsilon_F A_F + E_M \varepsilon_M A_M. \tag{5}$$

The volume fraction of fibers can be expressed by

$$V_F = \frac{v_F}{v_C} = \frac{A_F\, L}{A_C\, L} = \frac{A_F}{A_C}, \tag{6}$$

where v_F, v_C are volumes of fibers and the whole composite respectively. The volume fraction of matrix V_M could be expressed analogously considering the volume of matrix v_M. Then, considering V_M, V_F as volume fractions of matrix and fibers respectively, the rule of mixture can be stated

$$E_C = E_F V_F + E_M V_M. \tag{7}$$

Similarly, strength equation

$$\sigma_C = \sigma_F V_F + \sigma_M V_M \tag{8}$$

and the equation for ultimate strength

$$\sigma_{C,ult} = \sigma_{F,ult} V_F + \sigma_{M,ult} V_M, \tag{9}$$

where $\sigma_{M,ult}, \sigma_{F,ult}, \sigma_{C,ult}$ are the ultimate stresses obtained in matrix, fiber and composite respectively.

2.2 Expression for transversal loading

Let us consider the tensile load of F_C acting along the vertical direction perpendicular to that of the fiber direction (Lateral direction) as shown in the Fig. 2. The Fig. 2 shows the blue region which characterizes the carbon fiber orientation direction, and the white region is the Onyx region. Both the fibers and the matrix are experiencing the same force and stress along the loading direction, which leads to establish the relation analogously as for the longitudinal direction.

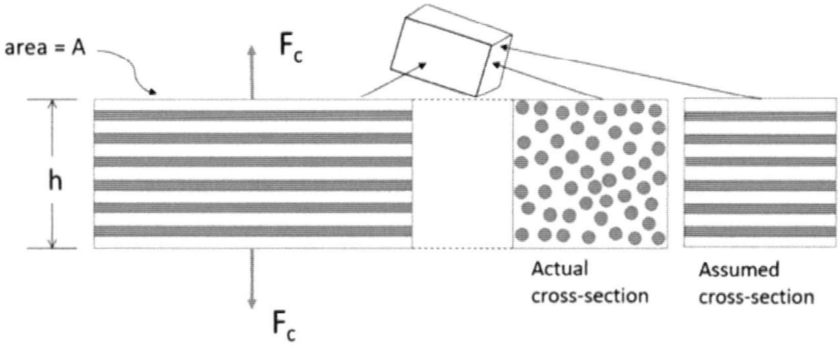

Fig. 2: Unidirectional fiber in Transversal direction.

Equilibrium of the forces:

$$\begin{aligned} F_C = F_F &= F_M, \\ \sigma_c = \sigma_F &= \sigma_M. \end{aligned} \tag{10}$$

Geometry of deformation: Total strain in the composite can be written as

$$\varepsilon_C = V_F \varepsilon_F + V_M \varepsilon_M. \tag{11}$$

The stress-strain relationships are given by

$$\begin{aligned} \sigma_C &= E_C \varepsilon_C, \\ \sigma_F &= E_F \varepsilon_F, \\ \sigma_M &= E_M \varepsilon_M, \end{aligned} \tag{12}$$

where $\varepsilon_M, \varepsilon_F, \varepsilon_C$ are strains obtained in Matrix, Fiber and composite respectively; $\sigma_M, \sigma_F, \sigma_C$ are stresses obtained in Matrix, Fiber and composite respectively; E_M, E_F, E_C is Young's modulus obtained in Matrix, Fiber and composite respectively; V_M, V_F, V_C are volume fractions obtained in Matrix, Fiber and composite respectively.

Combining the above Eqs. 12 with the Eq. 11, we get

$$\frac{\sigma_C}{E_C} = \frac{\sigma_F V_F}{E_F} + \frac{\sigma_M V_M}{E_M}. \tag{13}$$

Axial stress is the same in each cross-section

$$\sigma_C = \sigma_F = \sigma_M. \tag{14}$$

Therefore, the rule of mixture

$$E_C = \frac{E_F E_M}{V_M E_F + V_F E_M}. \tag{15}$$

The above Eqs. 7 and 15 are used to perform the analytical prediction of Young's Modulus particularly for unidirectional fibre reinforced composite.

3 Experiments in tensile mode

All mechanical tensile tests were performed using LabControl 100kN/1000Nm testing machine in accordance with ASTM D3039 standard [7]. Young's modulus obtained for the longitudinal and transversal direction of fibers was evaluated for different long fiber volume ratios, see Fig. 3.

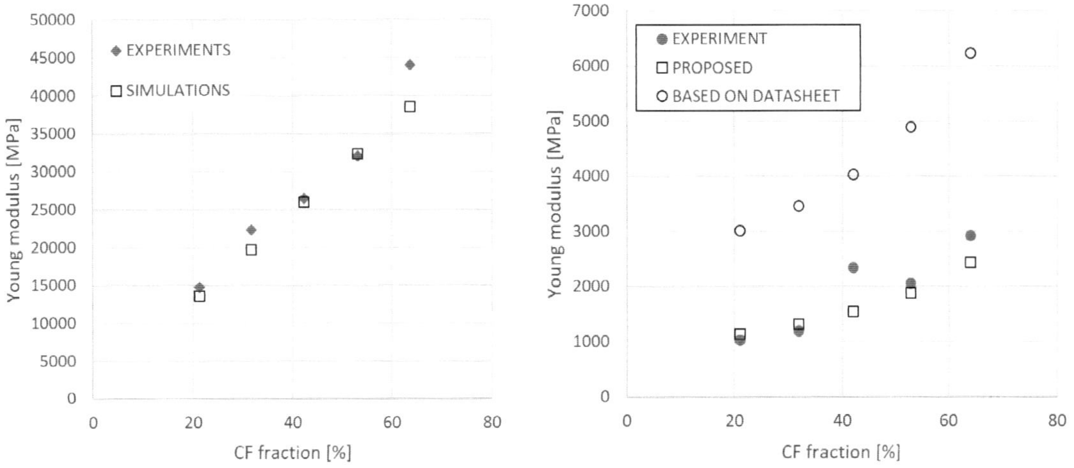

(a) Longitudinal

(b) Transverse

Fig. 3: Dependency of Young's modulus of Onyx+CF composite on the CF fraction.

Fig. 3 shows that the correlation is very good for all loading cases for proposed simulation in case of longitudinal loading. In case of transversal loading, the proposed and experimental correlations are good but the datasheet values are giving poor results. The proposed values from our experiment are reliable in case of transversal loading. The Young's modulus of fiber and matrix utilized in the prediction is 60,000 MPa and 900 MPa respectively.

4 Experiments on shear specimen

Based on the ASTM D7078 Standard [8], the shear testing specimen was designed, 3D printed and tested. Two contours carbon fibers were created around the perimeter in Eiger software. The shear-type specimen has been tested on Testometric 50CT-500 universal testing machine. All tests were conducted capturing full field data by Digital Image Correlation (DIC) system Mercury RT. Dantec Dynamics Electronic Speckle Pattern Interferometry (ESPI) system Q100 was used for verification on a shear-type specimen subjected to tensile load. The stiffness of specimen considering line with the length of 27.4 mm was 1864.2 N/mm as measured by ESPI and 1795 N/mm as measured by DIC. In other two loading cases, the stiffness of specimens made of onyx + two contour lines of CF was studied based on DIC data only.

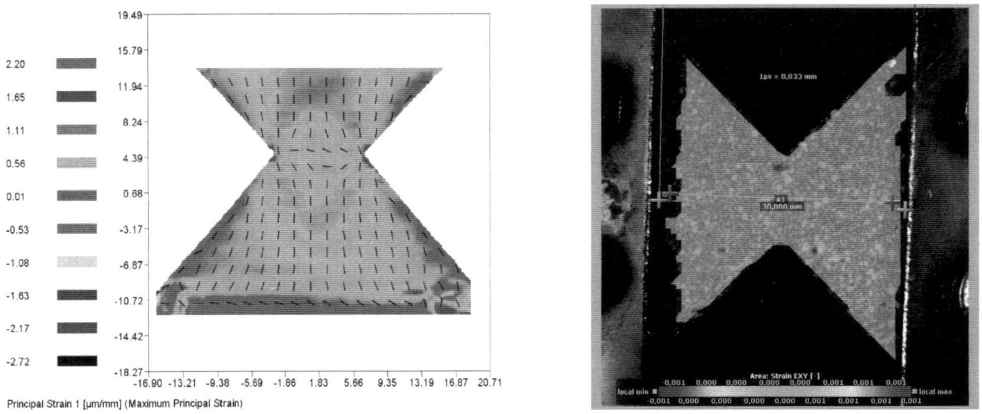

(a) ESPI – tensile load (F = 41 N)

(b) DIC – shear load (F = 22 N)

Fig. 4: Strain contours from full-field measurements on Onyx+CF composite.

5 Numerical modeling

FE modeling was performed in ANSYS 2020R1. The mesh prepared for the shear specimen is shown in the Fig. 4. In total, 4956 of solid185 elements were used for characterization of matrix and 216 of BEAM188 elements were used for long carbon fibers creation.

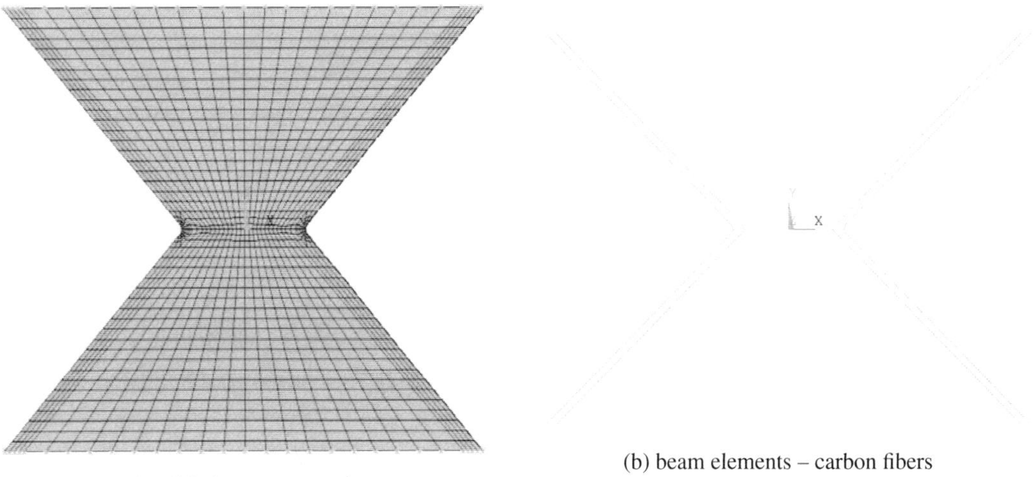

(a) solid elements – matrix

(b) beam elements – carbon fibers

Fig. 5: FE model of shear specimen characterizing Onyx+CF composite.

Specimen model was subjected to tensile, shear and combined loading using Ansys APDL macro. All nodes on the bottom area were fixed. Appropriate boundary conditions were applied on all nodes lying on the top area. Comparison of numerical and experimental results is done in Tab. 1.

Tab. 1: Comparison of experimental and numerical results of specimen's stiffness.

Loading mode	Experiment [N/mm]	FEM [N/mm]	Error [%]
tensile	1,795	1,890	5
shear	904	977	8
combination	2,354	2,474	5

6 Conclusion

All experiments realized on tensile specimens were used to identify the elastic properties of Onyx and carbon for subsequent Finite Element modeling. The most important parameter is Young's modulus. In order to reach good correlation of mixture rule with experiments for the longitudinal and transversal direction of fibers evaluated for different long fiber volume ratios its optimized value for matrix and fibers has been found: E_M = 900 MPa and E_F = 60,000 MPa.

Numerical results of shear specimen simulations for three loading modes are in good correlation with performed experiments. The developed computational strategy is based on solid elements usage for matrix and beam element application for long carbon fibers. Authors focused on linear elastic structural analysis, which is important for engineering applications. However, Onyx is a viscoplastic material and it is necessary to take it into account in the case of higher loads [4].

Acknowledgement

This research was supported by the project Innovative and additive manufacturing technology – new technological solutions for 3D printing of metals and composite materials, reg. no. CZ.02.1.01/0.0/17_049/0008407 financed by Structural Funds of Europe Union. It has been also created in connection with SGS2021/66 of specific research from Ministry of Education, Youth and Sports of the Czech Republic.

References

[1] F. Ning, W. Cong, J. Qiu, J. Wei, S. Wang, Additive manufacturing of carbon fiber reinforced thermoplastic composites using fused deposition modelling. Composites Part B 80 (2015) 369–378, www.elsevier.com/locate/compositesb.

[2] M. Saga, F. Barnik, M. Vasko, M. Handrik, P. Kopas, Identification of Physical Characteristic of Composite Materials Produced by Additive Technology from Perspective of Selected Mechanical Properties, Acta Physica Polonica A 138 (2020) 249–252, doi: 10.12693/APhysPolA.138.249.

[3] J.S.D.P. Pauldoss, Modeling of Mechanical Tests on Composite Specimens Produced by 3D Printing. Master thesis, Ostrava: VŠB - Technical University of Ostrava, Faculty of Mechanical Engineering, Department of Applied Mechanics, 2021, 68 p.

[4] R.A.J.A. Nicholas, Investigation of stress-strain behaviour in ONYX material and its modelling. Master thesis, Ostrava: VŠB Technical University of Ostrava, Faculty of Mechanical Engineering, Department of Applied Mechanics, 2020, 57 p.

[5] H. Al Abadi, H.-T. Thai, V. Paton-Cole, V. I. Patel. Elastic properties of 3D printed fibre-reinforced structures. Compos. Struct. 193 (2018) 8–18, doi: 10.1016/j.compstruct.2018.03.051.

[6] G.D. Goh, V. Dikshit, A.P. Nagalingam, G.L. Goh, S. Agawarla, S.L. Sing, J. Wei, W.Y. Yeong, Characterization of mechanical properties and fracture mode of additively manufactured carbon fiber and glass fiber reinforced thermoplastics. Mater. Des. 137 (2018) 79–89, doi: 10.1016/j.matdes.2017.10.021.

[7] Standard test method for tensile properties of polymer matrix composite materials. ASTM D3039/D 3039M; 2017.

[8] Standard Test Method for Shear Properties of Composite Materials by V-Notched Rail Shear Method. ASTM D7078 / D7078M; 2020.

Reduction of Gearing Pressure Pulsations and Noise Level in Gear Pumps using Non-standard Gearings and FEA Analysis

J.M. Hradecký[1,*], M. Dub[1]

[1] *Czech Technical University in Prague, Faculty of Mechanical Engineering, Department of Designing and Machine Components, Technická 4, 166 07 Prague 6, Czech Republic*

* *jakub.hradecky@gmail.com*

Abstract: In industry all over the world there is a need to make gear pumps (which are included in many industry units and machines) more silent and without unnecessary noise and pulsations they producing. Main part of the gear pump noise is caused by meshing gearing while rotating in the pump (causing fluid transport) in discontinuous run and causing pressure pulsations. This article reveals approaches leading to lower noise level using transformation of gears such as asymmetric gearing or continuous contact gearing, which both providing different tension scatter and reduce level of tension in stressed areas of the tooth. In this article are also mentioned some of the tools which can analyze gear noise using FEA and analyze pulsations in numerical models.

Keywords: asymmetric gearing; continuous contact gearing; mechanical analysis; gear meshing; silent gear pump.

1 Introduction

Gear pumps [1] use rotating gearing with two same gears to transport working fluid in gear space from the inlet to the outlet. So the work is not continuous which results into pressure pulsations between co-occupying teeth. These pulsations are the main source of noise radiating from the gear pump. Reducing pulsations with standard solutions such as increase of number of teeth, bigger pressure angel or helical gears does not have satisfying results because of the limitations of standard involute gears.

Asymmetric gears provide possibility to gain number of teeth and pressure angle to the satisfying level without losing its load capacity. There is also a possibility to use non-involute gearing, such as continuous contact gearing which provides very different tension in the tooth and nearly zero pulsations.

New development experiments for verification of the design modifications can be expensive and take a lot of time. There are programs such as EgeMATor or HYGESim which can simulate these experiments and help with further optimalizations. Gear noise can be also analyzed with FEA. Tension results can be transformed thanks to the Euler-Lagrange analysis to have noise level results.

2 Standard solutions

Standard solutions to prevent from pulsations and noise are increase of teeth on each gear, increase the value of a pressure angle and using helical gear. Unfortunately, it is not enough for desired noise reduction. It is possible to see how big the decrease of noise level is when standard solution is used in [2], where a modification for a wheel loader was made.

The article reveals a difference where gear pump (originally with nine-teeth gearing) was transformed to gear pump with twelve-teeth gearing. According to the measurement original level of noise was 87.6 dB. After an increase of number of teeth, new value was demonstrably lower by 1.7 dB to the value of 85.9 dB. Decrease of level of meshing pressure was also proven. For twelve-teeth gearing was value lower by 30%.

Interesting solution came from Jihostroj Velešín company called T3T [3]. Design of a gearing can be seen in Fig. 1. Gear width was divided into tree parts, where axis of each tooth is shifted by 30°. Thanks to this

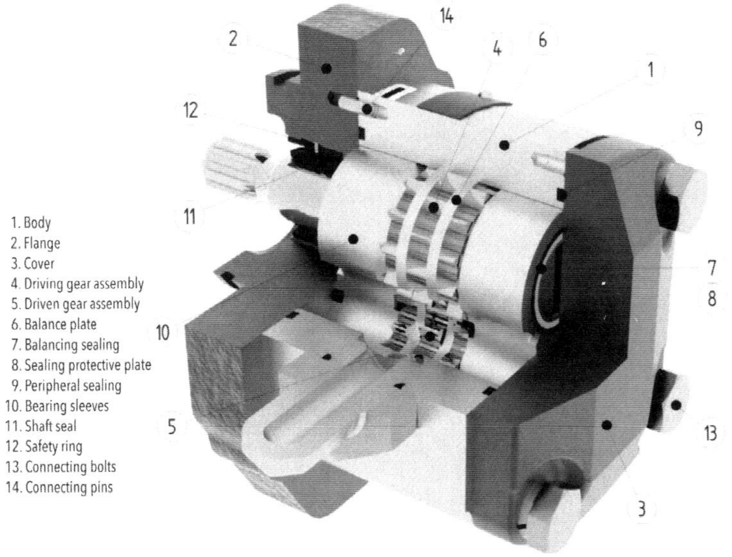

Fig. 1: Gear Pump Jihostroj T3T – lower pulsations [3].

the output flow goes more smoothly from the pump – exactly one third after another compared to the standard gearing. So, the pressure pulsations are partly eliminated. They have higher frequency but on the other hand much lower amplitude.

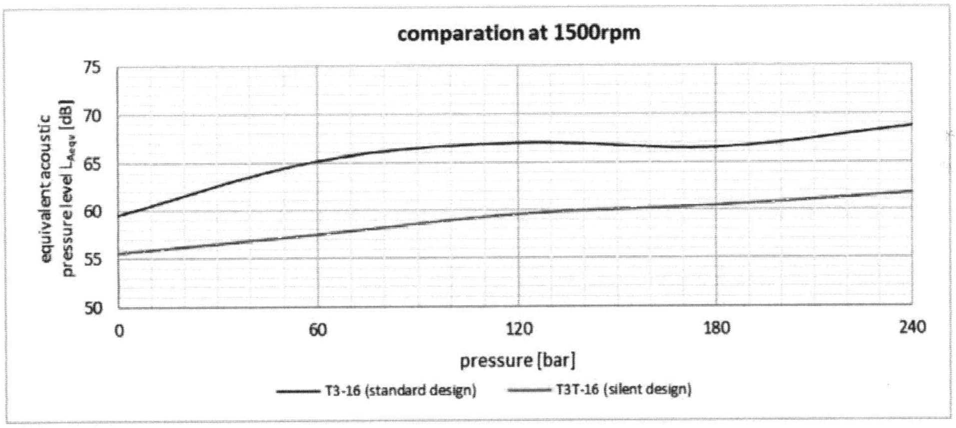

Fig. 2: Gear pump T3T noise reduction [3].

The result is noise reduction. In Fig. 2. there can be seen course of acoustic pressure in dB and output pressure from the pump, where there is clearly demonstrated a noise reduction roughly by 7 dB.

3 Asymmetric gearing

Increase of number of teeth on the gear leads to thinner teeth and so to the decrease of load capacity of each tooth. Asymmetric gearing is normally used to increase load capacity in applications where greater torque is needed (so a higher load capacity in dedendum of each tooth is provided). The main idea of using asymmetric gearing in gear pumps is to remain on the same level of load capacity (as it is on standard pump gearings) while increasing a number of teeth [4]. So, when number of teeth is increased, width of tooth dedendum is decreased. This decrease needs to be compensated to stay on the same level of load capacity per tooth. One of the solutions is asymmetric gearing which has different pressure angle on the working (meshing) side and cost side of the tooth.

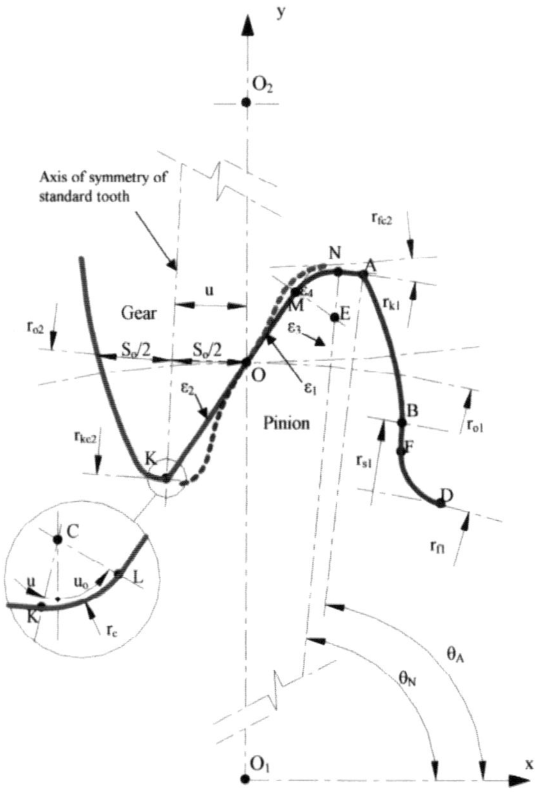

Fig. 3: Asymmetric gear geometry [4].

The level of asymmetry is expressed as k. For the gearing running clockwise and also anti-clockwise the asymmetry is not really great and the value of k is near 1. Tooth asymmetry leads to the different tension in tooth dedendum, which can be seen in Fig. 3.

About modification from standard to asymmetric gearing and their comparison is written for example in Optimized asymmetric gear tooth design [4]. There is a description of a modification to asymmetric gearing with standard pressure angle (20°) on meshing side of the tooth and a calculation which changes pressure angle on a coast side. The aim is to narrow width of an addendum and extend the width of a dedendum of a tooth.

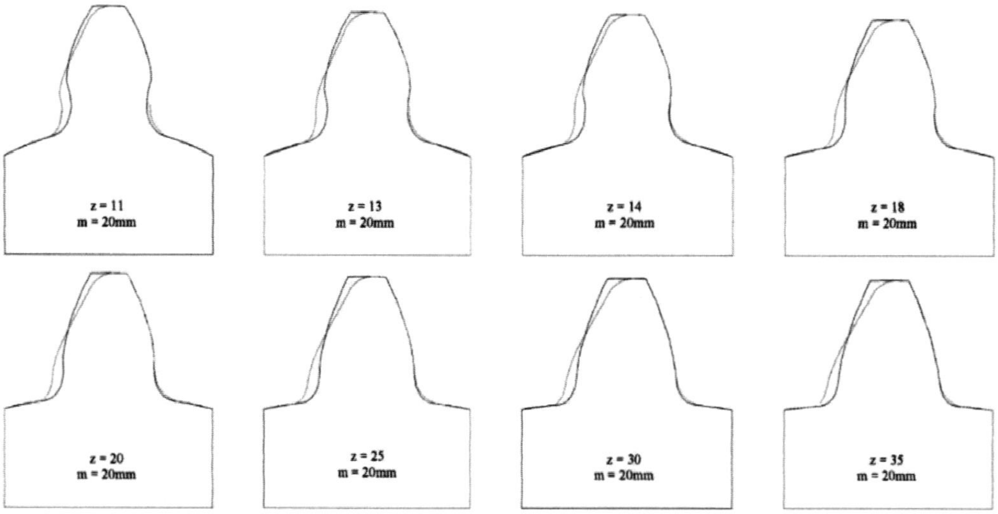

Fig. 4: Standard and asymmetric teeth comparison [4].

According to [5] for decrease of a bending tension in a tooth there should be small value of pressure angle on a meshing side. On the other hand, for decrease of meshing pressure, noise and vibrations there is a need to have bigger value of a pressure angle.

One of the advantages of asymmetric gearing in comparison to unconventional gearing is using standard production tools, assuming production of a gear is via cutting tools. In Fig. 4 there can be seen comparison of standard and asymmetric teeth for the same number of teeth in a gear and the same module.

The asymmetry causes changes in scatter tension in a tooth, which is shown in Fig. 5 with FEA analysis. It is clear, the areas of high tension are smaller in asymmetrical tooth, visual difference is best seen in dedendum of coast side of the tooth.

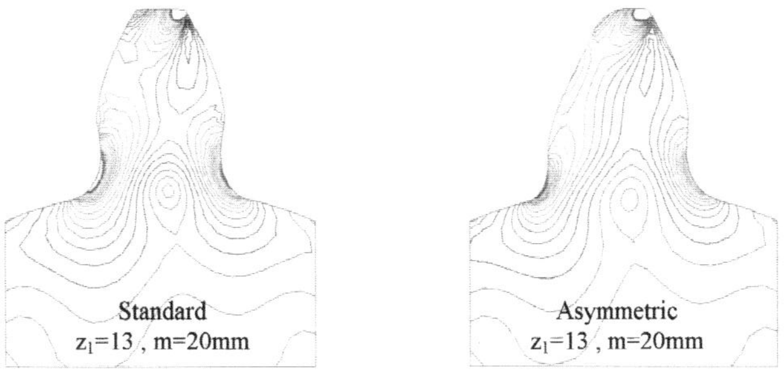

Fig. 5: Von-Misses stresses obtained by FEA [4].

Percentage decrease of maximum values of tension with asymmetric gears is available to see in next Fig. 6, where it is also described the reduction of tension is different for each number of teeth in a gear. The lower the number of teeth, the higher tension reduction.

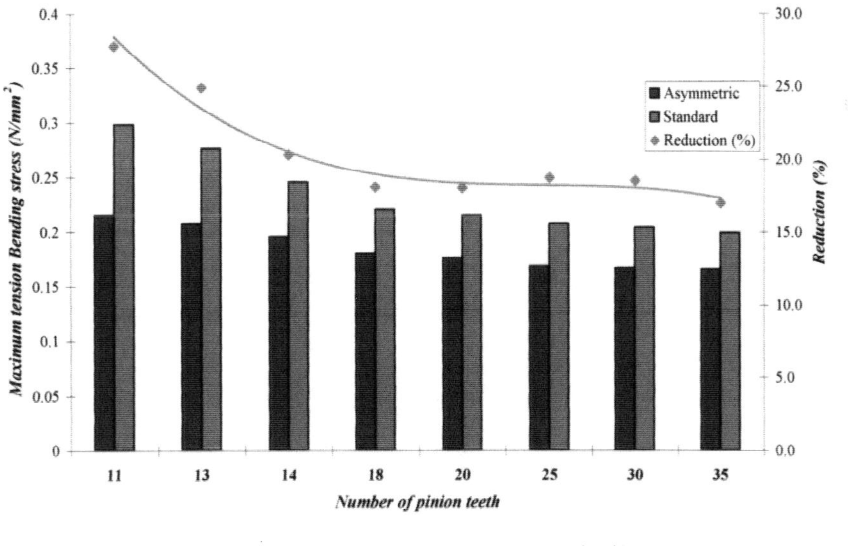

Fig. 6: Comparison of tension in a tooth [4].

So, thanks to using asymmetric gearing it is possible to stay on the same load capacity of a tooth while increase of number of teeth on a gearing (which requires decrease of width in tooth addendum). According to [6] it occurs to connection of gear advantages with smaller and bigger pressure angle. The tooth has advantages of wider dedendum and involute and at the same time reaches the height of the tooth higher than the standard one.

Asymmetric gearing in gear pumps, where bigger number of teeth is used, can be seen in Danfoss's products with marketing name Shhark. Thanks to the extreme asymmetry a nineteen-teeth gear pump was born from standard eleven-teeth pump (Fig. 7), which was consequently modified to the helical.

Danfoss company in their cataloque [7] declares the reduction of pulsations of 78%, how it is documented on flow-pulsation chart (Fig. 8). There is a comparison of a standard pump and Shhark shown. In their documentation there is also written a reduction of 10 dB of noise level thanks to the decrease of these pulsations.

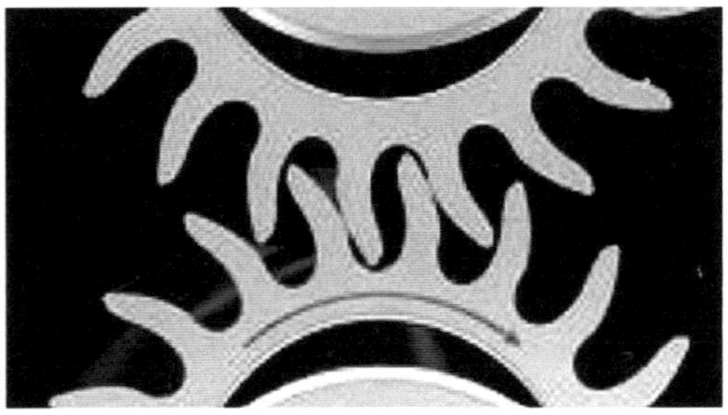

Fig. 7: Gearing – Shhark [7].

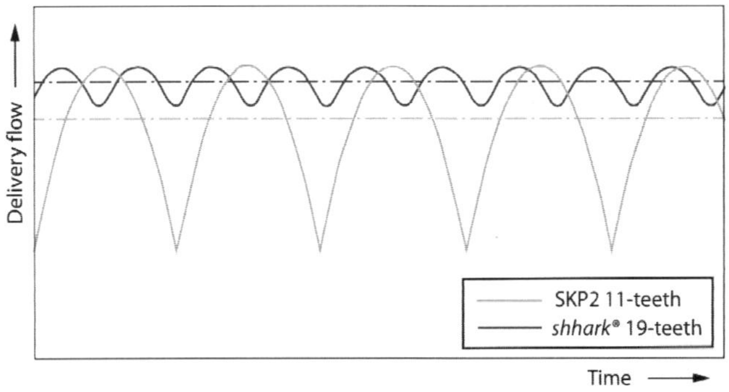

Fig. 8: Flow-pulsation comparison chart [7].

4 Continuous contact gearing

One of the newest phenomenon in silent gearing in gear pumps are continuous contact gears (Fig. 9). These pumps are offered for example by Settima [8] or Bosch Rexroth [9].

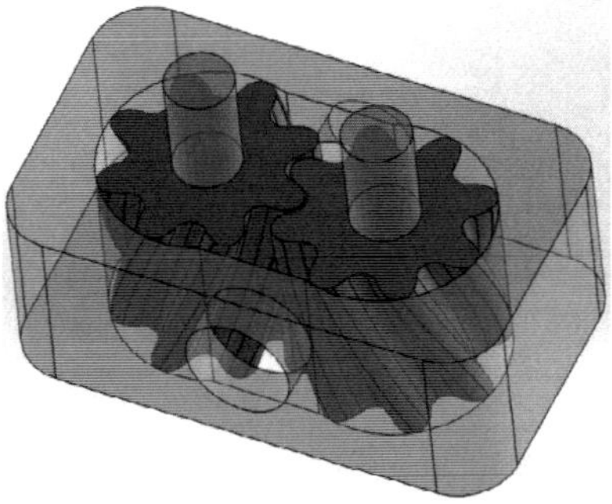

Fig. 9: Continuous contact gearing [10].

This type of gearing has its basis in Root-type blower [10], which is based on rotation of two cycloid elements. These elements are meshing in every point of their contact (Fig. 10).

Rotating elements are made of two arches. Top spline is possible to define using equations:

$$x_1 = \rho_r \sin\theta, \quad y_1 = a + \rho_r \cos\theta. \tag{1}$$

Base spline is compatible to Top spline an it is define as:

$$x_2 = \rho_r \sin(\theta - 2\phi) - a \sin 2\phi + 2r \sin\phi, \tag{2}$$

$$y_2 = \rho_r \cos(\theta - 2\phi) - a \cos 2\phi + 2r \sin\phi, \tag{3}$$

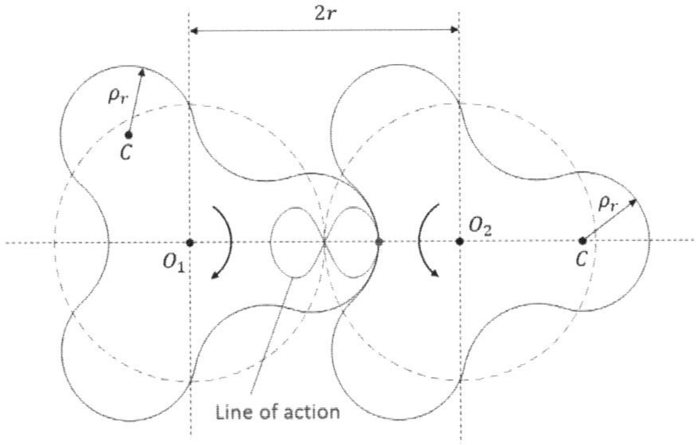

Fig. 10: Continuous gearing geometry [10].

where ρ is radius of an arch, a is the distance between middle of the arch and middle of the element and r is pitch radius. θ is a parameter of a gearing and ϕ is an angle defines element rotation. Two main parameters describing this gearing are number of teeth and ratio a/r. This ratio defines when the gearing is nearly flat or opposite – undercutted (Fig. 11). For preservation of convex-concave shape there is a need for the value of a ratio a/r to remain between 0.5 and 1.

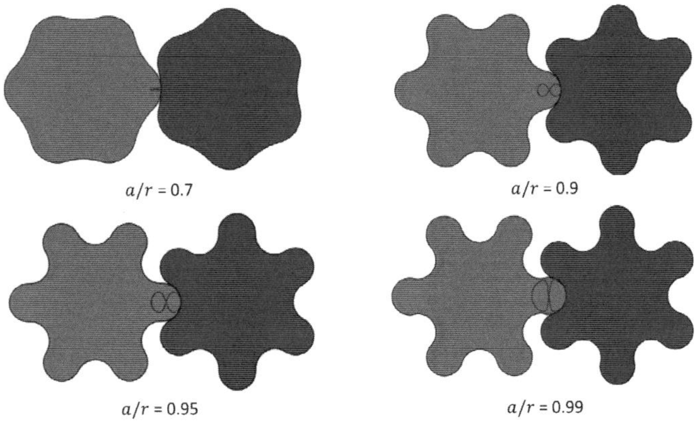

Fig. 11: Continuous contact gearing – different values of a/r [10].

With further modifications and implementation of the involute to the geometry of the gear (Fig. 12) there is a "low noise and pulsations" gearing provided, which combines advantages from involute gearing and gearing coming from Root-type blower. Top and base arch are the same, but the have involute spline connecting them.

Radius of these new arches is defined as:

$$\rho_r = r \cos \alpha \frac{\pi}{2N}, \tag{4}$$

where r is pitch radius, α is pressure angle and N is number of teeth. The position of ending points of the involute B and C is defined as:

$$\theta_B = \tan \alpha + \frac{\pi}{2N}, \tag{5}$$

$$\theta_C = \tan \alpha - \frac{\pi}{2N}. \tag{6}$$

For involute applies:

$$\mathrm{f_{inv}}(\theta) = r_b \begin{bmatrix} \cos(\phi - F) & -\sin(\phi - F) \\ \sin(\phi - F) & \cos(\phi - F) \end{bmatrix} \begin{bmatrix} \phi \\ 1 \end{bmatrix}, \tag{7}$$

where

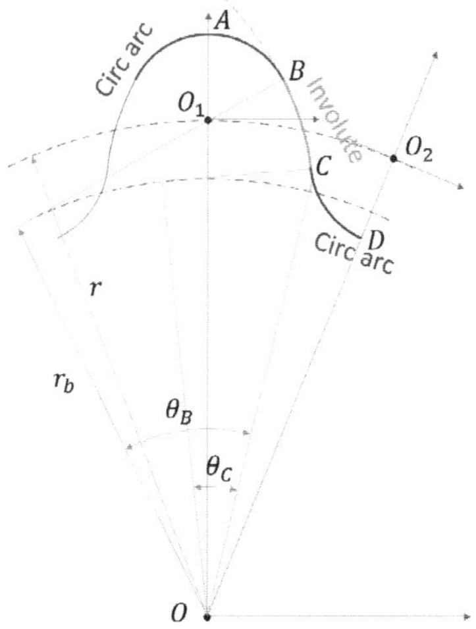

Fig. 12: Geometry of continuous contact gearing [10].

$$F = \tan\alpha - \alpha + \frac{\pi}{2N}. \tag{8}$$

So, arch radius is a function dependent on pressure angle, pitch radius and number of teeth. It is possible to define this geometry by these three parameters. For continuous meshing there is a need to keep right the condition a/r ¿ 0.5.

For even smoother meshing is recommended [8, 9, 10] helical gearing (Fig. 13).

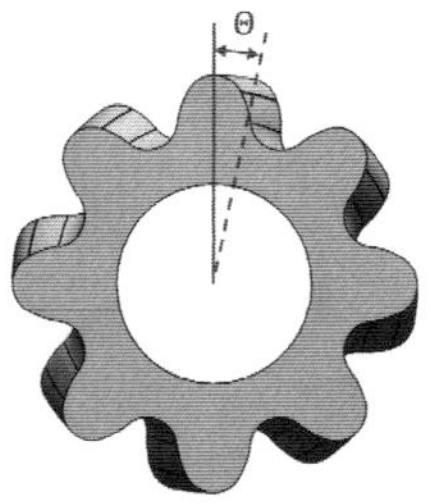

Fig. 13: Helical continuous contact gear [10].

Where helical angle θ is defined as:

$$\theta = \frac{H\tan\beta}{Nm_t\pi}2\pi, \tag{9}$$

where H is total width of a gear, m is module of a gear and β is angle of top arch. Even for helical gear there is a condition which needs to be consisted:

$$CR_{helix} = \frac{\theta}{2\pi/N} \geq 0.5. \tag{10}$$

The result of using helical continuous contact gearing is (according to [8]) decrease of gear pump noise level by 15 dB in comparison to standard gear pump. It also guarantees extremely low pulsations.

Map of noise level in relation to output pressure and rotation speed (Fig. 14), which is to be seen in [9], makes better image about noise level of gear pump based on working mode. It is clear with lower speed and lower output pressure, noise level is on its half values in comparison to standard pumps. Even on top working parameters the noise level values are clearly smaller than standard gear pumps.

Even when continual contact gearing is significantly more quiet, it has its own limits. For gear pumps including this type of gearing it is difficult to reach high volume efficiency. For this type of gearing there is a need for more complex axial balance system [10]. If there is a need to start with this new gearing, it must be

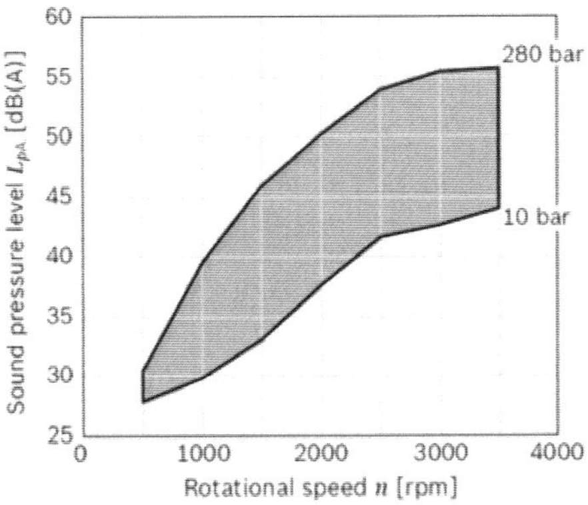

Fig. 14: Map of noise level in relation to output pressure and rotation speed [9].

counted with big investments into technology. Continual contact gearing cannot be made with standard gearing tools.

5 Prediction via mechanical/numerical analysis

There are many possibilities how to improve gear pumps and its gearing. But testing new variations for verification of its impact could be expensive and time taking, because for one gear geometry exist more than dozen possible geometric volumes of the gear pump (geometrical volume is defined via different width of a gearing). Nevertheless there are some suitable approaches, which generate sophisticated mathematical model. With this model there is a possibility to calculate the amount of noise level for every specific gear modification.

There is for example program called EgeMATor (External Gear Machine Multi Tool Simulator) [11], developed in Italian Naples and American Minnesota, which connects mechanical model, hydraulic model, cavitation calculations and more. In case of this program there is taken into account other aspects such as roughness of the surface and tolerances. Output results could be reactions in bearings, torque, meshing pressure of the gearing or flow pulsations in the output of the gear pump which affects final noise level. Decrease of pulsations using geometrical modifications from EgeMATor is declared (by developers) up to 61%. In the figure 15 is possible to see a comparison of pulsations between standard gear pump and its optimized version.

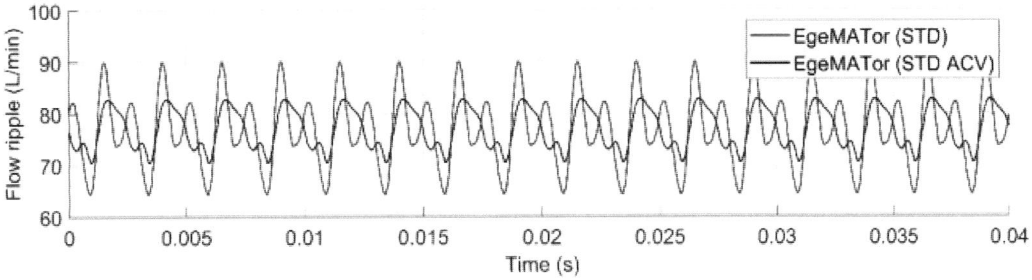

Fig. 15: Comparison of pulsations between standard gear pump and optimized one [11].

The accuracy of all the counting has not been verified yet in comparison with real prototype. Gear pump is still in stage of production.

From different approach there was developed program called HYGESim [12]. Its aim is not to compare pulsations (as in the case before), but level of noise in dB. Basically, it divides noise sources into three categories – mechanical born noise (friction, meshing gears), fluid born noise, air born noise (air contained in fluid). Based on these categories there are separated mathematical models made and consequently brought together into one acoustic model.

Using Ansys Virtual Lab a noise radiation in spheric area around the pump was created (Fig. 16). It

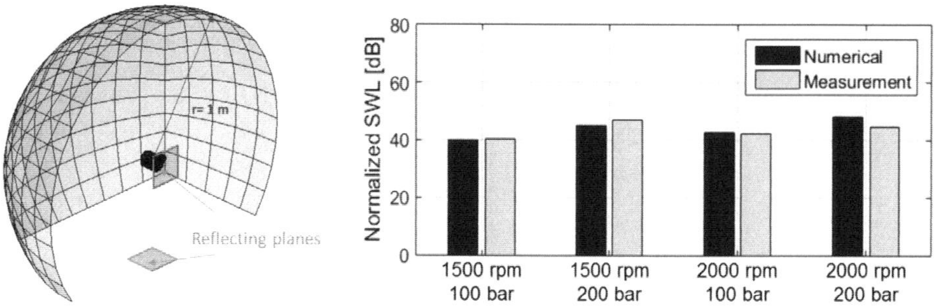

Fig. 16: Model of gear pump surroundig and comparison of noise level from program and real experiment [12].

simulates one meter radius around the pump. Noise level in this area is the result which was compared with real measures. Comparison is shown in Fig. 16.

So HYGESim analyses with pretty good accuracy noise level of a gear pump, but besides EgeMATor there is not a possibility to optimize geometry of a gear pump for a purpose of noise level decrease.

Noise level could be also investigated via FEA analysis. Results of Von-Misses stresses are transformed using Euler-Lagrange analysis which converts meshing tension to a level of noise as it can be seen in Fig. 17 [13].

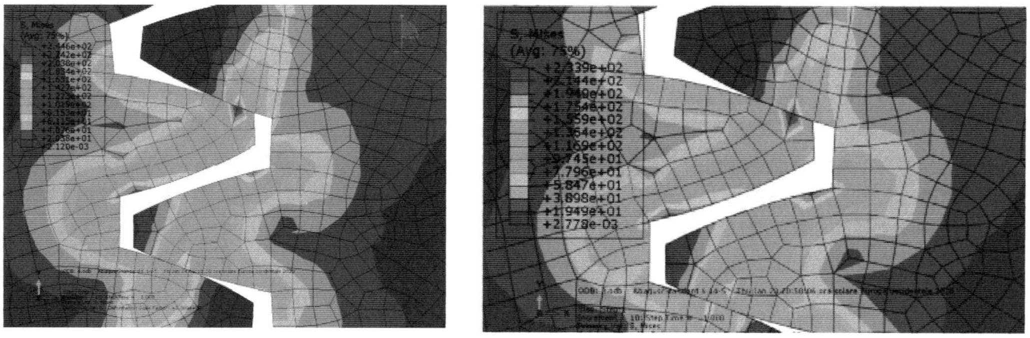

Fig. 17: FEA analysis of meshing gears [13].

In paper [13] it is possible to read about noise level comparisons between some models of gearing (Fig. 18). There are models including different materials or take or not take into account friction in calculations.

So based on FEA results there is a recalculation which reflects noise level of a gearing. With this method there is a possibility to create quick preliminary noise level analysis for different materials or friction variations.

6 Conclusion

Noise level of a gear pump is caused by discontinuous meshing of a gearing, and so discontinuous fluid transfer. For decrease of gear pump noise level there is a need to decrease pressure pulsations and more fluent fluid transfer. Standard possibilities, which are inadequate, are increase of number of teeth, bigger pressure angle or helical gearing.

Increase of number of teeth makes gear tooth thinner and vice versa. With standard involute gearing it is not possible to sufficiently overstep these limitations. Asymmetric gearing allows to maximize pressure angle (e.g. 45°) without making a tooth wider. When the cost side of the tooth is thinner, there is a possibility to increase number of teeth (e.g.: from 12 to 19 teeth). Thanks to this, pulsations happen to be eliminated, so the noise level of the pump also decreases.

Undoubtedly the most quiet are pumps including continuous contact gearing. Gearing is born from Root's blower. It includes not just arches in its geometry, but also partly involute. The result are negligible pulsations, rapid decrease of noise level and it also lower a frequency of noise level which is (for ear) on acceptable level.

The advantage of asymmetric gearing is possibility to produce it using standard tools aimed for involute gearing, for continuous contact gearing it isn't possible and there is a need for special tools. This factor con-

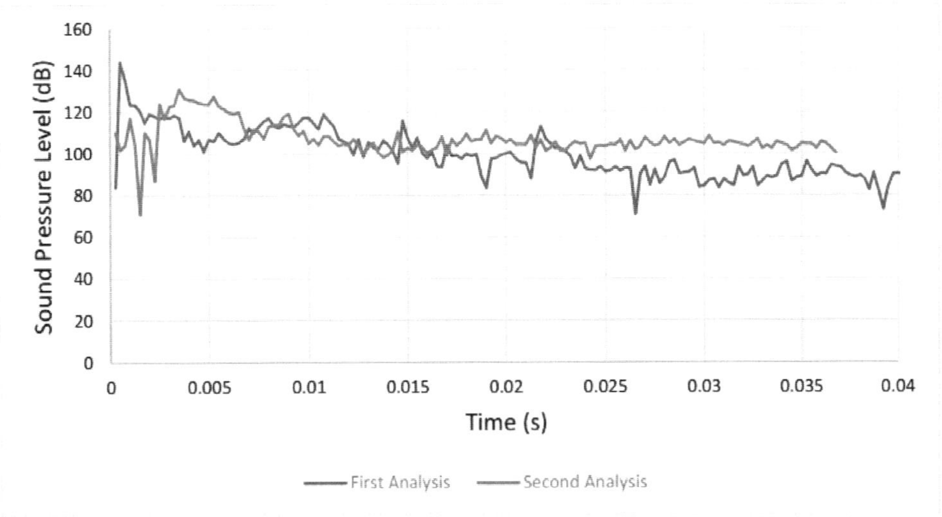

Fig. 18: Course of recounted noise level [13].

siderably influences final price of a gear pump. The question is, how great is a range of hydraulic aggregates, which are in direct contact with a human (where is a bigger need to lower the noise level) and how big range cover gear pump as a key component. And so, if it is useful for gear pump producers to create technological environment for brand new product.

Noise optimalization and design modifications requires following verifying test to prove their effectiveness. It can be financially and time taking. There exist programs as EgeMATor or HYGESim, which are capable to simulate efficiency of modifications or can be helpful with whole optimalization process. Numerical solution of noise level reduction ca be done also with FEA analysis and following recalculation for noise level.

Acknowledgement

This article was created in cooperation with Jihostroj Velešín a.s. where also quiet gear pump solutions are made and tested.

References

[1] J. Bláha, K. Brada, Příručka Čerpací Techniky, Praha, ČVUT, 1997.

[2] P. Huang, et al., Study on Noise Reduction of Gear Pumps of Wheel Loaders Based on the ICA Model, International Journal of Enviromental Research and Public Health. MDPI 13 (2019), doi: 10.3390/ijerph16060999.

[3] Retrieved from: https://www.jihostroj.com/en/hydraulics/gear-pumps-catalogue.html.

[4] G.K. Sfantos, V.A. Spitas, T.N. Constopoulos, Optimized asymmetric gear tooth design 23 (2003).

[5] M. Dymáček, Plastová ozubená kola a jejich analýza pomocí MKP, Brno, (2006).

[6] A. Mehr, Gear Solutions, retrieved from: https://gearsolutions.com/features/the-benefits-of-asymmetric-gears/.

[7] Retrieved from: https://assets.danfoss.com/documents/149364/BC315962412551en-000105.pdf.

[8] Retrieved from: https://www.settima.it/products/continuum-pumps/continuum.html.

[9] Retrieved from: https://www.boschrexroth.com/cs/cz/produkty/skupiny-produktu/mobilni-hydraulika/pumps/external-gear-pumps/azpj.

[10] X. Zhao, A. Vacca, Analysis of continuous-contact helical gear pumps through numerical modeling and experimental validation, Elsevier 26 (2018) 352–378, doi: 10.1016/j.ymssp.2018.02.043.

[11] G. Marinaro, E. Frosina, A. Senatore, A Numerica Analysis of an Innovative Flow Ripple Reduction Method for External Gear Pumps, Energies 22 (2021), doi: 10.3390/en14020471.

[12] S. Woo ,et al., Modeling Noise Sources and Propagation in External Gear Pumps, Energies 20 (2017), doi: 10.3390/e22111306.

[13] A. Liguori, et al., Noise Reduction in Spur Gear Systems, Entropy 17 (2020), doi: 10.3390/e22111306.

Variability of the Gyroid Structure for Biomedical Applications

A. Jíra[1,*], L. Řehounek[1], G. Javorská[1]

[1] *Czech Technical University in Prague, Faculty of Civil Engineering, Thákurova 7, 166 29 Prague 6, Czech Republic*

* *jira@fsv.cvut.cz*

Abstract: This work deals with gyroid structures and their high level of variability. By manipulating the basic gyroid equation, we can create various structures with either tubular or sheet character. Various specimens created by additive manufacturing are subsequently mechanically tested with the goal of comparing the effectiveness of individual geometrical solutions. The tests showed that gyroid structures offer much greater values of mechanical properties despite having the same porosity as other porous structures. Detailed tests of basic element cells also showed the importance of maintaining periodicity for conservation of isotropic properties of the basic element cell. Isotropy of the basic element cell is an important characteristic for subsequent numerical analyses utilizing the numerical homogenisation method.

Keywords: gyroid; 3D printing; osseointegration surface; gyroid variations.

1 Introduction

An optimal porous surface is the basis of every implant [1]. With the advancement of additive technologies, it is now possible to manufacture structures varying between the trabecular and sheet-based geometry [2, 3]. While a thicker porous structure in big implants does not pose a problem for greater trabecular beams, small implants only lend themselves to small diameters of beams and it is not uncommon to reach the lower precision limit of the machine, which poses quality risks [4]. A viable alternative is the gyroid structure. The gyroid structure is described by Eq. (1) and enables for a wide variety of the final shape and overall behavior.

$$F(x,y,z) = F_{sg}(x,y,z) = t = \sin(\bar{x})\cdot\cos(\bar{y}) + \sin(\bar{y})\cdot\cos(\bar{z}) + \sin(\bar{z})\cdot\cos(\bar{x}), \tag{1}$$

where $\bar{x}$, $\bar{y}$ a $\bar{z}$ are modified spatial coordinates so that $\bar{x} = 2\pi x/a$, $\bar{y} = 2\pi y/a$ a $\bar{z} = 2\pi z/a$. And parameter t determines the boundary curvature of the whole structure and therefore its overall character. A gyroid structure specified this way is called the Single gyroid ($F = F_{sg}$) and its applicability is very wide [5, 6].

Fig. 1 shows variants of the gyroid structure with different values of the parameter t. From this comparison, we can see that by roughly defining $t \in \langle 0.6; 1.2)$ and $t \in (-1.2; -0.6\rangle$, a tubular structure resembling beams is created. On the contrary, if we define $t \in (-0.6; 0.6)$, a sheet structure is created. The gyroid structure is defined for $-1.5 < t < 1.5$. If we set the parameter t between $1.2 < |t| < 1.5$ the structure will be defined, but various discontinuities and defects in continuity and smoothness can emerge [7].

The gyroid structure can be defined by two single gyroids with opposite orientations of surface curvature – the double gyroid. The structure is, therefore, represented twice and its volume is greater as compared to the single gyroid [5]. The basic equation for the double gyroid $F = F_{dg}$ can be written as following:

$$F(x,y,z) = F_{dg}(x,y,z) = t^2 = \big(\sin(\bar{x})\cdot\cos(\bar{y}) + \sin(\bar{y})\cdot\cos(\bar{z}) + \sin(\bar{z})\cdot\cos(\bar{x})\big)^2, \tag{2}$$

where t is a parameter influencing the surface curvature and is defined in the range of $t \in \langle -1.413; 1.413\rangle$ for the double gyroid.

2 Methodology

The influence of parameter t on the gyroid geometry is shown on Fig. 2. By defining $t \in \langle 0; 0.4 \rangle$, two sheet gyroids are created. When intersected with a homogeneous volume, a gyroid structure with a defined wall thickness is created. This is the second way of creating a volume. Parameter $t \in \langle 0.6; 1.2 \rangle$ creates a vastly more dense tubular structure than the single gyroid variant shown on Fig. 1.

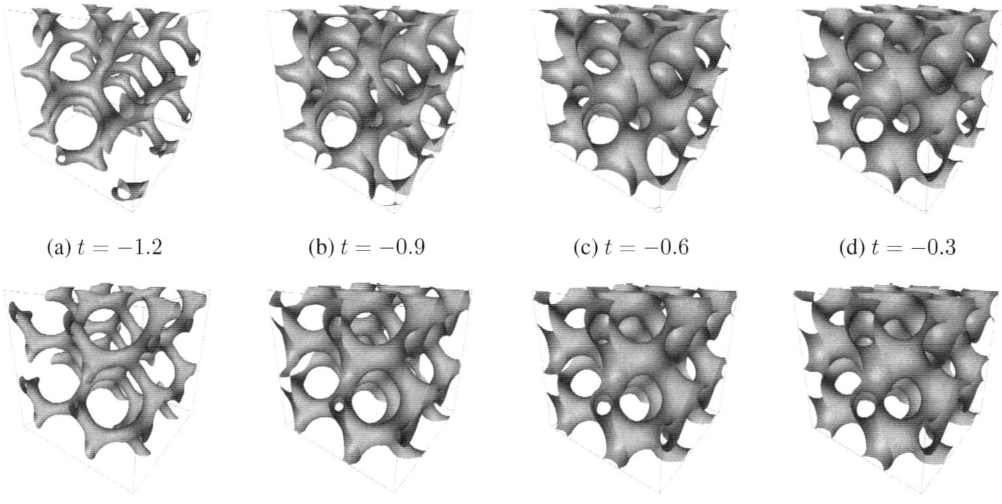

(a) $t = -1.2$ (b) $t = -0.9$ (c) $t = -0.6$ (d) $t = -0.3$

(e) $t = 1.2$ (f) $t = 0.9$ (g) $t = 0.6$ (h) $t = 0.3$

Fig. 1: Variants of the gyroid structure modelled according to Eq. (1) with different values of parameter t.

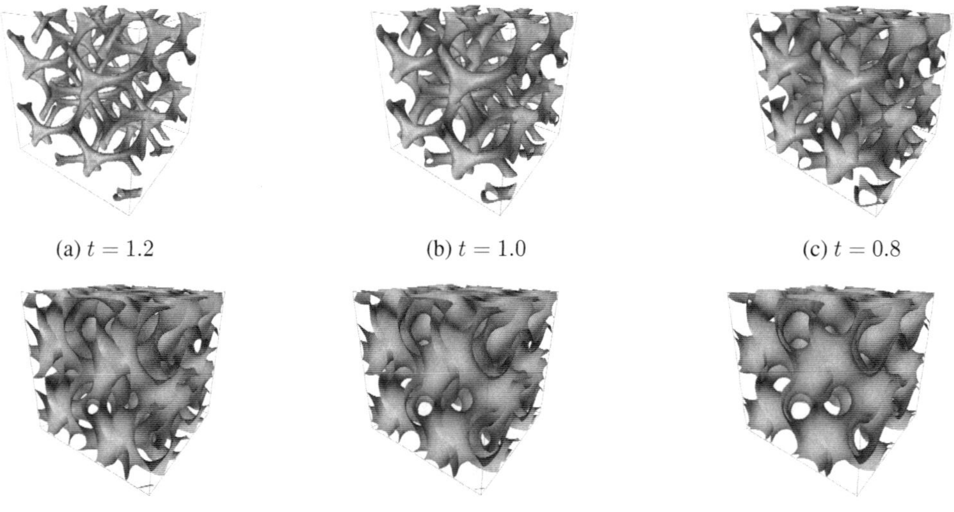

(a) $t = 1.2$ (b) $t = 1.0$ (c) $t = 0.8$

(d) $t = 0.3$ (e) $t = 0.4$ (f) $t = 0.2$

Fig. 2: Variants of the double gyroid structure modelled according to Eq. (2) with different values of parameter t.

Structures shown on Fig. 1 and 2 were generated in the MathMod software, version 11 [8] and the figures show a structure in a circumscribed cube of 2.5π edge length for demonstration purposes of periodically composed cells. The great variability of the structure gives rise to a question – how will these two variants (tubular and sheet) of the structure behave in comparison? This question can be answered by experiment.

The gyroid structure, which is defined by an implicit equation (Eq. (1)) enables for shaping of the structural character (tubular v. sheet) and influences porosity, an important characteristic for osseointegration of implants in bone cavities. [9].

The specimens were designed in Autodesk NetFabb and 3D-printed on the SLS Sinterit Lisa Pro printer. The dimensions of the porous parts were $25.12 \times 25.12 \times 25.12$ mm and a 2 mm thick load-distributing homogeneous plate was added on the top and bottom ends (Fig. 3). The porous part consisted of 4^3 cells with individual dimensions of $6.28 \times 6.28 \times 6.28$ mm. The porous structure is derived from Eq. (1) for the single gyroid. The sheet variant corresponds to a structure with $t = 0$ and explicitly defined wall thickness of 0.5 mm. The tubular version of the gyroid corresponds to parameter $t = 0.78$. Both variants modelled this way have the same porosity of $n = 0.75$.

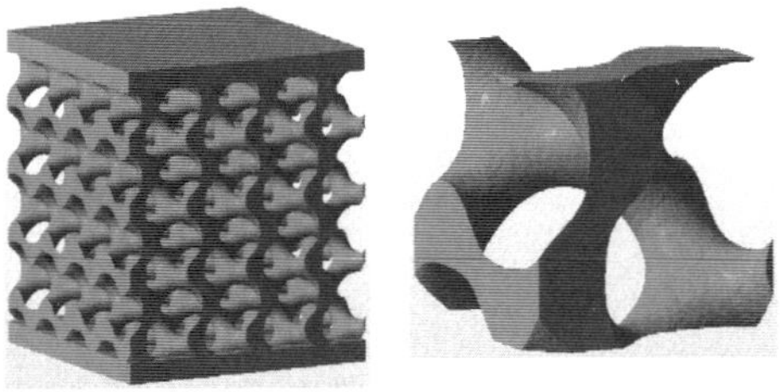

(a) tubular gyroid

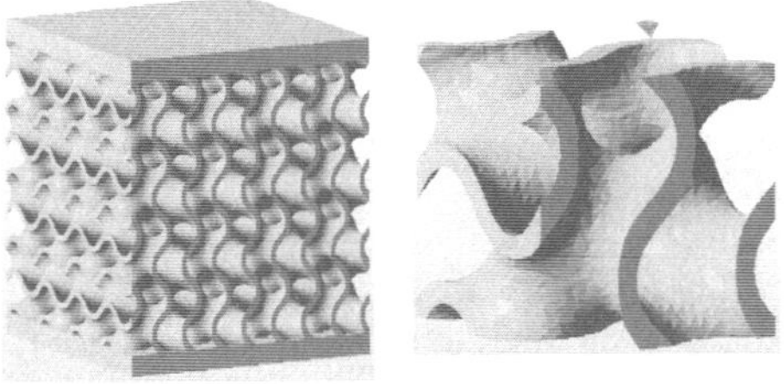

(b) sheet gyroid

Fig. 3: Geometrical models of test specimens and basic element cells – (a) tubular gyroid, (b) sheet gyroid [7].

Specimens shown on Fig. 3 were 3D-printed out of polyamide (PA12) and subsequently loaded on the LiTeM machine in the mode of controlled displacement with speed of 0.04 mm/s until failure.

3 Results and conclusion

The graphs on Fig. 4 show a significant difference in ultimate force and elastic modulus. The sheet variant of the gyroid structure reached an ultimate compressive strength of 4.43 ± 0.47 MPa, while the tubular variant reached only 1.75 ± 0.11 MPa. Similar differences can be observed in numerical analyses of the global modulus of the porous structure E, which was $E = 72.43 \pm 6.13$ MPa for the sheet variant, while only $E = 28.74 \pm 1.80$ MPa for the tubular variant. Therefore, the sheet variant reaches much better values of mechanical parameters at the same level of porosity and presents a more advantageous solution for load-bearing parts of implants.

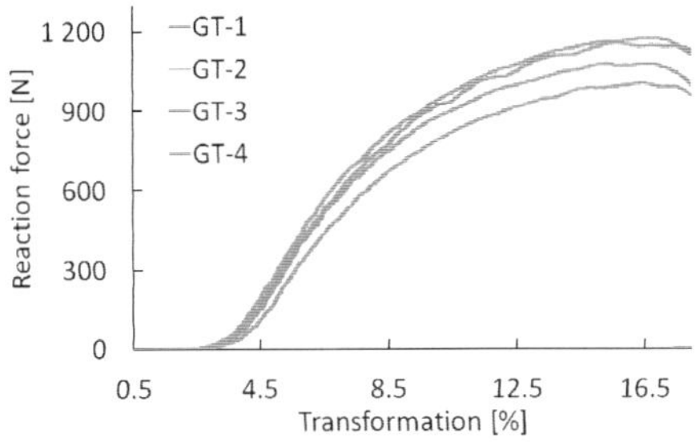

(a) tubular gyroid

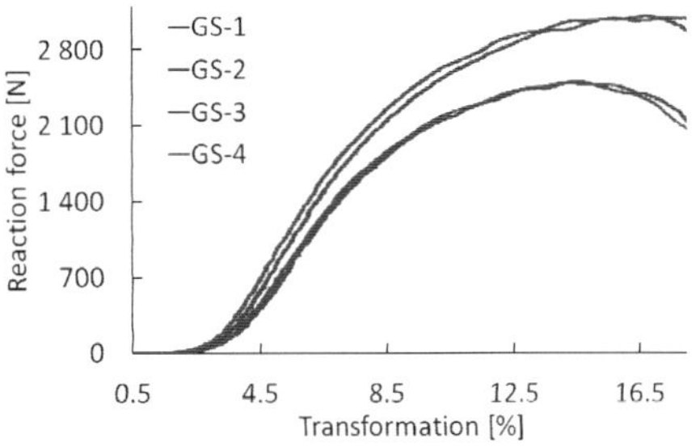

(b) sheet gyroid

Fig. 4: Force-displacement diagrams of tested structures manufactured via 3D printing from the PA12 material. (a) – tubular gyroid (GT), (b) – sheet gyroid (GS).

On Fig. 5, we can see sequential shots from the mechanical tests with differences in deformations of individual specimens. The choices of images on Fig. 5 are, beginning from left: test start, halfway through the test and at the end, following failure. The images show greater that the sheet gyroid structure has greater strength and stiffness as compared to the tubular variant. Both structures show formation of crack leading to total failure of the specimens. Different values of ultimate stress are shown on graphs on Fig. 4. They confirm the fact that the sheet variant of the organic structure is a viable alternative in regard to its greater strength.

In spite of being a TPMS (Triply Periodic Minimal Surface), the gyroid's periodicity can vary and depends on the section made in the structure. Periodicity is important for subsequent numerical analyses via FEM utiliz-

Fig. 5: Example of specimens deformation during the uniaxial compression test. Top row is the tubular gyroid variant, bottom row is the sheet gyroid.

Tab. 1: Parameters of the basic gyroid element cell for verification of symmetry and isotropy.

	cell dimensions [mm]	wall thickness [mm]	cell period in axial direction x	y	z
sample 1	$12,56 \times 12,56 \times 12,56$	0,5	$-\pi; +\pi$	$-\pi; +\pi$	$-\pi; +\pi$
sample 2	$12,56 \times 12,56 \times 12,56$	0,5	$-\pi; +\pi$	$-\frac{1}{2}\pi; +\frac{3}{2}\pi$	$-\pi; +\pi$

ing the numerical homogenisation method. For verifying the influence of periodicity on mechanical properties of the basic element cell, another static mechanical test was performed, using the same parameters as the former experiment. Specimens containing only one cell of the sheet single gyroid were made using 3D printing from the PA12 material (Tab. 1).

A simple comparison based on the mechanical tests of a single cell showed that a cell with the same period in all axes is isotropic. On the contrary, if we change the period by $\pi/2$ in the direction of one axis, the behavior changes to ortotropic (Fig. 6). With a specimen comprised of multiple periodical cells, we predict this influence of period offset will be eliminated and the overall response will be isotropic. However, for the purposes of using the gyroid cell for homogenisation and numerical FEM analyses, it is necessary to use a basic element cell that has the same period in all three axes.

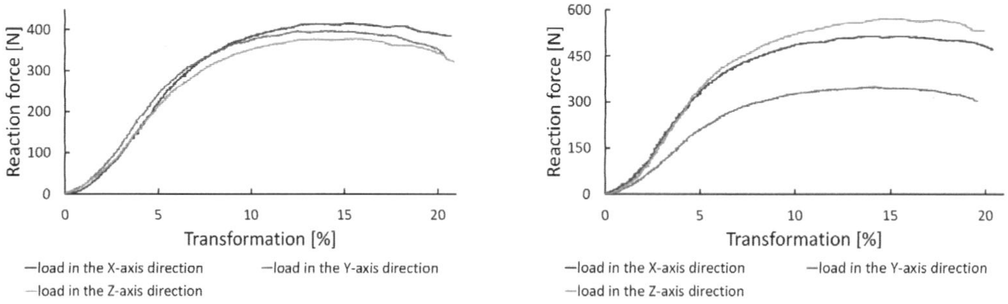

Fig. 6: A comparison of mechanical behavior of a single gyroid cell with the same period in all axes (a) and a gyroid cell with a period offset by $\pi/2$ in the direction of the y axis (b)

Acknowledgement

This work was carried out with the financial support of an internal CTU grant n. SGS 161 1612078A132.

References

[1] L. Řehounek, F. Denk and A. Jíra, Geometrical optimization of dental implants with regard to osseointegration, Acta Polytechnica CTU Proceeding 13 (2017) 97–101, doi: 10.14311/APP.2017.13.0097.

[2] P. Szymczyk, et al., The effect of geometry on mechanical properties of Ti6Al4V ELI scaffolds manufactured using additive manufacturing technology, Archives of Civil and Mechanical Engineering 20 (2020) 1–13, doi: 10.1007/s43452-020-0011-y.

[3] E. Gawronska and R. Dyja, Numerical Study of Geometry's Impact on the Thermal and Mechanical Properties of Periodic Surface Structures, Materials 14 (2021) 1–14, doi: doi.org/10.3390/ma14020427.

[4] A. Jíra, P. Hájková and L. Řehounek, Trabecular structures as efficient surface of dental implants, 57th International Scientific Conference on Experimental Stress Analysis, Luhačovice, Czech Republic (2019) 169–174.

[5] E. Yang, et al., Effect of geometry on the mechanical properties of Ti-6Al-4V Gyroid structures fabricated via SLM: A numerical study, Material and Design 184 (2019) 1–24, doi: 10.1016/j.matdes.2019.108165.

[6] M. Niinomi, Mechanical properties of biomedical titanium alloys, Materials Science and Engineering, 234 (1998) 231–236, doi: 10.1016/S0921-5093(97)00806-X.

[7] A. Jíra, et al,. Mechanical Properties of Porous Structures for Dental Implants: Experimental Study and Computational Homogenization, Materials 14 (2021) 1–25, doi: 10.3390/ma14164592.

[8] A. Taha, MathMod open source (GPL), Retrieved from: https://sourceforge.net/projects/mathmod/.

[9] M. Geetha, et al., Ti based biomaterials, the ultimate choice for orthopaedicimplants – A review, Progress in Materials Science, 54 (2009) 397–425, doi: 10.1016/j.pmatsci.2008.06.004.

Measurement of Screw Connection Parameters of Air Conditioning Systems

R. Jírová[1,*], L. Pešík[1]

[1] *Department of Design of Machine Parts and Mechanism, Faculty of Mechanical Engineering, Technical University of Liberec, Studentská 2, 461 17 Liberec, Czech Republic*
* *radka.jirova@tul.cz*

Abstract: In service, screw connections used in air conditioning systems require high reliability. Especially, a sealing function of the screw connection has to be appropriate to prevent coolant leaking. Therefore, the paper focuses on the measurement of screw connection parameters. The examined parameters related to the screw connection reliability were the friction coefficient in a thread of the screw connection and axial forces caused by the settlement process in screw connections.For this purpose, we designed a measurement device and analysed screw connections with different parameters.

Keywords: screw connection; force measurement; friction coefficient measurement; air conditioning system; sealing function.

1 Introduction

The reliability of screw connections of aluminium alloy bodies used in air conditioning systems depends on several factors. The principal factor is related to the friction in a thread of a screw and a connected body [1]. Two different materials are used for connection, steel and aluminium alloy. The coefficient of friction depends on a special surface layer on the screw thread, whereas thread lubrication is usually not performed during the assembling process. The lower coefficient of friction may cause a greater axial force at the prescribed tightening torque. Important is also a small variance of the friction coefficient values for the force ratios prediction in the screw connection. In addition to the friction coefficient that determines the magnitude of axial force in the connection, we need to ensure a constant magnitude of the force during the service and after a settlement of bodies. For this purpose, a disc spring in the form of a washer under the screw head was used in the screw connection of the air conditioning system [2, 3].

Both parameters, the friction coefficient and the constant force during the service of the connection, have to be carefully monitored during and after the installation [4]. This is the way to ensure sufficient reliability of the screw connection. For this purpose, we developed an original measurement device, which allows measuring both mentioned parameters.

2 Design of measurement device

The measurement device is based on a double reversing lever (Fig. 1).

2.1 Measurement of friction coefficient

Measurement of friction coefficient is ensured by the lever configuration, where at one end of the lever, the screw is tightened, and at the other end, the axial force in the screw is measured using a load cell.

By measuring the axial force and tightening torque, we found that thread lubrication is major for getting a higher value of axial force at the given value of tightening torque. In the measured case, we compared two different sliding surface layers on a screw thread.

The friction coefficient must be inspected before tightening of screw connection because it decides the reliability of the connection.

According to this measurement, we got statistical dispersion of axial forces at the given tightening torque.

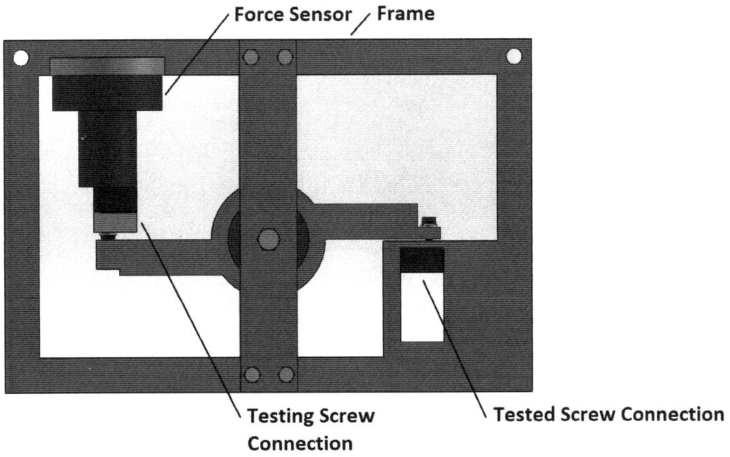

Fig. 1: Measurement device.

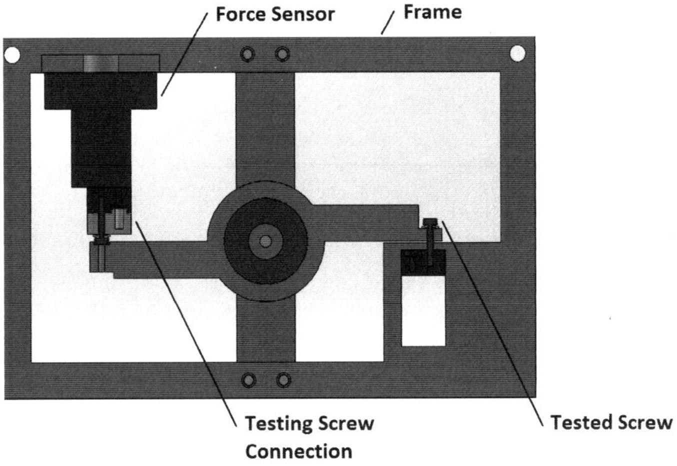

Fig. 2: Measurement of friction coefficient.

2.2 Measurement of axial force

By mounting the screw connection, also the embedding of connected bodies must be included to its reliability. For measuring the change of the axial force in the assembled screw connection caused by the settlement of bodies, the axial force is created at one end of the lever using an auxiliary screw connection, which is arranged in series with the load cell at the other end of the lever.

The measurement device is a relatively flexible construction in that the screw connection with simulated preload is mounted. In real conditions, the screw is preloaded by tightening torque, and connected bodies are pressed between the head of the screw and the body with thread. In the measured case, connected bodies are pressed similarly, but axial force appears through the measurement device.

For getting the preload loss due to embedding of connected bodies, the stiffness of the measurement device k_p and the stiffness of bodies pressing the connected bodies k_s must be known.

By the preload F_Q in the screw connection, according to Fig. 3, the deformation of connected bodies Δl_p and the deformation of the measurement device Δl_s appears. We may get a system stiffness k_{ps} through the measurement of overall deformation Δl_{ps}.

$$k_{ps} = \frac{F_Q}{\Delta l_{ps}} = \frac{F_Q}{\Delta l_p + \Delta l_s} \tag{1}$$

The overall stiffness is defined as

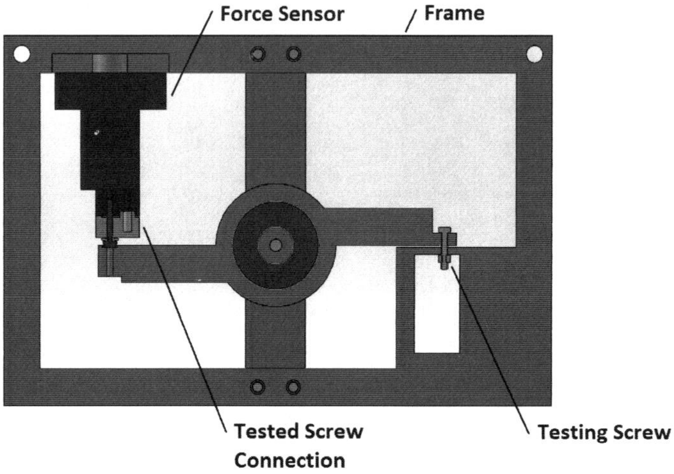

Fig. 3: Measurement of axial force.

$$\frac{1}{k_{ps}} = \frac{1}{k_p} + \frac{1}{k_s} \tag{2}$$

where stiffness k_p is

$$k_p = \frac{F_Q}{\Delta l_p} \tag{3}$$

and stiffness k_s

$$k_s = \frac{F_Q}{\Delta l_s} \tag{4}$$

Then, we used a strut with known stiffness k_r, which is higher by orders than stiffnesses k_p and k_s. Then the overall system stiffness changes to

$$\frac{1}{k_{rs}} = \frac{1}{k_r} + \frac{1}{k_s} \tag{5}$$

Through Eq. 5, we may get the stiffness of the measurement device k_s. Whereas

$$k_{rs} \cong k_s \tag{6}$$

Then, through Eq. 2, we may get the stiffness of bodies pressing the connected bodies k_s.

After mounting the screw connection with known stiffness k_b, the preload loss due to the embedding process is described in Fig. 4.

We may note that

$$\operatorname{tg} \alpha_p = k_p \tag{7}$$

and

$$\operatorname{tg} \alpha_b = k_b \tag{8}$$

We observed that lower screw stiffness generates lower preload loss due to the embedding of connected bodies.

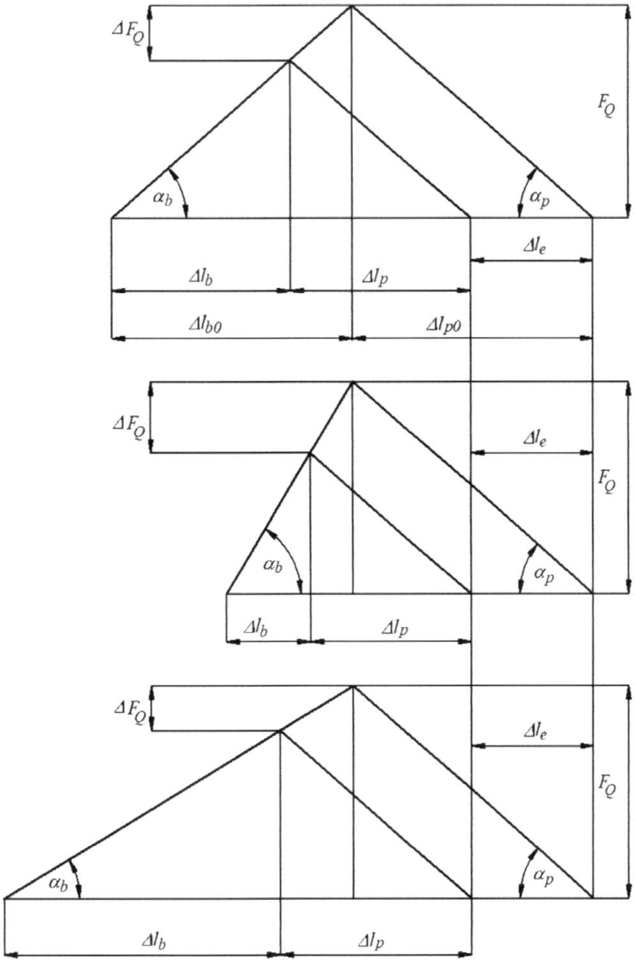

Fig. 4: Influence of screw stiffness on the preload loss due to embedding process.

3 Results

3.1 Friction coefficient

The friction coefficient influences the preload value of the screw connection. Two types of sliding surface layers of the screw were tested, Type A (Tab. 1) and Type B (Tab. 2), by the tightening torque of 12 kN. Then, the variation of preload was observed.

Tab. 1: Preload of screw connection – sliding surface layer Type A.

Specimen	1	2	3	4	5	6	7	8	9	10
Force [kN]	6.4	7.7	7.0	6.4	6.5	7.3	7.1	7.5	6.3	8.6

Tab. 2: Preload of screw connection – sliding surface layer Type B.

Specimen	1	2	3	4	5	6	7	8	9	10
Force [kN]	10.4	10.8	10.0	10.7	12.0	9.7	11.6	9.5	10.6	12.4

Results show that the sliding surface layer Type A of the screw reports higher friction than Type B layer. In the first case, we got the preload ca. 7.0 kN. In the second case, we reached nearly 11.0 kN.

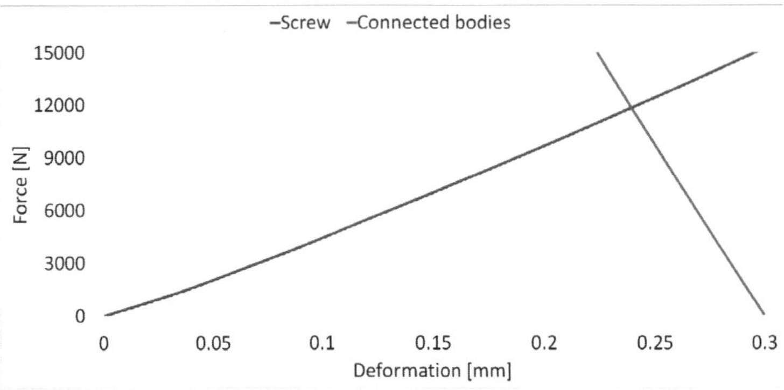

Fig. 5: Diagram of screw connection – after mounting.

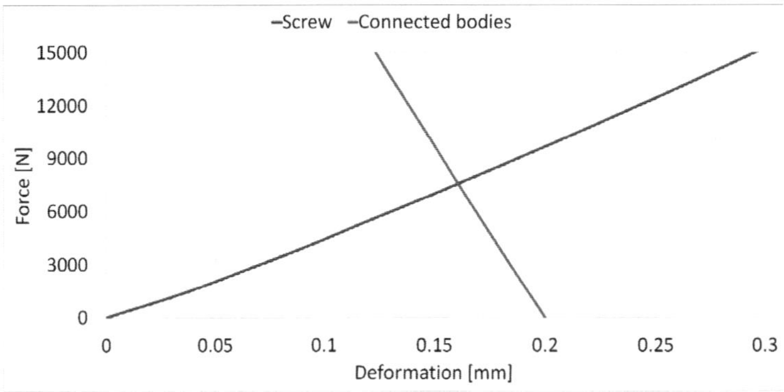

Fig. 6: Diagram of screw connection – after embedding process.

3.2 Axial force

By measuring axial force and deformation, we got diagrams of screw connection after mounting (Fig. 5) and after the embedding process (Fig. 6). In diagrams, the preload value is represented by an intersection of a screw line (red) and a connected bodies line (blue).

The initial condition of preload was 12 kN after mounting. Duo to embedding process, the preload decreased at 60% of the initial value.

4 Conclusion

The paper showed two possibilities of measurement of screw connection parameters in practice. Measurement results of the friction coefficient and the axial force in the screw connection are significant for the reliability and the quality of screw connections in air conditioning systems.

Regarding friction, the proper lubrication or sliding surface layer of thread is needed for sufficient axial force. Regarding the minimal preload loss due to the embedding of connected bodies, the low stiffness of the screw is demanded that may be reached by using the disc spring with degressive force-deformation characteristics as a washer under the screw head.

Acknowledgement

This work was supported by the Student Grant Competition of the Technical University of Liberec under project No. SGS-2019-5036.

References

[1] J.E. Shigley, Mechanical Engineering Design, McGraw-Hill Book Company, 1986.

[2] J. Homišin, et al., Súčasné trendy optimalizácie strojov a zariadení, Košice, C-PRESS, 2006.

[3] I. Mašín, P. Lepšík,Analytické a kreativní postupy v údržbě strojů a zařízení, Liberec, Technická univerzita v Liberci, 2015.

[4] E. Tomeh, Technická diagnostika. Vibrační diagnostika strojů a zařízení, Liberec, Technická univerzita v Liberci, 2015.

Composite Laminate Deflection during Low-velocity Impact

M. Kadlec[1,*], O. Vích[1], D. Novotný[1]

[1] *VZLÚ – Czech Aerospace Research Centre, Beranových 130, 199 05 Prague 9, Czech Republic*
[*] *kadlec@vzlu.cz*

Abstract: For the development of a composite structure damage detection system, it is important to create a reliable numerical predictive model of the damaged material. This paper is focused on experimental and numerical analysis of a drop weight impact test and the possibility of its extension by the presence of a laser interferometer for deflection measurment. The subject of the test was 1.58 mm thick composite laminate made of a commercially inaccessible carbon fibre–reinforced epoxy matrix. The impact test was simulated with the help of ABAQUS/Explicit 6.14 solver and compared with the experiment. Results of the laminate deflection were statistically assessed showing 4% difference between the model and the experiment.

Keywords: experiment; impact; deflection; simulation; composite.

1 Introduction

To detect in-service damage of aerospace structures, a structural health monitoring (SHM) system is being developed. It will be able to detect damage produced by structural ageing and impacts during the service life using sensors. Usually, the damage is monitored by periodic inspections, visually or with the utilization of non-destructive inspection (NDI) equipment. The SHM system will enable more frequent inspections using sensors [1, 2], with the final target of a continuous in-flight monitoring, with the potential cost reduction. The SHM system needs a definition of impact damage virtual morphology validated by experimental impact test results.

Collisions between fast-moving objects can lead to their complete, partial or barely visible damage (BVID). Since the change in the properties of the material from which they are made can have a significant relation to the load-cases for which they are designed [3, 4], it is very advantageous to have a computational model that corelates with reality. Due to the fact, that the properties of composite materials are influenced by the production technology, each material is moreover unique. Computational models require large number of parameters that need to be experimentally supported [5–7]. Verification of the response and the rate of damage to the material to impact by a standardized body plays a crucial role here. This work is mainly devoted to the comparison of the simulation with the experiment, which was performed using a drop weight machine, especially the verification of the possibility of its extension by the presence of a laser interferometer.

2 Experiment

Laminate samples were made from components, which properties shows Tab. 1. The composite layup was $[-45/+45/0/90/0/0]_S$ with thickness of 1.58 mm. The experiment was based on ASTM D7136 standard [8]. A total of 5 samples measuring 150 × 100 mm were cut from the plate using a diamond wheel and milled to the required dimensions. The sample was placed on a rectangular frame with a hole of 125 × 75 mm and fixed with 4 pressure holders with rubber heads. Using a pad machine impactor (Fig. 1a), the laminate was damaged in the middle of the sample by a spherical steel impactor with a diameter of 12.7 mm with an indicated energy of approximately 10 J. The impactor weight was 5,804 g and the corresponding impact velocity was approximately 1.8 m/s. Design of the impactor ensured that after the recoil, the punch stopped and there was no repeated impact.

The actual pre-impact punch speeds were measured using Hall probes. At the same time, the acceleration was measured using an accelerometer located inside the punch with the recording frequency 20 kHz. The data from the accelerometer were used to calculate the absorbed energy, which was approximately 8 J.

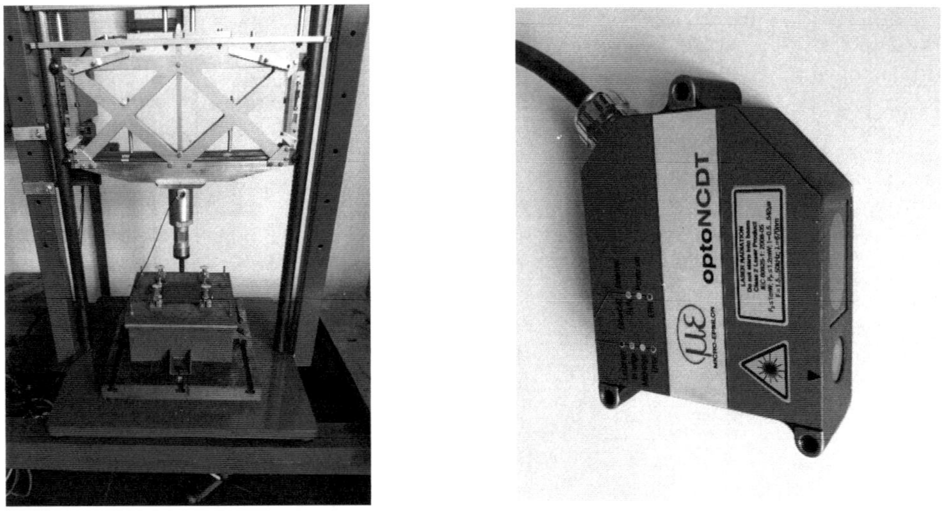

(a) Impact test setup

(b) Laser interferometer optoNCDT ILD 2300-20

Fig. 1: Experiment description.

Tab. 1: Properties of components.

Component	Material	Tensile modulus	Poisson ratio
[-]	[-]	[GPa]	[-]
Matrix	Epoxy	4.30	0.35
Fibre	Carbon	240.00	0.20

To measure the sample deflection under the impactor the laser interferometer was used (Fig. 1b), which works on the principle of optical triangulation. It utilizes a laser beam with a wavelength of 670 nm (max. Power 1 mW) and can be active in two modes. At the first mode the direct reflection of laser beam is used (angle of incidence of the beam is equal to the angle of reflection). In the second mode the diffuse reflection is used (laser impinges on the surface at right angles). The first mentioned mode/method was used in the described experiment. Laser beam angle of incidence was 15° deviated from the perpendicular direction of the surface (the set angle depends on the distance of the sensor from the surface).

The measuring range of the used sensor is 20 mm (40 to 60 mm from the front of the sensor) with a resolution of 0.3 μm. Sample frequency of the detector could be selected from the working range 1.5 to 49.1 kHz. The interferometer was connected to the measuring chain using a C-Box/2A converter, which simultaneously provides power to the sensor, time synchronization and allows setting of its parameters.

3 Model description

The role of the impact in the jig is a model case, where is bending between the support and a local deformation under the impactor, which is difficult to calculate. The task must be formulated very precisely already at the level of elastic parameters. The calculation of material parameters for the simulation was performed by use the values ??for epoxy matrix Hexcel 3501-6 and carbon fibre AS-4 (Tab. 1). Properties of fibres are considered as isotropic here. Elastic properties were calculated by using Halpin-Tsai relations. Strategy of other parameters calculation follows the methodology of Daniel and Ishai [9]. The parameters of efficiency, elasticity, strength, and damage evolution are shown in Tab. 2 and Tab. 3. The experimentally determined values ??were subjected to statistical analysis and compared with the results of simulations performed with the help of the ABAQUS/Explicit 6.14 solver.

In connection with its effective use, elements with a reduced number of integration points were used in the whole model. In case of the test fixture with rubber attachments C3D8R eight-node elements were used.

Tab. 2: Mechanical properties of laminate, Major properties.

Direction	Elasticity	Efficiency parameter	Laminate strength	Laminate damage evolution	Cohesive strength	Cohesive damage evolution
[-]	[GPa]	[1]	[GPa]	[J/m^2]	[GPa]	[J/m^2]
Longitudinal (11)	130.00	10^6	2.970	0.074	-	-
Transverse (22)	12.00	0.68	0.101	0.014	-	-
Transverse (33)	12.00	0.68	-	-	0.101	0.074
Shear (12)	4.99	1.00	0.101	-	-	-
Shear (13)	4.99	1.00	-	-	0.101	0.014
Shear (23)	4.64	1.00	-	-	0.101	0.014

Tab. 3: Mechanical properties of laminate, Poisson ratios.

Direction	Poisson ratio	Efficiency parameter
[-]	[1]	[1]
In-Plane (12)	0.27	10^6
Out-of-Plane (13)	0.27	10^6
Out-of-Plane (23)	0.43	0.31

For the instance of the impactor, which we consider to be very rigid and whose elastic deformation is not essential for the result, C3D4 tetrahedral elements were applied. The ply-by-ply method in connection with continuum-shell SC8R elements was used. This modelling technique does not take into account compressive and tensile deformations in the thickness direction of the laminate. For each layer, a minimum number of integration points was considered so that the local bending mode with the influence of transverse shear stresses was properly affected. The number of integration points must always be chosen with respect to the assumed course of the stress components as odd. Delamination of the layers is realized by cohesive contact. The failure of material was considered in the sense of Hashin's formulation.

4 Results

The time while impactor was in contact with the laminate sample was approximately 15 milliseconds. The values of applied energy, which were calculated from speed and maximum bending displacement during the impact test, are pointed in Tab. 4. On the impacted side, a hole could be observed without significant fibre breakage. Approximately 1 minute after the impact, the depth of the pit on the impacted side was measured using an indicator, which was in range from 0.82 to 1.26 mm. The back of the sample was severely damaged. The variety of failures such as fibre breakage, fibre bundle separation and delamination widening in the diagonal direction were observed.

To speed up the calculation, the velocity of the impactor on contact with the sample was increased to 5 m/s compared to the experiment 1.8 m/s. To keep the calculation and experiment energetically equivalent, the weight of the impactor was reduced at the expense of higher velocity. Because it was not an application of classical mass-scaling within an explicit solver, this adjustment did not affect the quality of the result (Fig. 2b). Comparison of experimental results and simulations show Tab. 4 and Fig. 3.

The deflection values obtained by the experimental were on average 4% higher than calculated by the simulation. F-test showed significantly different variances because the simulations were much more consistent. T-test confirmed a statistically significant difference of the means (p-value = 0.04).

(a) Specimen after impact

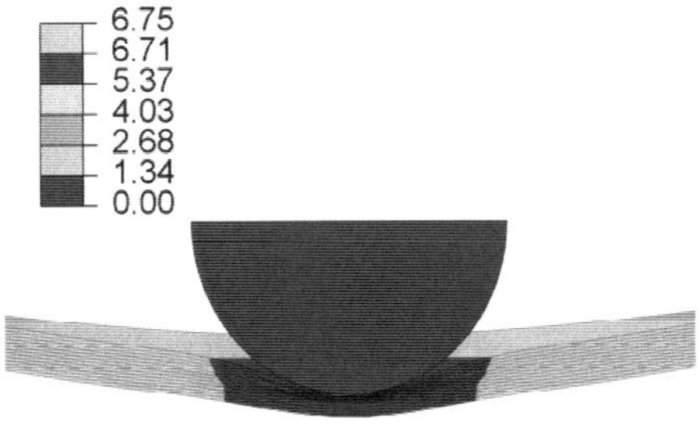

(b) Deflection under impactor [mm]

Fig. 2: Real and modelled damage.

Tab. 4: Vertical displacement under impactor.

Sample ID [1]	Energy [J]	Experiment [mm]	Simulation [mm]
1	10.42	7.13	6.71
2	10.30	6.90	6.68
3	10.80	6.76	6.79
4	10.22	7.27	6.66
5	10.50	6.89	6.72
Average	10.45	6.99	6.71
Std. dev.	0.22	0.21	0.05

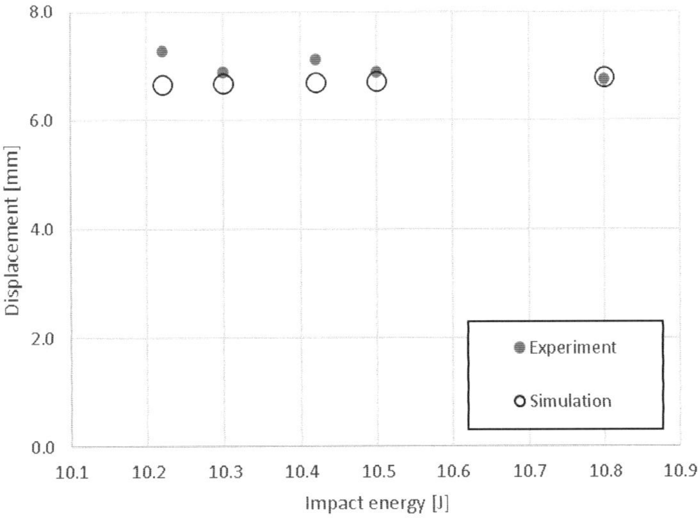

Fig. 3: Vertical displacement (laminate deflection) under impactor.

5 Conclusion

Use of a laser interferometer during the experiments in connection with the verification of simulation results has proven to be very effective. The results, which were subjected to statistical analysis, show a very small deviation of the model and the measurement with a difference 4%. The model can be used to analyse the internal damage of laminates and its effect on the deformation field of the sample at different types of loads.

Acknowledgement

Results incorporated in this paper received funding from Clean Sky 2 Joint Undertaking (JU) under grant agreement No 886703. The JU receives support from the European Union?s Horizon 2020 research an innovation programme and Clean Sky 2 JU member other than the Union.

References

[1] D. Kinet, et al., Fiber Bragg grating sensors toward structural health monitoring in composite materials: Challenges and solutions. Sensors 14 (2014) 7394–7419.

[2] R. Růžek, et al., Monitoring of compressive behaviour of stiffened composite panels using embedded fibre optic and strain gauge sensors, International Journal of Structural Integrity 8 (2017), 134–150.

[3] S. Sánchez-Sáez, et al., Compression after impact of thin composite laminates, Composites Science and Technology 65 (2005) 1911–1919.

[4] M. Kadlec, V. Kafka, Strain concentration during the compression of a carbon/epoxy composite after impact, International Journal of Structural Integrity 6 (2015) 279–289.

[5] V. Balasubramani, S. Rajendra Boopathy, R. Vasudevan, Numerical Analysis of Low Velocity Impact on Laminated Composite Plates, Procedia Engineering 64 (2013) 1089–1098.

[6] M.A. Hassan, S. Naderi, A.R. Bushroa, Low-velocity impact damage of woven fabric composites: Finite element simulation and experimental verification, Materials & Design 53 (2014), 706–718.

[7] K.K. Namala, P. Mahajan, N. Bhatnagar, Digital Image Correlation of Low-Velocity Impact on a Glass/Epoxy Composite, International Journal for Computational Methods in Engineering Science and Mechanics 15 (2014) 203–217.

[8] ASTM D7136M-20, Standard Test Method for Measuring the Damage Resistance of a Fibre-Reinforced Polymer Matrix Composite to a Drop-Weight Impact Event, ASTM International, West Conshohocken, PA, 2020.

[9] I. Daniel, O. Ishai, Engineering mechanics of composite materials, Oxford university press, 2006.

Torque Tightening Process Qualification

J. Kanaval[1,*], E. Cézová[1]

[1] *Department of Designing and Machine Components, Faculty of Mechanical Engineering, Czech Technical University in Prague, Technická 4, 166 07 Prague 6, Czech Republic*

* *jan.kanaval@fs.cvut.cz*

Abstract: In order to achieve quality products, certification of production processes is currently required in companies. It is possible to mention the procedure of screw tightening. To obtain a certificate, it is necessary to check in the company the technological assembly procedure and the equipment of the workplace with jigs and tools. It is necessary to check whether all provisions of regulations and standards are observed during the assembly of the product. In the case of screw connections, it is necessary to carry out experiments to verify the correctness of the tightening torque and to determine the Torque-angle curve. It is often required to assess the quality of threaded surfaces after disassembly of the product on a metallographic cut. This paper describes the certification procedure of tightening the main screw connection of a selected product. Without the granting of the certificate, it would not be possible to proceed with further assembly.

Keywords: prestressed screw connection; screw tightening; Torque-angle curve; process certification.

1 Introduction

The strength calculation of prestressed screw connections is usually performed according to *VDI 2230* [1]. The size of the *tightening factor* α_A is selected according to the selected method of mounting the screw connection and for the considered scattering of the coefficients of friction, the required range of tightening torque $M_{k\,\min}$ to $M_{k\,\max}$ is then calculated. The *Torque-angle curve* is an indicator of the correct tightening process of the screw connection. Certification of the screw connection assembly process is performed according to the standard *EN 14831: 2005 - Fasteners - Tightening performance - Torque/Angle simplified test method* [2].

The whole certification process was implemented in the following steps [3]:

- visual inspection of the tightened connection,
- check of the technological process of tightening the screw connection,
- verification of the correctness of the tightening torque determination,
- check the equipment of the workplace with jigs and tools,
- testing of untightening torque values and statistic evaluation,
- experimental determination of the Torque-angle curve,
- inspection of threaded surfaces after disassembly of a screw connection from a metallographic cut.

2 Certification of the screw tightening process

In the following subchapters, a brief description of all implemented activities performed within the certification of the tightening process of the screw connection is given. Used screws M6×16 are made of stainless steel A2-80. A threaded hole is made in the aluminum alloy.

2.1 Visual inspection of the tightened connection

To perform the measurements tightening and untightening torque were delivered 5 pieces assembled products (cover and housing). Tested samples meet the requirements according to the standard EN 14831: 2005. During and after the tests there were not found any visible damages or wrongly performed tightening.

2.2 Check of the technological process of tightening the screw connection

The cover of Pulse Generator assembled according to the specified technological procedure (assembly instruction). The prescribed tightening torque (10–11) Nm corresponds to the EN 14831: 2005 standard.

Fig. 1: Assessed product with screwed cover.

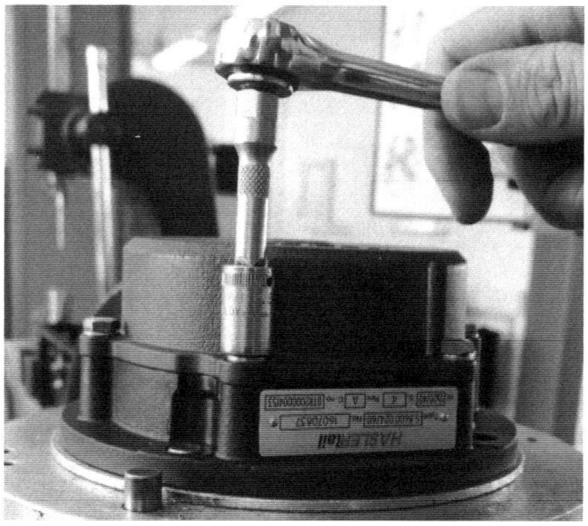

Fig. 2: Tightening the screw with a torque wrench.

The following assembly procedure is available at the workplace (see Fig. 1 and Fig. 2), prepared in the form of a clear table. Below are the individual operations for assembling the cover:

- insert the O-ring to the groove on the cover,
- before you mounted both parts again together after the EOL-Test, check that O-ring is still in the correct position and that the both parts are clean,
- apply a slight layer of grease Klüber Petamo GHY 133N inside the thread of the housing,
- fixing the two parts cover and housing with screws M6×16, sealing screws M6×16 and high tension spring rings together,
- tight them with a torque of (10–11) Nm according the assembly drawing,
- insert the sealing wire through the hole of the screw and twist it clockwise with using the special pliers,
- put the label and insert the wire through the hole of the second screw,
- twist counter clockwise the sealing wire with special pliers,
- when it's twisted, insert the seal through the wires and pull the wires far apart as its shown on picture below,
- when the seal is prepared, then use sealing pliers to press it. Cut the exceeded wires after seal pressing,
- bend the sealing wire towards the cover.

2.3 Verification of the correctness of the tightening torque determination

Tab. 1: Tightening torques for screws according to EN ISO 4014 and EN ISO 4017 depending on the coefficients of friction.

Diameter nominal Md [mm]	Thread pitch P [mm]	0.06	0.08	0.10	0.12	0.14	0.16	0.18
M6	1.00	6.23	7.02	7.80	8.6	9.4	10.1	10.9
M8	1.25	13.96	15.92	17.88	19.8	21.8	23.8	25.7
M10	1.50	30.75	34.00	38.57	42.5	46.4	50.3	54.2
M12	1.75	52.16	58.94	65.72	72.5	79.3	86.0	92.8
M14	2.00	78.26	89.11	99.96	110.8	121.7	132.5	143.4
M16	2.00	127.06	144.43	161.81	179.2	196.6	214.0	231.3

As an example, the following table shows the values of the tightening torques for the coefficients of friction 0.06 to 0.18. These tightening torque values have been calculated according to EN 14831: 2005 for bolts according to EN ISO 4014 [4] and EN ISO 4017 [5] (see Tab. 1). See 1st row – diameter nominal M6 and coefficients of friction 0.16 and 0.18. The specified tightening torque is determined correctly.

2.4 Check the equipment of the workplace with jigs and tools

Tightening of the screws crosswise is always performed with a calibrated Torque wrench 10.5 Nm.

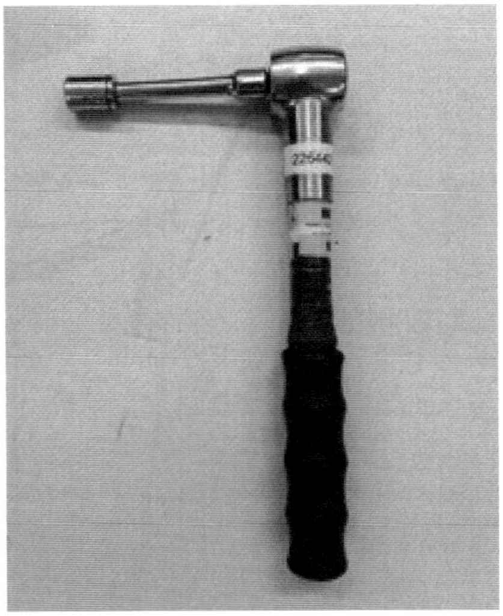

Fig. 3: Calibrated Torque wrench 10.5 Nm (side view).

Fig. 4: Calibrated Torque wrench 10.5 Nm.

The mounting fixture used is shown in Fig. 1.

2.5 Testing of untightening torque values and statistic evaluation

A simple test stand was designed to verify the tightening torque. Tightening and untightening torque were measured using a calibrated torque sensor HBM T20WN/20 Nm (see Fig. 5 and Fig. 6).

Fig. 5: A simple test stand for tightening and untightening torque measurement of the bolts were realized.

Fig. 6: Calibrated Torque wrench 10.5 Nm.

The experimentally determined values of tightening and loosening torques of the bolts were statistically evaluated (30 measurements were performed). Selection average value of the experimentally determined untightening torque of bolt joints is 9.289 Nm and selection standard deviation of the experimentally determined untightening torque of bolt joints is 1.533 Nm.

2.6 Experimental determination of the Torque-angle curve

Maximum and minimum values of tightening torques and axial forces in the M6 screw connection were calculated according to the standard EN 14831: 2005 (Fasteners – Tightening performance – Torque/Angle simplified test method) (see Tab. 2).

Tab. 2: Maximum and minimum values of tightening torques and clamping forces in the M6 screw connection according to the Standard EN 14831: 2005.

Thread parameters of the M6 screw		
Basic major diameter	d [mm]	6.000
Thread pitch	P [mm]	1.000
Minor diameter	d_3 [mm]	4.773
Pitch diameter	d_2[mm]	5.350
Minimum outer diameter of the seating face	$d_{w\,\min}$ [mm]	8.880
Minimum hole diameter	$d_{n\,\min}$ [mm]	6.400
Minimum friction coefficient in threads	$\mu_{\min}$ [1]	0.120
Maximum friction coefficient in threads	$\mu_{\max}$ [1]	0.180
Calculated values of tightening torques and clamping forces		
Preliminary torque	T_p [Nm]	3.000
Minimum clamping force	$F_{p,\min}$ [kN]	2.135
Maximum clamping force	$F_{p,\max}$ [kN]	3.031
Testing force	F_{test} [kN]	8.000
Minimum testing force	$F_{test,\min}$ [kN]	7.756
Maximum testing force	$F_{test,\max}$ [kN]	8.631
Minimum torque	$T_{\min}$ [Nm]	8.544
Maximum torque	$T_{\max}$ [Nm]	10.870

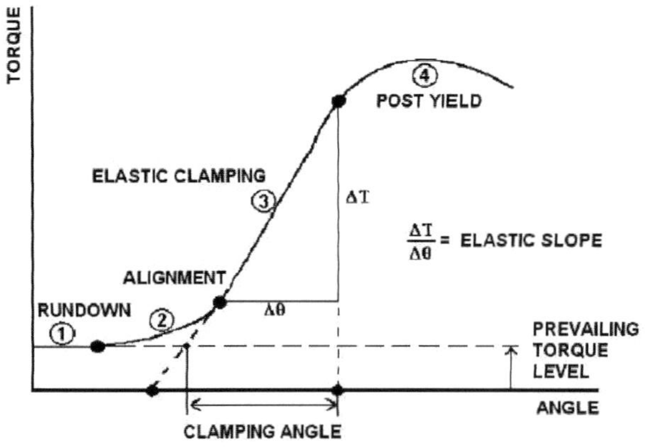

Fig. 7: Torque-angle curve [6].

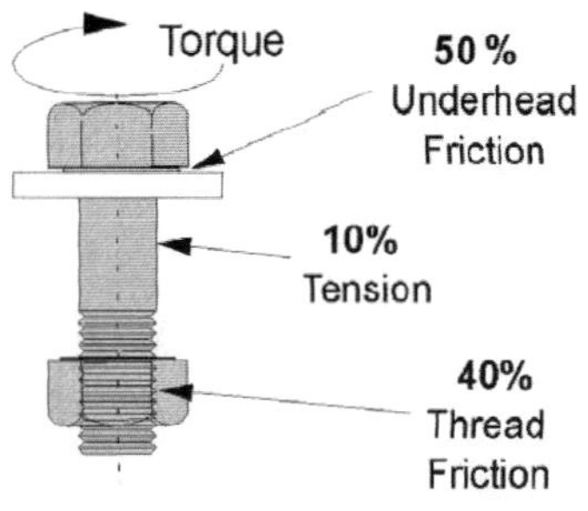

Fig. 8: Energy distribution when tightening the screw joint (standard thread pitch) [6].

The Torque-angle curve is a reliable indicator of the correct tightening of the screw connection (see Fig. 7 and Fig. 8). The implementation of the experiments was controlled by a PC, the collection and evaluation of measured data was carried out by a control program created using the software "LabVIEW FDS" from the company "National Instruments". Experimental determination of Torque-angle curves was measured on six M6 screw joints on the sample No. 3.

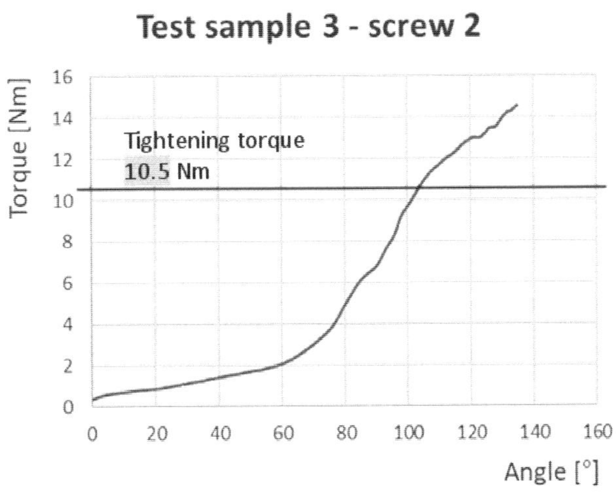

Fig. 9: Experimentally determined Torque-angle curves (for test sample 3, screw 2).

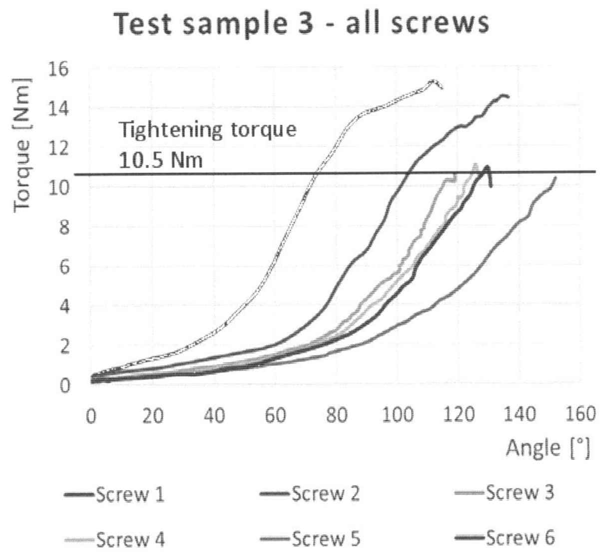

Fig. 10: Experimentally determined Torque-angle curves (for test sample 3, all screws).

From the experimentally determined Torque-angle curves, it is evident that from the torque of about 3 Nm, the Torque-angle curve dependence is steeper linear in accordance with the theory of tightening bolted joints assembly up to a tightening torque of 10.5 Nm (see Fig. 9). In the case of lower tightening torques, the contact surfaces of the joint are seated and the lock washer is compressed.

The measured curves show a mutual displacement in the direction of the horizontal axis, which was probably caused by a different angle of rotation before started the actual process of the screw joint tightening when turning by hand with a torque sensor HBM T20WN/20 Nm (see Fig. 10). This fact must always be taken into account when evaluating the measured Torque-angle curves.

There are presented the linear parts of all measured Torque-angle curves after adjusting the source data in Fig. 11. For each line (shown by the dashed line), its equation in parametric form and the reliability value R^2 are given in Fig. 11.

Experimentally evaluated values of untightening and tightening torques of the assessed bolted joints prove that the bolted joints are tightened with the correct specified tightening torque and that is used very good thread locking system (washer) too.

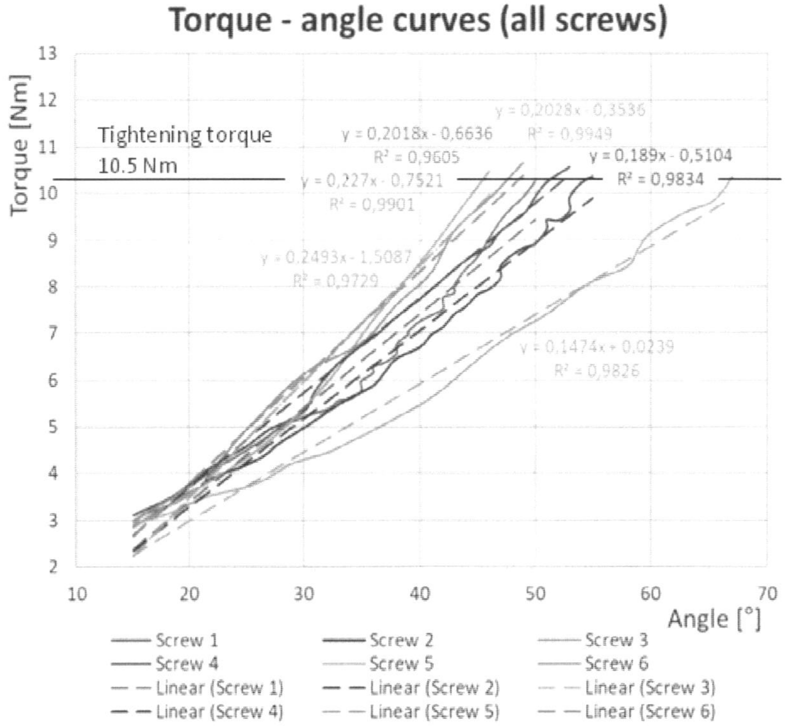

Fig. 11: Experimentally determined Torque-angle curves after adjusting the source data (for test sample 3, all screws).

2.7 Inspection of threaded surfaces after disassembly of a screw connection from a metallographic cut

After disassembly of the screw joints, metallographic samples of the screw and internal thread in aluminum alloy were made in order to assess the state of the threaded surfaces.

For the purpose of metallographic analysis, delivered part was cross-sectioned in the area of screw joint using LECO MSX255 sectioning machine. The samples of screw and threaded hole were mounted into fenol resin using LECO MX400 mounting press. Obtained metallographic specimens (see Fig. 12) were grinded up to grit P4000 and polished by 0.08 μm colloidal silica suspension (threaded hole) and 0.05 μm Al_2O_3 polishing suspension (screw) using LECO GPX300 grinding and polishing machine. Due to the etching nature of colloidal silica suspension, the microstructure of cross-sectioned threaded hole was revealed.

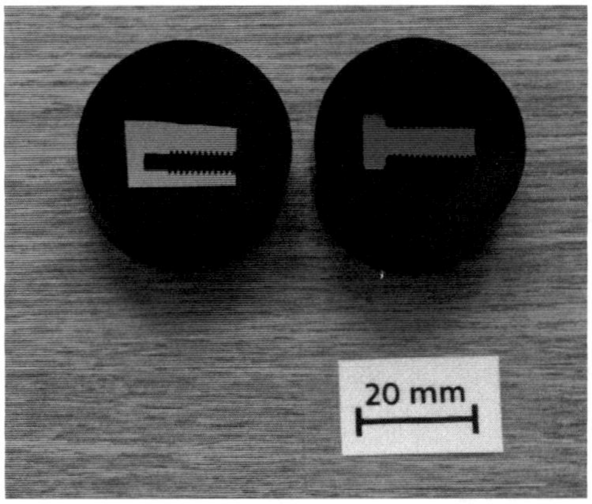

Fig. 12: Metallographic specimens, threaded hole and screw.

Fig. 13: Digital high-resolution light microscope OLYMPUS DSX 1000.

Metallographic analysis was performed using high-resolution light microscope OLYMPUS DSX 1000 (see Fig. 13). The brightfield illumination method (BF) was used for documentation of thread profiles of both specimens (threaded hole and screw).

There are presented details of photographs of the external thread of the M6 screw and the internal thread of the M6 hole after disassembly of the screw connection in Fig. 14 to Fig. 15.

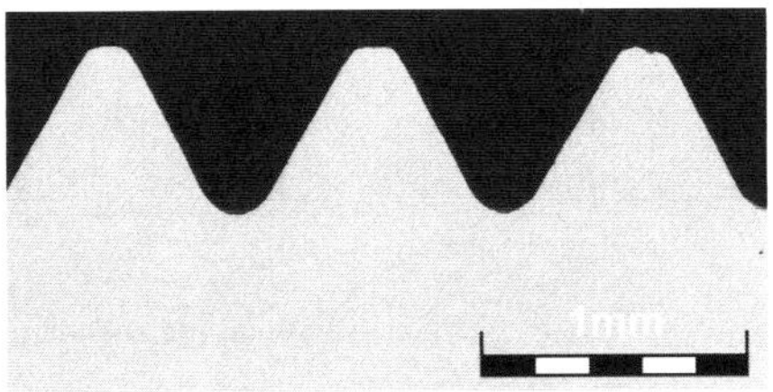

Fig. 14: Part of the external thread of the M6 screw, detail.

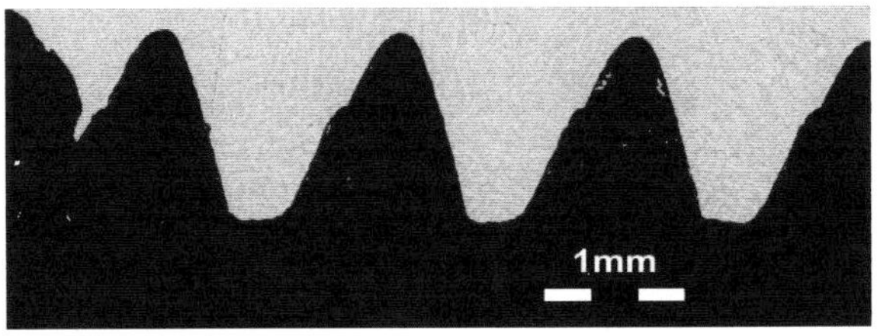

Fig. 15: Part of the internal thread of the M6 threaded hole, detail.

There are visible small deformations of the threaded surface of the internal thread M6 in the aluminum alloy in the Fig. 15, while the external thread of the steel screw M6 shows no changes (see Fig. 14).

3 Conclusion

To perform the measurements untightening torque were delivered 5 pieces assembled products. Tested samples meet the requirements according to the standard EN 14831: 2005. During and after the tests there were not found any visible damages or wrongly performed tightening.

The cover of product assembled according to the specified technological procedure (assembly instruction). The prescribed tightening torque (10–11) Nm corresponds to the EN 14831: 2005 standard. A simple test stand was designed to verify the tightening torque. Tightening and untightening torque were measured using a calibrated torque sensor HBM T20WN/20 Nm. A simple test stand was designed for experimental determination of the Torque-angle curve. Tightening torques and wrench rotation angles were measured using a calibrated torque sensor HBM T20WN/20 Nm.

The experimentally determined values of tightening and loosening torques of the bolts were statistically evaluated. Selection average value of the experimentally determined untightening torque of bolt joints is 9.289 Nm and selection standard deviation of the experimentally determined untightening torque of bolt joints is 1.533 Nm.

Checking the tightening torque of bolt joints by the "untightening and tightening method":

1. untightening torque 9.56 Nm / tightening torque 10.66 Nm (row was O.K.) / untightening torque 6.71 Nm;

2. untightening torque 9.50 Nm / tightening torque 10.67 Nm (row was O.K.) / untightening torque 6.92 Nm.

From the experimentally determined Torque-angle curves, it is evident that from the torque of about 3 Nm, the Torque-angle curve dependence is steeper linear in accordance with the theory of tightening bolted joints assembly up to a tightening torque of 10.5 Nm.

The measured curves show a mutual displacement in the direction of the horizontal axis, which was probably caused by a different angle of rotation before started the actual process of the screw joint tightening when turning by hand with a torque sensor HBM T20WN/20 Nm. This fact must always be taken into account when evaluating the measured Torque-angle curves. In the case of lower tightening torques, the contact surfaces of the joint are seated and the lock washer is compressed.

Experimentally evaluated values of untightening and tightening torques of the assessed bolted joints prove that the bolted joints are tightened with the correct specified tightening torque and that is used very good thread locking system (washer).

There are visible small deformations of the threaded surface of the internal thread M6 in the aluminum alloy, while the external thread of the steel screw M6 shows no changes. Based on all performed analyzes, the process of tightening screw joints was certified.

Acknowledgement

Thanks to my colleagues Ing. Marek Šťádler from Department of Designing and Machine Elements for the preparation of measuring stand and Ing. Vladimír Mára from Department of Materials Engineering for performing metallographic cuts and photos.

References

[1] Guideline VDI 2230, Systematic calculation of high duty bolted joints, Part 1, Joints with one cylindrical bolt, 2003.

[2] Standard EN 14831: 2005 - Fasteners - Tightening performance - Torque/Angle simplified test method.

[3] J. Kanaval, Special Process Qualification “Torque Tightening”, Department of Designing and Machine Elements; Faculty of Mechanical Engineering; Czech Technical University in Prague, Prague, 2020.

[4] Standard EN ISO 4014: 2011-06, Hexagon head bolts - Product grades A and B.

[5] Standard EN ISO 4017: 2015-05, Fasteners - Hexagon head screws - Product grades A and B.

[6] R. Shoberg, Engineering fundamentals of threaded fastener design and analysis. MI 48335, USA: RS Technologies, a Division of PCB Load & Torque, Inc. 2010.

Tension and Compression Optimization of Integrated Loop Technology Joint

B. Kropík[1,*], T. Ponížil[2], A. Malá[1], T. Zámečníková[1],

N. Schmidová[1], K. Doubrava[1], M. Dvořák[1], H. Chlup[1], T. Mareš[1]

[1] *Department of Mechanics, Biomechanics and Mechatronics, Faculty of Mechanical Engineering, Czech Technical University in Prague. Technická 4, 160 00 Prague 6, Czech Republic*

[2] *Compo Tech PLUS, spol. s r. o., Nová 1316, 342 01 Sušice, Czech Republic*

* *bohumil.kropik@fs.cvut.cz*

Abstract: This contribution deals with the experimental part of development of an integrated loop technology joint from first idea to improved design of joint according to tension and compression tensile test. Paper is focused on the experimental part of development with support of FEM analysis outputs. Initial experiments and simulations led to the change of production technology of integrated loop technology joints. Then different variants of new specimens were made by change of section layers positions to achieve the highest tension and compression strength. These new specimens were subjected to detailed experimental investigation with use of digital image correlation, acoustic emission, strain measurement and video recording and the optimal variant was found. Final variant was subjected to additional tests in other loading scenarios which will be published separately.

Keywords: optimization; CFRP; integrated joint; filament winding; integrated loop technology.

1 Introduction

Current situation in the field of joining techniques of composites (whether with other materials or with each other) is still poorly dependent on automation. Most widely expanded connection techniques of composites, such as manual lamination or bonding, are highly dependent on human work, which also affects the quality and repeatability of these joints. There are some automated techniques such as wound pin joints of metal inserts in composite tubes, but this is hard to achieve with composite to composite joint without any additional inserts. At this point it is the turn of the Integrated Loop Technology (ILT) joint which allows the connection of composite tubes to any other tube with possibility of automated production. Members of our team have been working on this for a long time. More complex view on Finite Element Method (FEM) analysis of the first generation of specimens with ILT joints is shown in reference [1]. First generation of specimens with ILT joints is discussed in reference [2]. As a demonstration of application could be shown the bicycle frame main triangle of mountain bicycle used for enduro races presented last year in reference [3] which was made with use of ILT joints. Some basic measurements of the first generation of specimens using Acoustic Emission (AE) are discussed in reference [4]. Filament wound tube samples made of carbon fibers with the ILT joints were optimized according to the analysis and experiment in the form of tensile and compressive tests. Specimens were monitored with a wide range of monitoring methods such as Digital Image Correlation (DIC), AE, Strain measurement and video monitoring during the experimental part of optimization to identify their behavior and failures.

2 Second generation of specimens

According to the results of experiments of first generation of specimens a change was made in the technology of production process of ILT joints to avoid the fiber-free area near the joint. Technology was adjusted by moving of the pins, so the winding of tube was ended behind the problematic area. New specimen is shown in the Fig. 1.

Fig. 1: New specimen with ILT joint.

New set of specimens was made by variation of fiber areas shown in Fig. 2 and Fig. 3. The relevant part of area is always highlighted in red. Overview of specimen variants is in Tab. 1. Used fiber areas are as follows:

- SN ("XY") – the area of fibers forming the main part of the tube. The "XY" represents number in brackets which means the ± angle of laid fibers to the longitudinal axis of tube. This area is shown in Fig. 2a.
- A"X" – the area of fibers forming the ILT joint. The "X" represents number and means that these fibers are wrapped around the joined tube (1) or not (2). These fibers are laid on the sides and oriented at angle of 0°. These areas are shown in Fig. 2b and Fig. 3b.
- Hoop – the area of fibers forming the main part of the tube oriented at angle of nearly 90°. This area is shown in Fig. 3a.

(a) SN ("XY") area of specimen (b) A1 area of specimen

Fig. 2: Fiber area types of new specimens.

In addition to change in production technology, the section A1 with wrapped fibers around the joined tube and the usage of composite tube as joined part are major changes between generations of specimens. All specimens consist of five layers which are chosen from these four described areas.

Tab. 1: Overview of new specimens and their composition.

Specimen type	First layer	Second layer	Third layer	Fourth layer	Fifth layer
SN_02	SN (45)	A2	Hoop	SN (0)	SN (45)
SN_07	Hoop	A1	SN (45)	SN (0)	SN (45)
SN_08	Hoop	A2	SN (45)	SN (0)	SN (45)
SN_09	Hoop	A1	A1	SN (45)	SN (45)

Due to the amount of new specimen variants, the time-consuming nature of specimen production and costs,

(a) Hoop area of specimen

(b) A2 area of specimen

Fig. 3: Fiber area types of new specimens.

a detailed FEM analysis was performed first. Based on the analysis, specimen types SN_02, SN_07 and SN_08 were chosen for more detailed analysis and tensile test. Specimen SN_09 was designed later based on the requirement of simpler production and according to obtained experimental results and it was also subjected to detail analysis and tensile test. Chosen specimens allow us to experimentally assess the influence of wrapping of fibers around tube in A section and to compare influence of swapping the fiber areas.

3 Experimental procedure

Chosen specimens were subjected to the tensile and compressive test using electromechanical universal loading machine TIRA test 2300. Number of specimens, according to the individual variants and loading scenario, are given in Tab. 2. Based on results of first generation of specimens, it was necessary to verify that the fiber-free area near the joint is not affecting the strength of designed joint anymore. Therefore, two specimens were selected in each variant and loading scenario for Digital Image Correlation (DIC) method monitoring. The specimens were also equipped with 1-axis linear strain gauge which was used for comparison of specimens and optimization of FEM analysis model. Mechanical loading of all specimens was recorded also on camera and monitored by AE method. All specimens were loaded with displacement at loading rate of 2 mm/min. All measured forces presented below ale related to the maximal tension force achieved during experiments and all mechanical strains are related to the maximal tension strain recorded during experiments given as normalized strain.

Tab. 2: Number of chosen specimens divided according to different types of loading scenarios.

Specimen type	Tension	Compression
SN_02	2	3
SN_07	2	4
SN_08	2	4
SN_09	2	2

3.1 DIC method

Configuration of DIC method setup was chosen based on previous experiences as 3D with two cameras capturing the same area of specimen as close to the ILT joint as the measured assembly allows. Configuration is shown in Fig. 4.

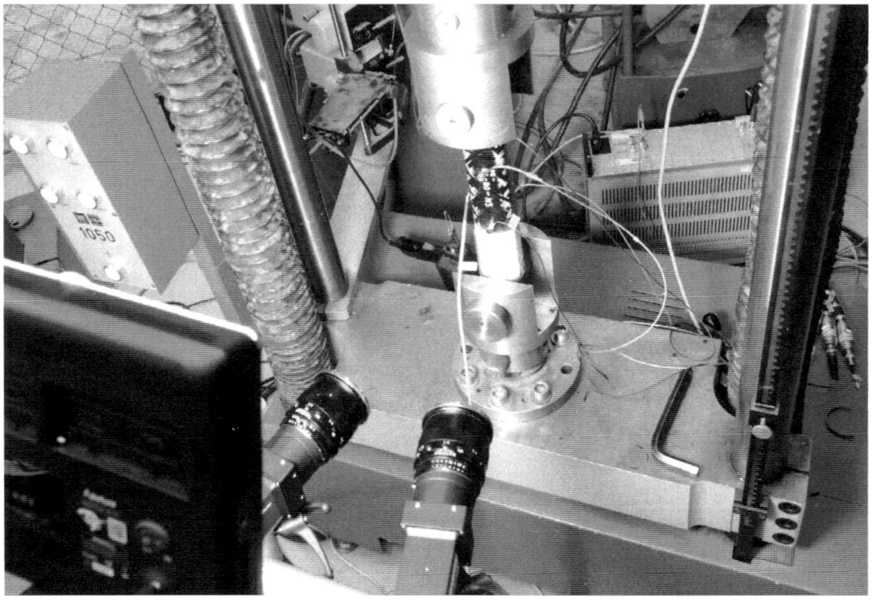

Fig. 4: Configuration of DIC method setup.

As mentioned before, two of each variant of specimens were captured by DIC in tension and two in compression with focus on previous fiber-free area of first generation of specimens. Below is presented result as Z axis displacement, (direction of view of the DIC images) which proved before to be a reliable indicator of potentially problematic area. Captures of all specimens both in tension and in compression right before failure showed that there is no more problem at this area. Specimens SN_02 and SN_08 and tension specimens of SN_07 variant showed quite uniform displacement in Z axis as is shown on example of SN_02 tension specimen in Fig. 5a. On the other hand, the SN_07 specimens loaded in compression showed different result as they suffered sudden loss of stability of composite tube before failure of ILT joints. Example of DIC result of compression specimen SN_07 is shown in Fig. 5b. The last variant of specimens SN_09 was not subjected to DIC measurements.

3.2 AE method

There were two configurations of AE method based on previous experience which was given by some issues of AE sensors during experimental measurements. Sensors were always located on the wider area of tube surface as shown in Fig. 6. One of these configurations was used on each specimen except the SN_09 tension specimens:

- Four AE sensors divided by two on both sides of specimen and located 40 mm outside the center of tube. One upper located and one lower located (Fig. 6 – right side).
- Three AE sensors on one side of specimen located in the center of specimen and 40 mm outside the center of specimen tube, also one upper located and one lower located (Fig. 6 – left side).

The number of counts was evaluated from recorded data in dependence to the measured force. Examples of most common results could be seen in Fig. 7a section for SN_07 tension loading scenario and in Fig. 7b section for SN_07 compression loading scenario. The first one in Fig. 7a shows change in increase of counts which indicates increasing of local failures of specimen and might be used for some failure prediction. On the other hand, the second one in Fig. 7b does not show significant change in increase of counts from the first recorded. Computed Tomography (CT) analysis of future specimens is planned so the comparison of AE measuring with this analysis could be useful.

(a) SN_02 specimen – tension

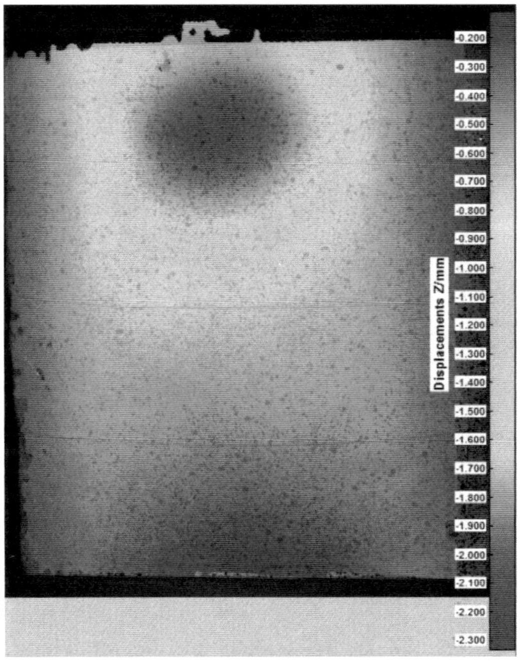

(b) SN_07 specimen – compression

Fig. 5: Z axis direction displacement of chosen specimens measured by DIC method right before final failure.

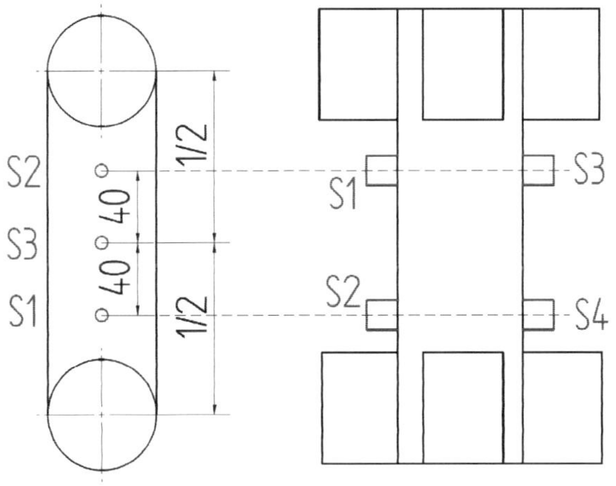

Fig. 6: Configuration of AE method setup.

3.3 Strain measurement

Due to the complexity of FEM analysis and the fact that the FE model was much more rigid than the measured data of displacement, the strain measurement was done on all specimens. Therefore 1-axis linear strain gauge was installed in the middle of narrower surface area of tube, oriented in the direction of applied loading as could be seen in Fig. 8.

Results of experimental measurements given as dependence between related force and normalized strain are shown in Fig. 9. Section of Fig. 9a is focused on tension loading scenario and section of Fig. 9b is focused on compression loading scenario. There could be seen that the differences of measured strains between specimen types are not much different. Still the SN_09 specimens seem to be slightly more rigid than the other types which is quite expected given the composition of these specimens. Based on measured experimental strain values the FE analysis was modified to contain strain gauges and the results were much closer to the measured ones than with comparison of displacements.

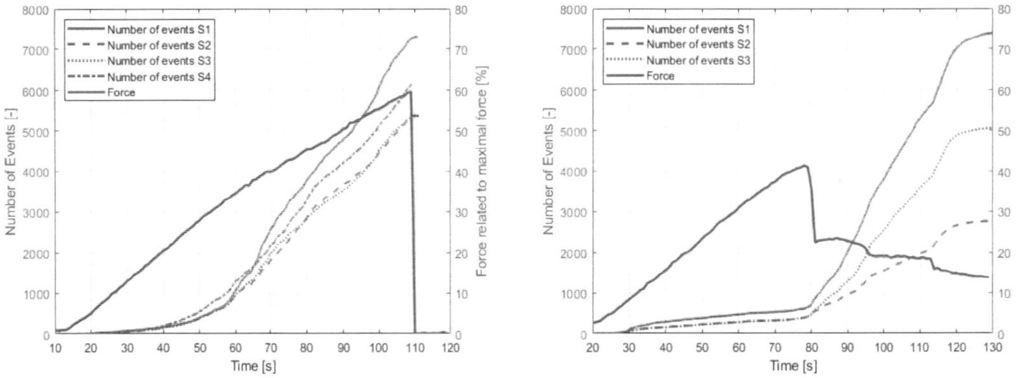

(a) SN_07 specimen – tension (b) SN_07 specimen – compression

Fig. 7: Example of evaluated number of counts of AE method measuring compared with measured force presented on SN_07 specimens

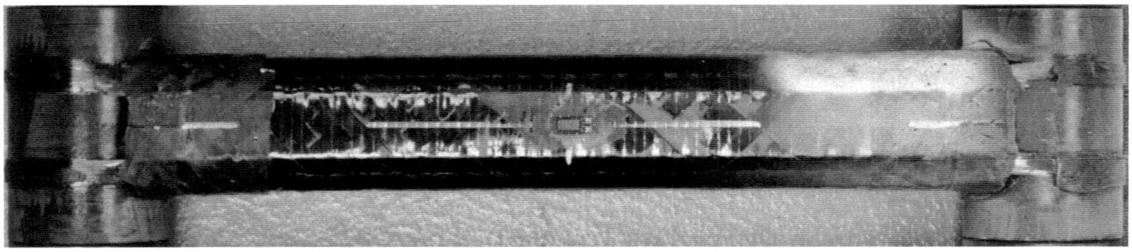

Fig. 8: One of measured specimens with installed strain gauge and sprayed grid used for DIC.

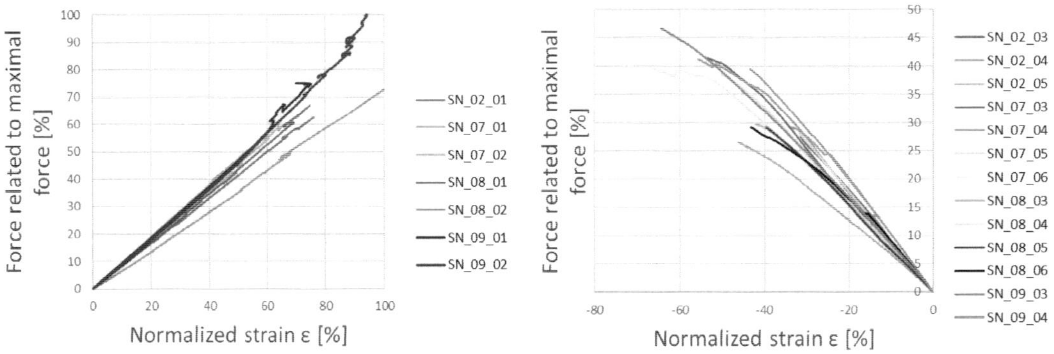

(a) Tension (b) Compression

Fig. 9: Result of experimentally measured strains given as dependence between related force and normalized strain divided according to the loading scenario.

3.4 Video recording

Video recording was performed for all specimens. The camera position was given by the position of DIC setup since they couldn't take up the same side of the specimens. Therefore, camera was placed from side of loading machine taking the narrower area of tube with whole ILT joints with pins inside. This composition gives us an idea of behaviour and failure of specimens and shows us some possible problems. The first problematic behaviour observable by eye on the recordings was fact, that the non-isotropicity of composite specimens given by their composition and production process could affect the results. For example, one of specimen loaded in tension partially failed when one side of wound fibers of single ILT joint broke. Even so, the specimen lasted another almost 15 % of tension loading before complete failure. Some recordings of specimens showed that the movable part of loading machine visibly moved fast with the failure of specimen. Upon closer look, it was observed that the specimen was rotated off-axis from the axis of loading. This phenomenon together with the difference in displacements between FE analysis and experimental measurements led to another measurement.

Based on the above observations, high resolution camera recording was prepared with markers on pins inside fixtures to compare measured displacement given by machine and displacement acquired by Edge Detection Method (EDM) and to detect possible problematic behaviour of machine during loading. Experiment was recorded by one camera as a 2D configuration for specimens SN_09 once in tension and compression. This measuring showed that during the compression loading of specimen the lower fixture (static one) most probably rotated. Also, we evaluated displacement of this measuring and the comparison between displacement recorded by machine and this method is show in Tab. 3. Although the displacement evaluated from this experiment is still containing some clearances for example between pins and ILT joints the result is much more rigid and thus closer to the FE analysis results.

Tab. 3: Comparison of displacement measured by loading machine and evaluated from EDM measurement.

Displacement of SN_09	Loading machine [mm]	EDM [mm]	Difference [%]
Tension	5.3	3.8	39
Compression	1.9	1.7	12

3.5 Overview of experimental results

Most of the partial results have already been shown in sections devoted to experimental methods. Measured dependence of displacement of loading machine on related force is given in Fig. 10. Section Fig. 10a deals with the tension loading scenario and could be seen that the SN_09 specimens are there slightly more rigid than the others. Also measured data of compression loading scenario in Fig. 10b are presented. The beginning of the signals represents the clearances between the pin and ILT joint which was different for different specimens. Comparison of average maximal forces achieved during experiments is shown in Tab. 4.

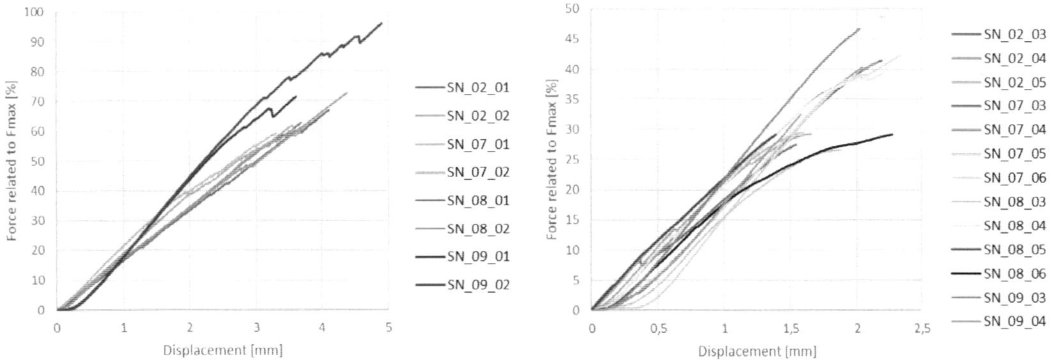

(a) Tensile test (b) Compressive test

Fig. 10: Tensile test results given as dependence between related force and displacement.

Tab. 4: Experimental results given as average maximal force achieved during loading [%].

Specimen type	Tension	Compression
SN_02	62	29
SN_07	61	41
SN_08	70	31
SN_09	87	43

As could be seen from results, the influence of wrapping the fibers around tube of ILT joint proved to be useful which could be seen on the SN_07 compression results. Also, the change in composition of specimens proved to be useful as could be seen on the SN_08 tension results. By combining of these findings with the technology demand was the SN_09 created and proved to be the most optimal variant. The sudden stability loss seen in compression of SN_07 specimens might be given by the fact that in all other experiments was the failed part the ILT joint and it happened by lower level of force. The SN_09 specimens did not show this type of failure.

4 Conclusion

Development of ILT joint according to the results of experimental measurements and monitoring was done. Experimental measurement was monitored by several different methods. DIC showed us that the previous fiber-free area was solved by the change of manufacturing technology and it pointed to a potential problem of sudden loss of stability of composite tubes by one variant of specimens. AE monitoring gave us dependence of counts on force during loading. Based on this result, it was decided to compare the measurement of AE with the CT of the next set of specimens. Strain measurement showed expected results in behaviour between variants of specimens. Video recording showed some potential problems of experimental setup and evaluation of additional special measurement by EDM method gave us more precise estimation of displacement. All mentioned results were important for making a reliable FE analysis model. The most suitable design appears to be SN_09 which was chosen for additional testing with different loading scenarios, such as bending and torque tests. Also, some specimens of this variant were tested in tension and compression again to extend the statistics and to do the mentioned comparison of CT results with AE monitoring which will be published separately.

Acknowledgement

This work has been supported by project No. SGS21/151/OHK2/3T/12 of the Grant Agency of the Czech Technical University in Prague and by project No. TJ02000252 of the Technological Agency of the Czech Republic.

References

[1] A. Malá, et al., Finite element analysis of composite tubes with integrated loop connections, Proceedings of computational mechanics 2019 (2019) 114–116, https://dspace5.zcu.cz/bitstream/11025/36298/1/Mal%C3%A1.pdf.

[2] A. Malá, et al., Analýza tuhosti a pevnosti vinutých trubek s integrovanými spoji, Mechanika kompozitních materiálů a konstrukcí 2020 - sborník (2020) 15–19, ISBN 978-80-01-06767-3.

[3] B. Kropík, et al., Identification of MTB enduro bicycle frame beaviour, Book of full papers (2020) 252–262, https://ean.vsb.cz/files/Book_of_Full_Papers.pdf.

[4] B. Kropík, et al., Detection and Monitoring of Failures in CFRP Specimens with Integrated Joints by Acoustic Emission Method, 27th Workshop of Applied Mechanics - Proceedings (2019) 20–23, ISBN 978-80-01-06680-5.

Evaluation of Mechanical Properties of Ribs Fusion by Three-point Bending Experiment

K. Kubášová[1,*], R. Sedláček[1], T. Suchý[1,2], L. Vištejnová[3],

M. Štěpánková[4], P. Klein[2], M. Bartoš[5,6]

[1] *Czech Technical University in Prague, Faculty of Mechanical Engineering, Department of Mechanics, Biomechanics and Mechatronics, Technická 4, 166 00 Prague 6, Czech Republic*

[2] *Department of Composites and Carbon Materials, Institute of Rock Structure and Mechanics, Czech Academy of Sciences, V Holešovičkách 41, 182 09 Prague 8, Czech Republic*

[3] *Charles University, Medical Faculty in Pilsen, Biomedical Center, alej Svobody 1655/76, 301 00 Pilsen, Czech Republic*

[4] *Klinik für Thoraxchirurgie, Fachkliniken Wangen, Am Vogelherd 14, 88239 Wangen im Allgäu, Germany*

[5] *Institute of Dental Medicine, First Faculty of Medicine, Charles University, General University Hospital in Prague, Kateřinská 32, 121 08 Prague 2, Czech Republic*

[6] *Institute of Anatomy, First Faculty of Medicine, Charles University, U Nemocnice 3, 128 00 Prague 2, Czech Republic*

[*] *kristyna.kubasova@fs.cvut.cz*

Abstract: The aim of the study is to compare the effect of surgical therapy of rib fractures with conservative treatment by means of biomechanical analysis of healed fractures. The surgical therapy is based on a metallic plate fixation of the rib fractures whereas the conservative treatment is based on spontaneous bone healing leading to fibrocartilage callus formation. Adult minipigs with terminated growth were used as the animal model and 6 fixed and 9 unfixed ribs were analyzed by means of the maximum bending moment, bending strength (maximum bending stress), and external bending stiffness of intact and broken (fixed and unfixed) ribs. Other evaluated parameters are maximum force, bending stiffness, maximum deflection, and energy. The experiment showed higher values of maximum bending moment and external bending stiffness (structural parameters) for unfixed ribs compared to fixed ribs. Conversely, the strength (material parameter) of unfixed ribs reached lower values than that of fixed ones.

Keywords: mechanical properties; ribs; three-point bending; experiment.

1 Introduction

Although many experts adhere to conservative treatments, there are several modern fixation devices for the treatment of the rib segment. This work deals with the influence of fixation with a metal plate on the healing of rib fractures. The prerequisite is an improvement in the stability of the chest wall, breathing, and a shorter recovery time [1, 2].

During the three-point bending experiment, many parameters can be measured that reflect the mechanical behavior of various technical and biological materials. In any case, it is important to realize what each parameter represents. Our task was to evaluate the structural and material parameters of the fusion of broken ribs. The structural parameters reflect the force transmitted by this fusion, and the material properties describe the quality of the evaluated material [3, 4].

2 Methods and materials

The study was approved by the Animal Welfare Advisory Committee of the Ministry of Education, Youth and Sports of the Czech Republic (approval ID MSMT_31280_2018_2). An animal model of an adult minipig (>9 months, >50 kg) with terminated skeletal growth was used in the project. Each minipig underwent total right-side osteotomy of 6 ribs under general anesthesia. 3 minipigs received surgical therapy (fractures fixation by metallic plate) and 3 minipigs remained without fixation. The experiment was terminated after 8 weeks and the ribs were extracted for biomechanical, micro-CT, and histological analysis. The time of termination was selected on previous experiments when the ribs were analyzed after 3 weeks and the fragments were connected by the callus composed of fibrous tissue, cartilage with minor of mineralized bone tissue, i.e. the healing process was not in its later stages when the majority of callus is composed of mineralized bone tissue. Furthermore, the physiological process of fracture healing takes several weeks and 8 weeks of healing represents the time period when the most of callus is composed of mineralized bone tissue, but still not healed completely [5]. For biomechanical analysis 6 fixed, 9 unfixed, and 39 intact ribs were applied. The main goal was to compare the mechanical parameters of fixed and unfixed ribs relative to intact ribs.

The samples were loaded on a special fixture for three-point bending at a constant feed of 5 mm/min. All samples were tested without metallic plates. Force, displacement, and time data were recorded during the experiment with a sampling frequency 40 Hz. The data recorded from the test were processed into graphical force-deflection dependences (see Fig. 1). The evaluated parameters were maximum force, strength, external bending stiffness, maximum bending moment, bending stiffness, maximum deflection (corresponding to maximum force), and energy. Using these parameters, we try to evaluate the structural and material properties of healed bone [4, 6].

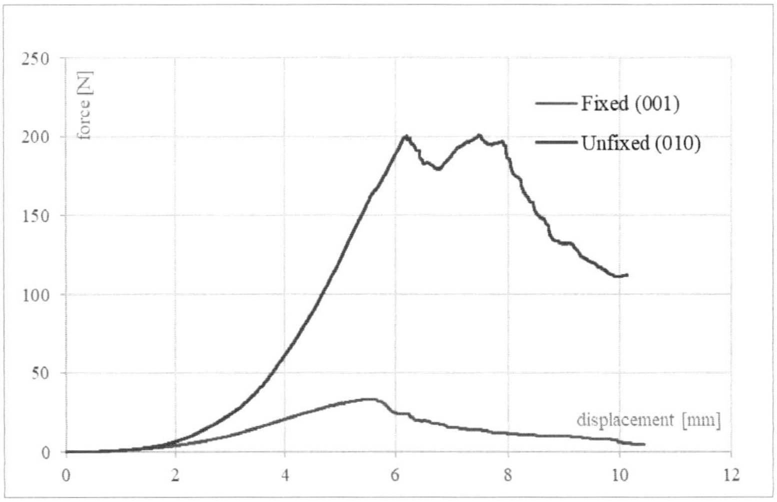

Fig. 1: Graphical force-deflection dependence of fixed and unfixed ribs.

A fibrocartilage callus is formed around the fracture of the unfixed ribs (see Fig. 2), its large size distorts the evaluated parameters, which depend on the cross-section. For these evaluated parameters to be comparable, it is necessary to compare similarly large ribs. The broken ribs (fixed and unfixed) were divided according to the size of the bending cross-section modulus into the intervals of intact ribs [3].

For further evaluation, 3 parameters were selected, 2 structural parameters – maximum bending moment, external bending stiffness, and 1 material parameter – strength. To enable the comparison of fixed and unfixed ribs, these parameters were related to the arithmetic averages of intact ribs of the same size for each parameter.

The division into groups was performed using Scott's rule, verification of normality using Q-Q plots, statistical evaluation using the Mann-Whitney nonparametric test, and effect size using the Cohen's delta.

(a) intact rib (b) rib after fixation (c) rib without fixation

Fig. 2: The loading of the ribs (intact, fixed, and unfixed) on a special fixture for three-point bending and force sensor (MTS Mini Bionix 858.02).

3 Results

It is evident (Tab. 1, Fig. 3) that the structural parameters of unfixed ribs are higher than those of fixed ribs. The fixed ribs reached only less than 10 % of the maximum bending moment of the intact ribs, however, the unfixed ribs reached more than 68% of the original bending moment of the intact ribs. The external bending stiffness of the fixed ribs increased by only over 3 %, while that of the unfixed ribs reached almost 10 %. The trend is opposite for the material parameter, the strength of fixed ribs reached more than 8 % of the strength of intact ribs, which is more than that of unfixed ribs, whose strength reached only about 4 % of the original (of the intact ribs) strength.

Tab. 1: Evaluated parameters of fixed and unfixed ribs related to intact ribs.

	STRUCTURAL PARAMETERS		MATERIAL PARAMETER
	Maximum bending moment M_{o_max} [%]	External bending stiffness S [%]	Strength σ_{max} [%]
Fixed ribs	9.95	3.37	8.20
Unfixed ribs	68.33	9.43	4.16

In the statistical comparison of the maximum bending moment and the external bending stiffness for fixed and unfixed ribs, there was a statistically significant increase (according to the Mann-Whitney test) for unfixed ribs compared to fixed ribs. The external bending stiffness increased by 180 % and the maximum bending moment even by 587 %, and the effect size reached (according to Cohen's delta) a very high significance. The opposite trend was observed for the strength when it was 49 % lower for unfixed ribs compared to fixed. However, this fact is not statistically proven due to the high data variance. A summary of the statistical evaluation is given in the table below (Tab. 2).

Tab. 2: Summary of statistical comparison of fixed and unfixed ribs.

	STATISTICALLY SIGNIFICANCE				EFFECT SIZE		
Relative parameters	Test	p-value	Significance	Test force	The ratio of arithmetic means	Cohen's delta	
Maximum bending moment	M-W	0.0018	proven	99.8 %	+ 587 %	2.88	Very high level
External bending stiffness		0.0056	proven	95.4 %	+ 180 %	2.14	Very high level
Strength		0.2629	not proven	19.4 %	– 49 %	0.57	Medium level

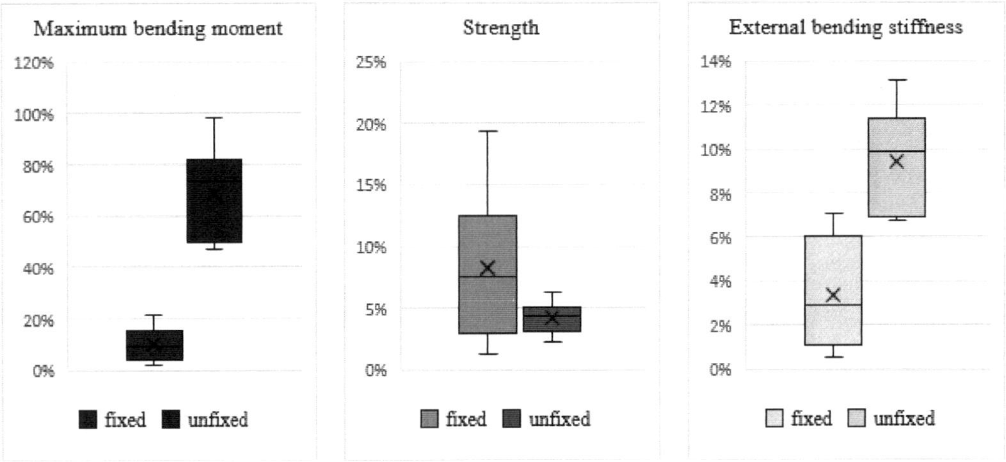

Fig. 3: Boxplots of evaluated relative parameters.

4 Conclusion

A three-point bending experiment was performed on a total of 54 rib samples, of which 6 were fixed and 9 were unfixed. The values of fixed and unfixed ribs were related to the values of intact ribs of the same size for parameters of maximum bending moment, external bending stiffness (structural parameters), and strength (material parameter). The obtained data show that the unfixed ribs reached the values of intact ribs better compared to fixed ribs in the structural parameters supported by statistical significance and high effect size. Conversely, the fixed ribs reached the values of intact ribs better compared to unfixed ribs in the material parameter, however without statistical significance and the effect size of medium level.

The biggest problem is the formation of a fibrocartilage callus around the unfixed rib, causing an uneven fracture and obstructing breathing. Further research will address the evaluation of these mechanical parameters depending on tissue healing using micro-CT and histology.

Acknowledgement

The project was supported by Charles University Grant Agency, grant number 1468218, and the SGS, with grant number SGS19/155/OHK2/3T/12.

References

[1] P. L. Althausen, et al., Early surgical stabilization of flail chest with locked plate fixation, Journal of Orthopaedic Trauma, vol. 25 (2011) 641–647, doi: 10.1097/BOT.0b013e318234d479.

[2] B.D. Solberg, et al., Treatment of Chest Wall Implosion Injuries Without Thoracotomy: Technique and Clinical Outcomes, The Journal of trauma, vol. 67 (2009) 8–13, doi: 10.1097/TA.0b013e3181a8b3be.

[3] D.T. Davy, J.F. Connolly, The biomechanical behavior of healing canine radii and ribs, Journal of Biomechanics, vol. 15 (1982) 235–247, doi: 10.1016/0021-9290(82)90170-1.

[4] M.M. Murach, et al., Rib Geometry Explains Variation in Dynamic Structural Response: Potential Implications for Frontal Impact Fracture Risk, Annals of Biomedical Engineering, vol. 45 (2017) 2159–2173, doi: 10.1007/s10439-017-1850-4.

[5] J. Lineau, et al., Initial vascularization and tissue differentiation are influenced by fixation stability, vol. 23(3) (2005) 639–45, doi: 10.1016/j.orthres.2004.09.006.

[6] C.G. Ambrose, et al., Mechanical properties of infant bone, Bone, vol. 113 (2018) 151–160, doi: 10.1016/j.bone.2018.05.015.

Numerical Simulation of the Low-velocity Impact on the Woven Composite Plate by Usage of Hyperplastic Material Model with Damage

T. Mandys[1,*], T. Kroupa[1], V. Laš[1]

[1] *University of West Bohemia in Pilsen, NTIS – New Technologies for Information Society, Technická 8, 301 00 Pilsen, Czech Republic*

* *tmandys@ntis.zcu.cz*

Abstract: The contribution is aimed to the numerical modeling of low-velocity impact and validation of the hyperplastic material model with damage designed by authors for modeling of woven fabric reinforced composites and implemented into commercial FEM software Abaqus. The composite plates were subjected to the low- velocity impact by usage of the drop-test machine with two types of impactor differed by the weight. The experimental data of the tested composite plate in form of deflections at three selected measurement points and the contact force – time dependencies were compared for particular impact velocities with the numerical simulations. The visual comparison of the occurrence of damage on the composite plates was performed and compared with corresponding damage results obtained from the computational model.

Keywords: low-velocity impact; drop test; damage; elastoplasticity; woven composite.

1 Introduction

The composite materials experienced an increased application in various manufacturing industries in the last time such as transport or renewable energy industry due to its specific advantages as decreased weight and high stiffness in comparison with conventional materials. The main disadvantage of the composite materials and its applications is its susceptibility to damage and subsequent failure, which may occur during the life-time of the composite construction. A special chapter of the loading states on the composite constructions are the impact events with the foreign object, such as the bird strike, fall of the different objects such as cell phones or screwdrivers during the inspection or maintenance, or the hits of the small objects such as the flying stones from the tires. The impact event is generally classified as low-velocity, medium velocity and high-velocity impacts [1]. The low-velocity impacts, when the energy of the impactor is fully absorbed by the material of the impacted body, appear to be the most dangerous for the safety and the further life-time of the composite structure [2]. The typical for the low-velocity impacts is causing of internal damages, which causes substantial strength reduction due to cracking of the matrix, fiber splitting or yarn locking [3]. Therefore it is necessary to be able to predict initiation of the damage and its propagation using the numerical simulations in the design part of the composite structure and to include the common cases of low-velocity impacts into the expected loading with regard to the specific application.

2 Tested material and material model

The user defined hyperplastic material model with damage was used for description of the non-linear behavior of the woven fabric reinforced composites [4]. The proposed orthotropic material model is based on thermodynamic principles in accordance of the first and second law of thermodynamics. The model assumes the finite strain theory and considers the additive decomposition of the total logarithmic strain tensor to the elastic and the plastic part [5]. The description of the damage states in the material model is realized by proposed degradation functions that are dependent on the relevant elastic strains. The equations of the damage evolution are applied in the material model in accordance with the principles of continuum damage mechanics [6]. The author's material model is implemented into commercial FEM software Abaqus 6.13 using UMAT and VUMAT subroutines that allows the usage both implicit and explicit solver. The subroutines are written in Fortran code.

The tested woven composite material was made from three layers of fiberglass fabric with plain weave pattern (density 812 g/m^2) and epoxy resin with designation MGS 385 LR. The total thickness of tested woven composite structure was 0.9 mm. Experimental compressive and cyclic static test were performed on the material specimens in accordance of ASTM standards. The mathematical optimization was used in the identification process of material parameters of tested composite. The material parameters are summarized in Tab. 1 and Tab. 2 [4].

Tab. 1: Elastic and strain/hardening curve parameters.

E_{11} [GPa]	E_{22} [GPa]	E_{33} [GPa]	G_{12} [GPa]	G_{13} [GPa]	G_{23} [GPa]	ν_{12} [-]	ν_{23} [-]	ν_{31} [-]	R_0 [MPa]	K_R [MPa]	n_R [-]
25.30	25.79	8.00	4.50	4.50	2.75	0.337	0.337	0.280	35	265	0.4

Tab. 2: Material damage parameters, units [-].

ε^0_{11T}	ε^0_{11C}	ε^0_{22T}	ε^0_{22C}	ε^0_{33T}	ε^0_{33C}	γ^0_{12}	γ^0_{23}	γ^0_{13}
0.0030	0.0007	0.0040	0.0076	0.0	0.0076	0.0048	0.0	0.0
ε^R_{11T}	ε^R_{11C}	ε^R_{22T}	ε^R_{22C}	ε^R_{33T}	ε^R_{33C}	γ^R_{12}	γ^R_{23}	γ^R_{13}
0.0084	0.0500	0.0060	0.0500	0.0001	0.0500	0.0102	0.0001	0.0001
ε^F_{11T}	ε^F_{11C}	ε^F_{22T}	ε^F_{22C}	ε^F_{33T}	ε^F_{33C}	γ^F_{12}	γ^F_{23}	γ^F_{13}
0.0125	0.2000	0.0137	0.2000	0.0233	0.2000	0.0391	0.0350	0.0350
a_{11T}	a_{11C}	a_{22T}	a_{22C}	a_{33T}	a_{33C}	a_{44}	a_{55}	a_{66}
0.0125	0.2000	0.0137	0.2000	0.0233	0.2000	0.50	0.50	0.50
b_{11T}	b_{11C}	b_{22T}	b_{22C}	b_{33T}	b_{33C}	b_{44}	b_{55}	b_{66}
0.0	10.0	0.0	10.0	100.0	10.0	100.0	100.0	0.8550
K^0_{11T}	K^0_{11C}	K^0_{22T}	K^0_{22C}	K^0_{33T}	K^0_{33C}	K_{44}	K_{55}	K_{66}
1.0	1.0	1.0	1.0	0.0	1.0	0.0	0.0	0.76
D^U_{11T}	D^U_{11C}	D^U_{22T}	D^U_{22C}	D^U_{33T}	D^U_{33C}	D^U_{44}	D^U_{55}	D^U_{66}
0.1290	0.7500	0.1210	0.75	1.0	0.75	1.0	1.0	1.0
D^U_{12}	D^U_{13}	D^U_{23}	$D^{\max}_{ijT}$	$D^{\max}_{ijC}$				
1.0	1.0	1.0	1.0	0.75				

3 Experiment – low velocity impact

A drop-test machine designed by authors was used for the realization of impact testing of the square specimens made from tested woven composite with dimensions 190 $\times$ 190 mm. The composite specimens were supported by the steel stand along the all edges. The overlap between the plate's edges and the steel stand was 20 mm at each side. The geometry of the tested square plate and its placement on the steel stand is shown in Fig. 1. The three laser sensors OptoNCDT 2200 were used for the measuring of the response of the tested body during the impact event. The position of the selected points for the laser sensors placement is shown in Fig. 1. Due to the width of the vertical linear guides of the drop-machine and compact dimensions of the plate the laser sensors had to be placed under steel stand and square plate. The impact point was in the middle of the plate. The linear guides caused the clearance of impactor in approximate value 3 mm in the horizontal plane that could shift the impact point off the centre of the composite plate. Two types of impactors with different weights (0.243 kg and 2.335 kg) were used for the impact testing. The impactors were guided by the vertical linear guides of the drop-test machine and were accelerated only by the gravity.

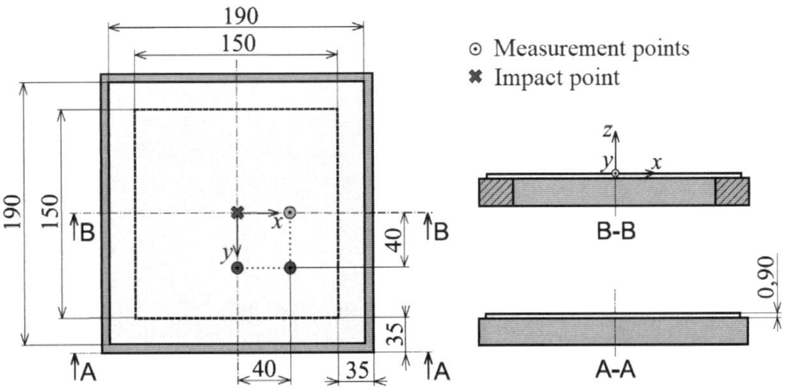

Fig. 1: The geometry of the tested woven composite square plate and the location of the laser sensors in the measurement point during the impact testing.

Fig. 2: Experimental apparatus with the woven composite plate and 0.243 g weight impactor.

The impact velocities were varied in the range from 1 to 4 m/s in case of 0.243 kg weight impactor and from 2 to 4 m/s in case of 2.335 kg weight impactor respectively. The step in the ranges of the impact velocities was for the both impactors the same, 1 m/s. The electro-mechanical brake of the impactor based on the principle of the optical gate was always activated during the impact after the first contact event of impactor with the tested composite plate. That avoided the other contact events of impactor with the composite plate and minimized the possibility of the accumulation of the damage in the structure or the occurrence of the secondary damage of the tested plate. The contact force between the impactor and the tested woven composite plate was recorded during the impact events by the force sensor Kistler 9712 B, that was built into the head (spherical shape of the head with radius r = 15 mm) of the impactor. The sampling rates for the laser sensors and the force sensor was 10 kHz. The experimental apparatus for the case of the 0.243 g weight impactor is presented in Fig. 2. The impact events were recorded by the high speed camera Olympus i-Speed 2 with the frame rate 2,000 fr/s. The records were used to other analysis of the behavior of the woven composite plates and investigation of the occurrence of the damage during the loading.

4 Numerical simulations and the results

The numerical simulations of the impact event were performed in the FEM software Abaqus using the authors defined hyperelastic material model with damage, explicit solver based on central difference scheme for time integration and finite strain theory. The computational model was created as a full contact problem of three bodies – composite plate, impactor and steel stand. The friction between bodies has been neglected. The complex geometry of the both types of impactors was simplified and only the spherical head of impactor with added mass to reach the total weighs were modeled. The attached cable from force sensor in the impactor head has been neglected.

The position of the laser sensors in the measuring points is invariable in the plane of the composite plate in the coordinate systems during the impact events, while the location of the nodes of the computational mesh is changed due to deformation of the composite plate during the impact. Therefore the values in form of deflections (z location) of the composite plate at the selected measuring points were interpolated from the space displacements obtained from the neighborhood nodes of computational mesh. The reason of the interpolations of the deflections in the measuring points of the laser sensors and the differences between interpolated deflection and the deflection from the corresponding node in the initial loading state is for the 2D state illustrated in Fig. 3. The 3D bilinear interpolation method was used; the interpolations and the results were post processed in computing platform Matlab [7].

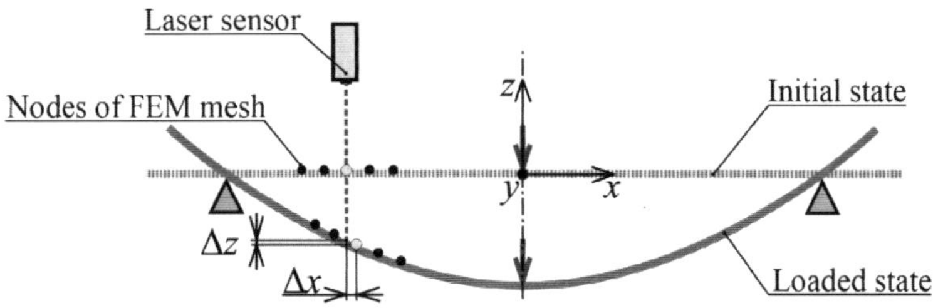

Fig. 3: The principle of the interpolation of the deflection in the measuring points from the space displacements of the neighboring nodes of computational mesh, 2D state of the situation.

The results from numerical simulations and the experiments were for all cases of the impact events compared for the range $t \in \langle 0, 30 \rangle$ ms, when the time $t = 0$ corresponds to the beginning of impact event. The comparison of deflections from the selected measurement points and contact force dependencies on the time is shown for the case of 0.243 kg weight impactor and particular impact velocities in Fig. 4–7. The same scale factor of compared values was used for all impact velocities with the corresponding weight type of impactor. The occurrences of damage were not visually observed on the composite plates in case of impact events with 0.243 kg weight impactor both in the case of numerical simulations and experiment too.

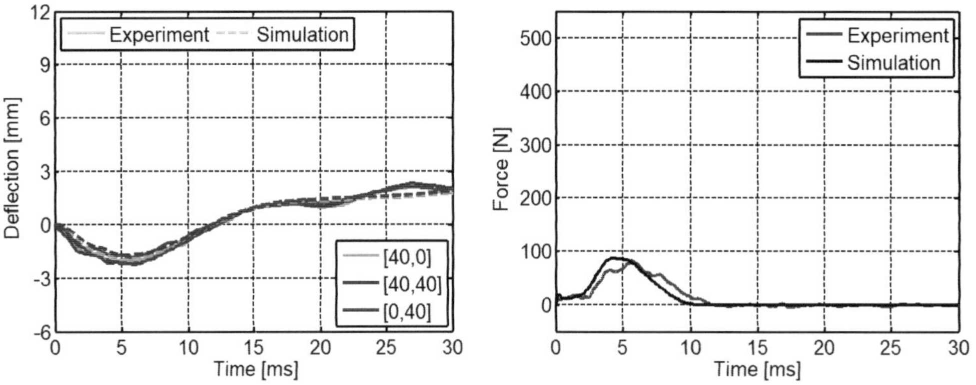

Fig. 4: The comparison of the deflections in measurement points (left) and contact force (right) between experiment and numerical simulation for 0.243 kg weight impactor and impact velocity 1.0 m/s.

The deflections at three measuring points and contact force comparisons between numerical simulations and experiment are shown for the case of 2.335 kg weight impactor and tested range of velocities in Fig. 8–10. The damage occurred during the impact testing only in the case of 2.335 kg weight impactor and only for impact velocity 4.0 m/s both in the experiment and simulation. The comparison of the occurrence of the damage on the composite plate is shown in Fig. 11. It is illustrated in case of numerical simulation on the composite plate using the contour plot of damage in the shear plane 12 (D_66) and damage in material direction 2 in tension

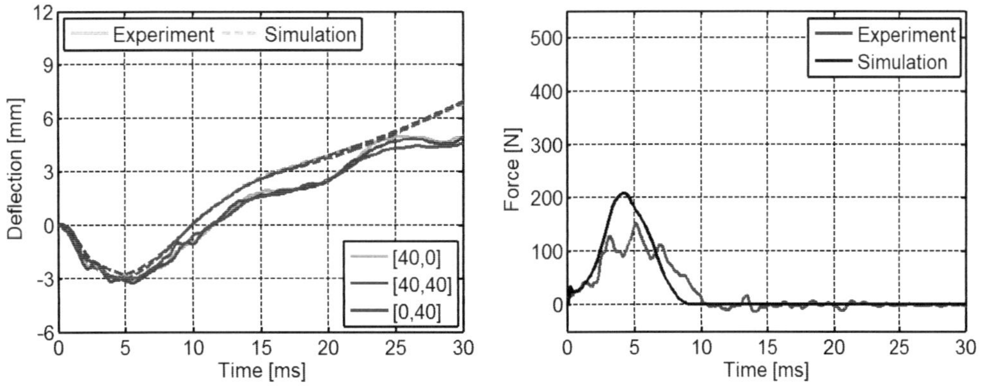

Fig. 5: The comparison of the deflections in measurement points (left) and contact force (right) between experiment and numerical simulation for 0.243 kg weight impactor and impact velocity 2.0 m/s.

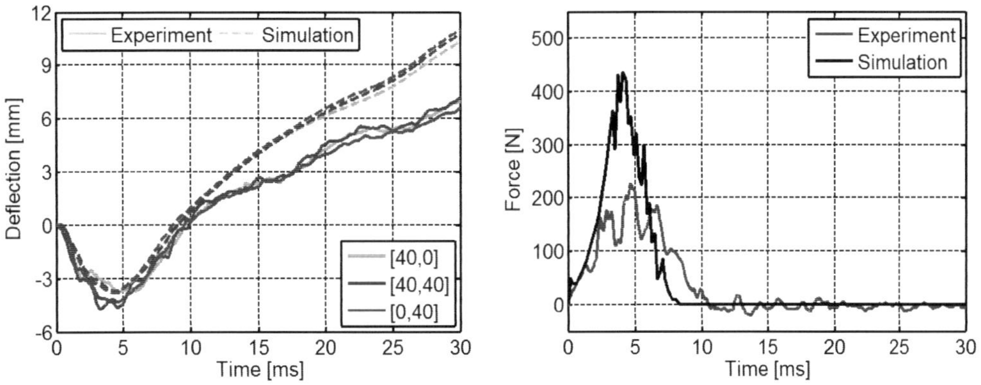

Fig. 6: The comparison of the deflections in measurement points (left) and contact force (right) between experiment and numerical simulation for 0.243 kg weight impactor and impact velocity 3.0 m/s.

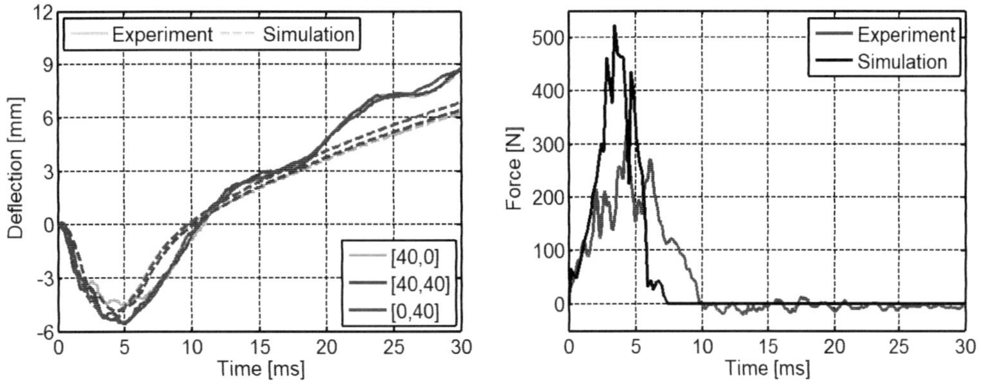

Fig. 7: The comparison of the deflections in measurement points (left) and contact force (right) between experiment and numerical simulation for 0.243 kg weight impactor and impact velocity 4.0 m/s.

(D_{22}) witch occurred locally in the impact point of the plate. The damage is in the case of experiment evident in the detail picture with the blue border in the Fig. 11.

The comparison of the behavior of the composite plate during the impact testing between the numerical computational model and the pictures obtained from high speed camera during the experiment is shown for the 2.335 kg weight impactor and impact velocity 4.0 m/s in Fig. 12. The maximal deflection is achieved in the time t = 15 ms after the start of impact event at t = 0 ms.

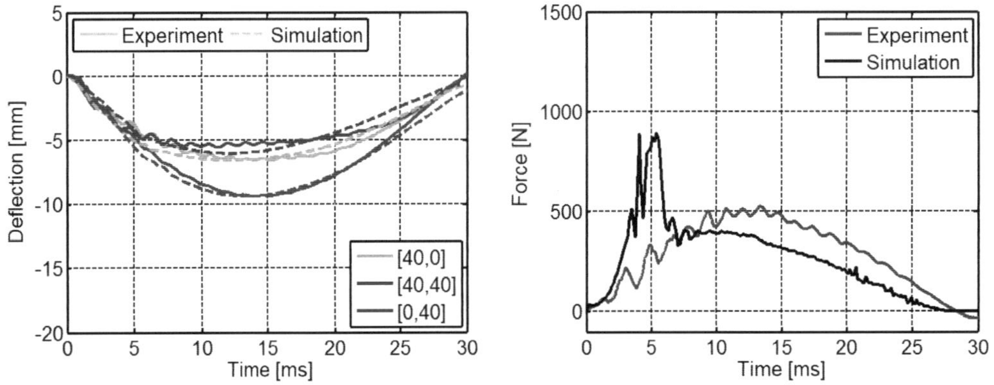

Fig. 8: The comparison of the deflections in measurement points (left) and contact force (right) between experiment and numerical simulation for 2.335 kg weight impactor and impact velocity 2.0 m/s.

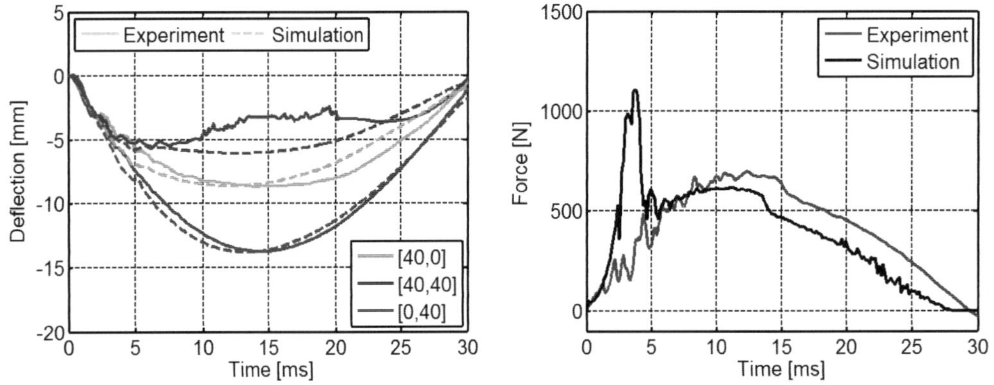

Fig. 9: The comparison of the deflections in measurement points (left) and contact force (right) between experiment and numerical simulation for 2.335 kg weight impactor and impact velocity 3.0 m/s.

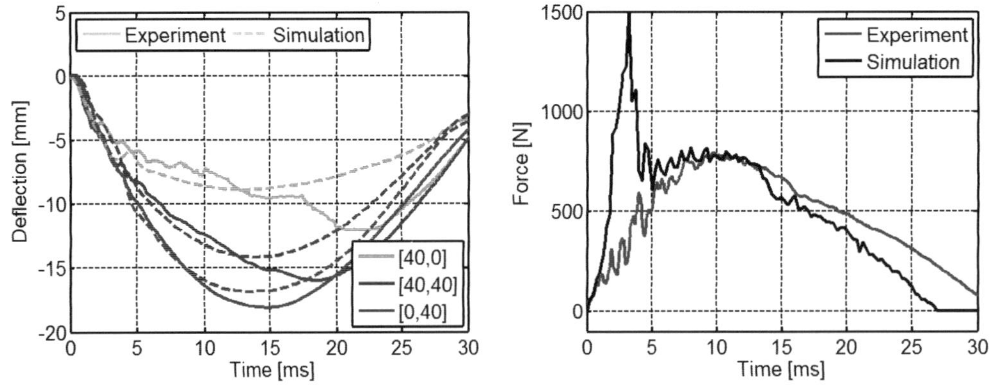

Fig. 10: The comparison of the deflections in measurement points (left) and contact force (right) between experiment and numerical simulation for 2.335 kg weight impactor and impact velocity 4.0 m/s.

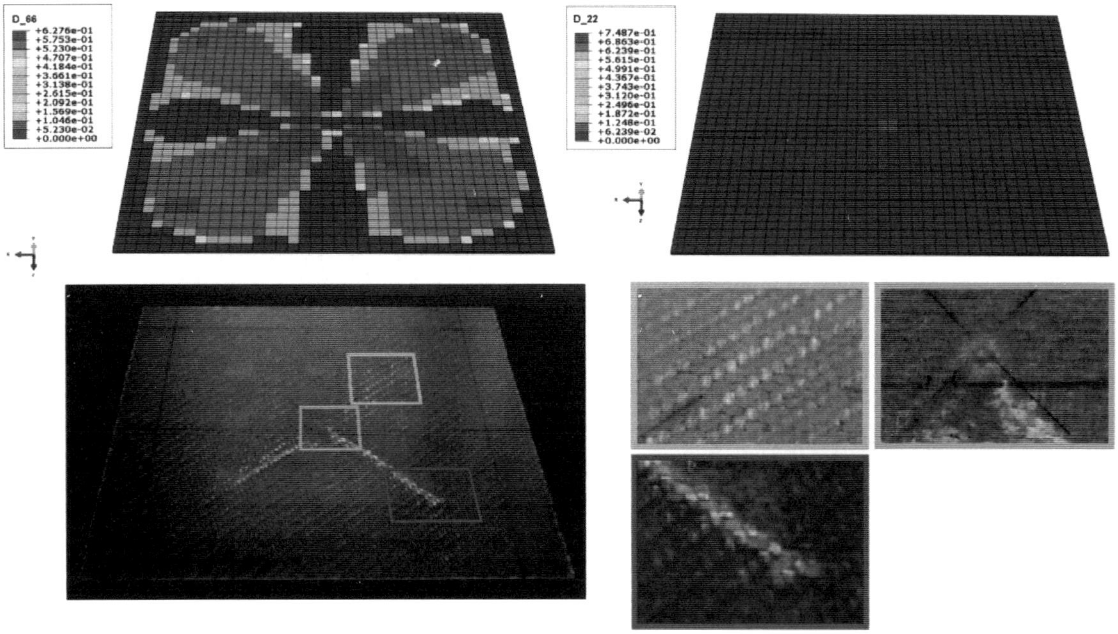

Fig. 11: The comparison of damage of the composite plate subjected to low-velocity impact with 2.335 kg weight impactor and impact velocity 4.0 m/s between numerical simulation and experiment.

5 Conclusions

User defined hyperplastic material model with damage implemented into commercial finite element method software Abaqus was verified using the experiment in the case of modeling of the dynamic events. The response of the composite square plate to low-velocity impact was experimentally investigated using the drop-test machine and two types of impactor with different weights. The deflection-time dependencies, contact force-time dependencies and the occurrence of the damage were compared. Even though the comparison of the deflection-time dependencies from the experiment and numerical simulations shown the good agreement, in the case of low-weight impactor the increasing deviations on the impact velocity in the secondary movements of the plate, i.e. after unloading of the impactor, are evident. The comparison of the contact force-time dependencies shows agreement between numerical and experimental results. However the experimental data did not show the increase of the maximum of the first peak of the contact force-time dependencies in the contrast to the numerical results. These deviations of the contact force dependencies can be caused by neglected friction and damping in contact between spherical head of impactor and composite plate in the computational model, possibly during the experiment there may have been excessive damping of the impactor due to clearance of the impactor in the horizontal plane in the linear guides of the drop-test machine. The visual comparison of the visible occurrence of damage on the composite plate shows the quantitative match with the numerical simulations. However, the damage model predicts the failure, or rupture that was not reached in the experiment.

Acknowledgement

The work was supported by the project SGS-2019-009 of the Czech Ministry of Education, Youth and Sports.

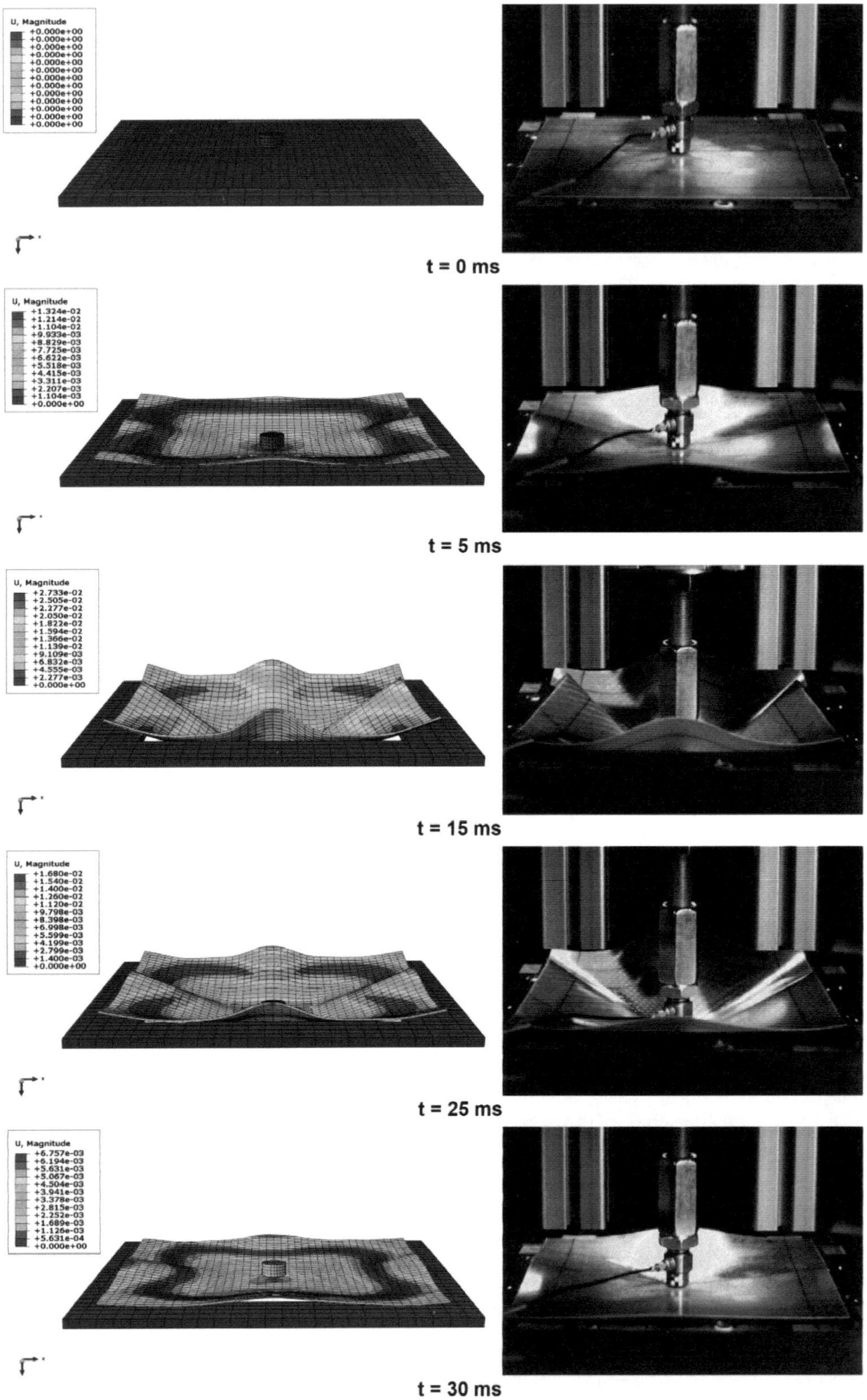

Fig. 12: The comparison of the behavior of the composite plate during the impact testing between the numerical computational model and the pictures obtained from high speed camera during the experiment for the 2.335kg weight impactor and impact velocity 4.0 m/s. The total displacement is presented in the case of numerical simulations.

References

[1] V. Balasubramani, S. Rejandra Boopathy, R. Vasudevan, Numerical Analysis of Low Velocity Impact on Laminated Composite Plates, Procedia Engineering 64 (2013), pp. 1089–1098.

[2] Y. Pai, K. Dayananda Pai, M. Vijaya Kini, A review on low velocity impact study of hybrid polymer composites, MaterialsToday: Proceedings, doi: 10.1016/j.matpr.2021.05.390.

[3] S. Abrate, B. Castanié, D.S. Yapa Rajapakse, Dynamic failure of Composite and Sandwich Structures, Dortrecht, Netherlands, Springer, 2013.

[4] T. Mandys, T. Kroupa, V. Laš, L. Lobovský, Hyperplastic material model with damage for woven composites, Composite Science and Technology, 183 (2019), doi: 10.1016/j.compscitech.2019.107790.

[5] M.Á. Caminero, F.J. Montáns, K.J. Bathe, Modeling large strain anisotropic elasto-plasticity with logarithmic strain and stress measure, Computers & Structures 89 (2011), pp. 826–843, doi: 10.1016/j.compstruc.2011.02.011.

[6] S. Murakami, Continuum Damage Mechanics – A continuum Mechanics Approach to the Analysis of Damage and Fracture, Springer Science + Business Media, 2012.

[7] S. Olivík, Ukázka možností interpolace dat v softwaru Matlab, 2006, https://mat.fsv.cvut.cz/komisevstez/13/prispevky/olivik.pdf.

Minimally Invasive Fixation Techniques of Sacral Bone Fractures

M. Marešová[1], L. Lobovský[1,*], T. Mandys[1], M. Salášek[1,2],

J. Hluchá[1], T. Pavelka[2], J. Křen[1]

[1] *NTIS - New Technologies for the Information Society, Faculty of Applied Sciences, University of West Bohemia, Univerzitní 8, 301 00 Plzeň, Czech Republic*

[2] *Clinic for Orthopaedics and Traumatology of Locomotive Organs, University Hospital, alej Svobody 8, 304 60 Plzeň, Czech Republic*

[*] *lobo@kme.zcu.cz*

Abstract: The study focuses on minimally invasive fixation techniques of sacral bone fractures by means of the transiliac internal fixator (TIFI). Several implementations of TIFI for unilateral transforaminal fracture treatment are studied both experimentally and computationally. The comparison of fixation techniques is performed on the basis of the calculated stiffness ratio, which is determined from the displacement of the sacrum and the applied external load. The results of the experimental measurements are compared with numerical simulations while a very good agreement is achieved.

Keywords: biomechanics; pelvis; injury; osteosynthesis; transiliac internal fixator, iliosacral screw.

1 Introduction

Pelvic bone fractures account for about 2 % of all fractures of the human musculoskeletal system. In patients without osteoporosis, most common causes of pelvic fractures are impacts during traffic accidents or falls from large heights. Unstable fractures in pelvic area are often associated with polytraumatic injuries and a high risk of post-operative complications. A selection of the most appropriate fixation technique is one of the key factors for successful medical treatment and minimization of potential negative consequences of the fracture. Either external or internal fixators may be applied [1–3].

The presented study focuses solely on a treatment of a unilateral transforaminal fracture which belongs to group C, subgroup 1 of Tile's classification (i.e. C1 type fracture) [4]. No other affiliated injuries are considered. In the following, minimally invasive fixation techniques are of interest. A special attention is paid to several types of treatment by the transiliac internal fixator (TIFI) [5] and a possible combination of TIFI with the illiosacral screw (ISS) [6].

The TIFI fixator consists of two polyaxial screws, a single polyaxial screw is inserted in the wing of each ilium, while the two screws are interconnected with a subfascial transverse connector rod, Fig. 1 (left), thus forming a stable metal structure which constrains relative motion of the iliac bones [7]. The head of the polyaxial screw is stored in a ball housing which allows an attachment of the transverse connector rod at an arbitrary spatial angle. This simplifies positioning of the TIFI fixation during surgical treatment.

The ISS fixator is a long threaded screw, Fig. 1 (right), which ties the sacral bone fragments together. In most cases, the ISS is inserted through ala of ilium into the first sacral vertebra (S1) [6].

In the following, four fixation techniques are analysed: the classic TIFI, the supraacetabular TIFI, the dual TIFI and the combination of supraacetabular TIFI with ISS (TIFI+ISS). When the classically positioned TIFI is utilised, the polyaxial screws are inserted near the posterior superior iliac spine parallelly to the superior gluteal line [8]. During the supraacetabular insertion of TIFI, the polyaxial screws are inserted within the supraacetabular corridor, i.e. inferior to the classic TIFI [9]. The dual TIFI employs both the classic TIFI and the supraacetabular TIFI in order to fix the sacral fracture [10]. The TIFI+ISS technique combines the supraacetabular TIFI with the ISS, which is inserted into the S1 vertebra.

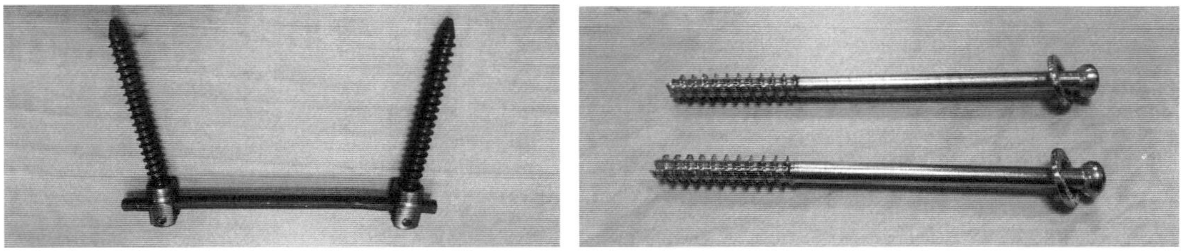

Fig. 1: Transiliac internal fixator (left), iliosacral screws (right).

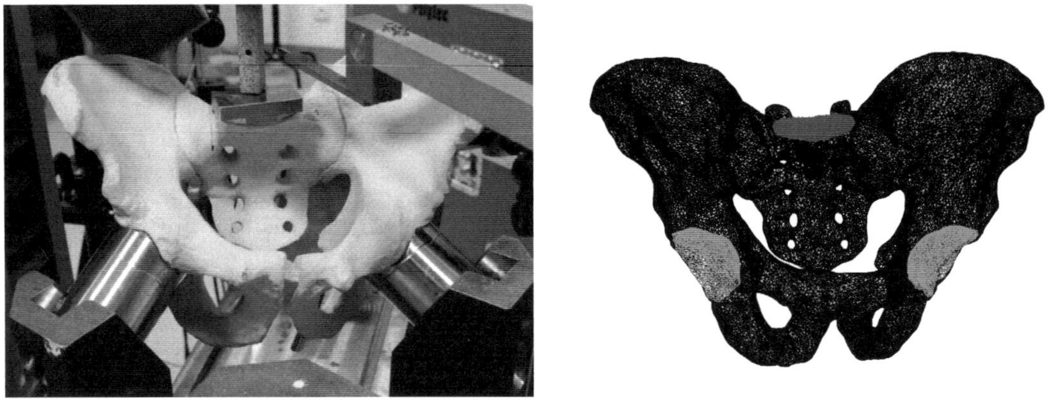

Fig. 2: Orthopaedic model of pelvic structure used during the experimental campaign (left) and FE computational mesh used for numerical simulations (right).

The selected types of fixation techniques are examined for a simplified test case which can be easily controlled within both experimental and computational modelling. This involves a male pelvic ring fixed in both acetabula so that no translations nor rotations in the acetabular region are allowed. The pelvic ring is positioned so that the sacral base is horizontal. The sacral base is loaded by a vertical force which magnitude corresponds to the physiological loading by an upper body weight of an average adult male.

2 Experimental analysis

The experiments were performed on artificial orthopedic models of an adult male pelvis [11]. These models were made of solid polyurethane foam and were mounted on a steel stand which enabled fixing the model in acetabula on steel femoral heads. The femoral heads were manufactured as three-dimensional negatives of acetabula. Thus it was possible to constrain displacements and rotations in the acetabular regions. In order to asses the performance of the selected fixation techniques, the treated pelvic models were subjected to compressive loading in mechanical testing machine, Fig. 2 (left).

On the contrary to cadaveric samples, application of the orthopaedic models enhanced repeatibility of the performed tests as the variations between the models were small. The material properties of the polyurethane foam were determined using a serie of dedicated tensile and compressive test measurements on cuboidal samples of material cut out of the pelvic models. The resulting Young modulus and Poisson ratio of the foam are presented in Tab. 1.

The solid foam, which the pelvic models were made of, was a homogenous isotropic material. Its mechanical response was almost linear while the viscoelastic effects were negligible. This significantly simplified development and verification of the computational model.

The stiffness of the treated pelvic structure was assessed based on the load-displacement curves recorded at the sacral base during the full scale pelvic tests. This data were recorded by the load cell and the extensometer of the mechanical testing machine while the three-dimensional motion of the fractured bone was analysed by digital image correlation of a set of simultaneously captured images. The quality of the pelvic ring fixation was assessed based on ratio between the stiffness of the treated pelvic structure and the stiffness of the intact pelvis. In addition a relative motion of selected points on the fractured bone parts during loading was investigated.

Tab. 1: Materials parameters.

	Young modulus [GPa]	Poisson ratio [-]
Solid foam	0.194	0.20
Titanium	120	0.33

3 Computational analysis

The experimental data were utilised for validation of the numerical computations based on the finite element (FE) method. The computational mesh is displayed in Fig. 2 (right). The geometry of the computational model was created from CT images of an orthopedic pelvic model [12]. The CT scans were performed in the transversal direction with a distance of 1.2 mm between two subsequent CT images. The computational model was simplified so that the pelvic ring was considered as a single continuous body, i.e. the cartilage and joints between pelvic bones were neglected.

The computational mesh of the model was created within the HyperMesh software package [13] and consisted of about 300 000 volume elements. Specifically, the elements were tetrahedrons with a characteristic edge length of 2.44 mm. Computational simulations were performed in the finite element solver Abaqus [14].

For the purposes of numerical analysis, a homogeneous isotropic linearly elastic model was used for both the polyurethane solid foam and the titanium fixators. The applied material parameters are given in Tab. 1.

The two sets of boundary conditions are highlighted in Fig. 2 (right). The red colour highlights the sacral base region where the static load was applied in cranio-caudal direction. The blue colour depicts the acetabula which were fixed in space, i.e. rotations and displacements in all directions were constrained.

The linear unilateral transforaminal fracture was considered. The direction and the size of the fracture were created in accordance with the experiments. The width of the fracture line was 0.7 mm. A surface to surface contact problem was considered between the surfaces of the fractured bone parts with a friction coefficient of 0.8.

The computational models of the individual fixator screws have been simplified by omitting their threads. Each screw was created as a three-dimensional cylindrical body with volumetric tetrahedral elements and constrained to the surface of the insertion hole within the pelvic bone. The screws and the cylindrical connector rod (also created using volumetric tetrahedral elements) were tied using a constraint.

4 Results and discussion

In order to assess the efficiency of each fixation technique, the stiffness of the entire pelvic structure k was determined for each model as

$$k = \frac{F}{u^B}, \tag{1}$$

where F was the external loading force acting in the cranio-caudal direction (i.e. perpendicular to the sacral base) and u^B was the displacement of the sacral base along the axis parallel to the direction of loading. In the experiments, the stiffness of the fractured pelvis model treated with a selected fixation technique was always compared to the stiffness of the original intact model by a stiffness ratio r_k. The same ratio was also determined during the numerical analysis. It was defined as

$$r_k = \frac{k_F}{k_I}, \tag{2}$$

where k_F corresponded to the stiffness of the fractured pelvis model with a fixation and k_I was the stiffness of the intact pelvis model.

In ordert to assess the stiffness of the intact pelvis, both experimental measurements and computational simulations were performed for the load of $F = 300\,\mathrm{N}$. The computationally predicted vertical displacement of the sacral base subjected to this loading was $u_I^B = 0.419\,\mathrm{mm}$. Thus the resulting stiffness of the intact model was $k_I = 714.6\,\mathrm{N\cdot mm^{-1}}$. This value was subsequently used as a reference value for the stiffness ratio

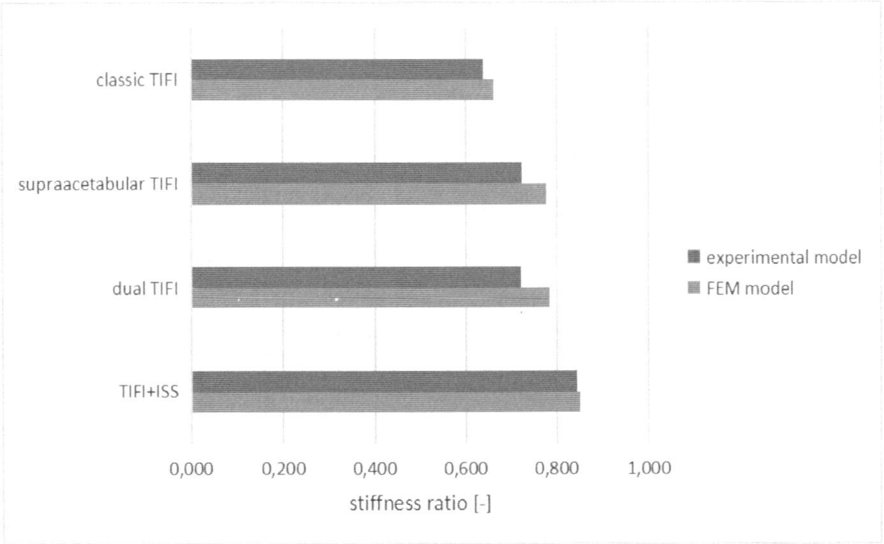

Fig. 3: Comparison of the resulting stiffness ratios obtained experimentally and computationally.

Tab. 2: Resulting stiffness ratios of both experimental and FEM model under load F = 500 N.

	stiffness ratio experiment [-]	stiffness ratio FEM [-]
classic TIFI	0.639	0.660
dual TIFI	0.720	0.785
supraacetabular TIFI	0.722	0.777
TIFI+ISS	0.842	0.849

calculations for all individual fixators within the computational study. The stiffness of the pelvic model treated with a selected fixator was determined using a maximal static load of 500 N.

Within the experimental part of the study, each fixator was tested using a new orthopaedic model of pelvis. Thus the stiffness of the intact pelvis had to be determined for every single pelvic model separately. Then, the fracture was created and a studied fixation technique applied.

The resulting values of stiffness ratio obtained from both experimental tests and numerical simulations are listed in Tab. 2 and graphically visualised in Fig. 3. The results were in good agreement while the computational data predicted slighlty larger values of the stiffness ratio than the measurements. The best results were achieved for the combination of TIFI and ISS which provided the stiffness ratio of about 85 %. Furthermore, results for the dual TIFI and for the supraacetabular TIFI were about identical. Thus applying the dual fixation did not bring any further improvement in the studied case. The computational values indicated the stiffness ratio of roughly 78 % for both techniques. The worst values were achieved by the classic TIFI which provided the stiffness ratio of 66 % within the FE analysis and only 64 % in the experiments.

The maximum deviation of the FE model from the experimental data is about 9 % for the model with dual TIFI fixation. On the contrary, the experimental and computational results differ in less than 1 % for the combination of TIFI and ISS.

5 Conclusion

Within this study, both the experimental and the computational model of the pelvic structure were developed and used to analyze several TIFI techniques. For the studied case of the linear vertical transforaminal sacral fracture without comminuted zone, experimental as well as computational results indicated that supraacetabular insertion of TIFI fixator and dual application of TIFI was superior to the classical TIFI technique and provided significantly higher stiffness of the lateral pelvic segment. However, enhancing the supraacetabularly inserted

TIFI with the second TIFI fixator (forming the dual TIFI) did not provide higher stability of the pelvic structure. For the studied case, the best results in terms of structural stiffness were achieved when the supraacetabular TIFI was combined with the ISS inserted into S1 vertebra. This restored the stiffness of the pelvic structure to almost 85 % of its original value.

Acknowledgement

This publication is supported by the European Regional Development Fund-Project "Application of Modern Technologies in Medicine and Industry" No. CZ.02.1.01/0.0/0.0/17_048/0007280.

References

[1] J. M. Matta, P. Tornetta, Internal Fixation of Unstable Pelvic Ring Injuries, Clin Orthop Relat Res. 329 (1996) 129–140, doi: 10.1097/00003086-199608000-00016.

[2] M. Scaglione et al., External fixation in pelvic fractures. Musculoskelet Surg 94 (2010) 63–70, doi: 10.1007/s12306-010-0084-5.

[3] S. Wu et al., Minimally invasive internal fixation for unstable pelvic ring fractures: a retrospective study of 27 cases, J Orthop Surg Res 16 (2021) 350, doi: 10.1186/s13018-021-02387-5.

[4] Fakler, J. K. M., Stahel, P. F., Lundy, D. W., Classification of Pelvic Ring Injuries, In W. R. Smith, B. H. Ziran, S. J. Morgan (Eds.), Fractures of the Pelvis and Acetabulum (2007) 11–26.

[5] T. Dienstknecht et al.,Biomechanical analysis of transiliac internal fixator. International orthopaedics 35(12) (2011) 1863–1868, doi: 10.1007/s00264-011-1251-5.

[6] M. A. Giráldez-Sánchez et al., Percutaneous iliosacral fixation in external rotational pelvic fractures. A biomechanical analysis, Injury 46(2) (2015) 327–332, doi: 10.1016/j.injury.2014.10.058.

[7] J. M. Vigdorchik et al., A biomechanical study of standard posterior pelvic ring fixation versus a posterior pedicle screw construct, Injury 46 (2015) 1491–1496, doi: 10.1016/j.injury.2015.04.038.

[8] B. Füchtmeier et al., Die minimal-invasive Stabilisierung des dorsalen Beckenrings mit dem transiliakalen Fixateur interne (TIFI) - operative Technik und erste klinische Ergebnisse. Unfallchirurg 107 (2004) 1142–1151, doi: 10.1007/s00113-004-0824-9.

[9] P. Korovessis, M. Stamatakis, A. Baikousis, Posterior stabilization of unstable sacroiliac injuries with the Texas Scottish Rite Hospital spinal instrumentation, Orthopedics 23(4) (2000) 323–327.

[10] C. Tempelaere, C. Vincent, C. Court, Percutaneous posterior fixation for unstable pelvic ring fractures, Orthop Traumatol Surg Res. 103(8) (2017) 1169–1171, doi: 10.1016/j.otsr.2017.07.024.

[11] Retrieved from: https://www.sawbones.com/full-male-pelvis-large-solid-foam-w-sacrum-1301.html.

[12] Retrieved from: http://sfepy.org/dicom2fem/.

[13] Retrieved from: https://www.altair.com/hyperworks/.

[14] Retrieved from: https://www.3ds.com/products-services/simulia/products/abaqus/.

Experimental Analysis of Vehicle Response

J. Melcer[1,*]

[1] *University of Žilina, Faculty of Civil Engineering, Department of Structural Mechanics, Univerzitná 8215/1, 010 26 Žilina, Slovak Republic*

* *jozef.melcer@uniza.sk*

Abstract: The present article deals with the experimental monitoring of the vehicle response when driving along the road and the subsequent processing of the measured signals. Describes the used technique, the properties of the sensors and the methodology of measurement. It shows the possibilities of statistical analysis of the obtained signals, correlation dependences and frequency composition of vibrations. Makes conclusions for possible applications in engineering practice.

Keywords: experiment; vehicle; dynamic response; analysis; signal processing.

1 Introduction

The behavior of the vehicle when driving on the road can be modeled numerically or monitored experimentally. The experiment represents the most objective possibility of assessing objective reality, therefore its results have more weight than the results of numerical solutions. As experimental technique develops, so experimental procedures are also improved, especially the possibilities of processing measured signals. Vehicle response monitoring has also been done in the past [1–7]. It was done by mechanical and civil engineers, each from their own point of view. Civil engineers are interested in the results related to the road load, driving comfort and the ability to verify the calculation models of vehicles.

2 Experimental measurements and instrumentation

The subject of the experiment was the measurement of vertical accelerations at selected points of the Tatra T815 truck at different speeds of its movement along the road. The accelerometers were located on the right side of the vehicle on the front and rear axles and at the center of gravity of the vehicle's sprung mass, Fig. 1.

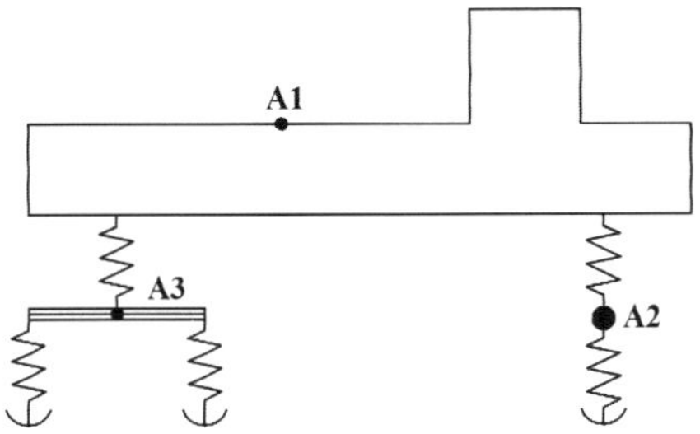

Fig. 1: Location of accelerometers on the vehicle.

The measuring line consists of the following components: sensor, amplifier, A/D interface, measuring center and data printer, Fig. 2. The signal transmission to the measuring center can be via cables (the center is located in the vehicle) or it is possible to use wireless transmission (the center is located outside the vehicle).

A piezoelectric miniature accelerometer Brüel & Kjær Deltatron 4508 B (Fig. 3) with the following parameters was used as a sensor: frequency range 0.4 – 8,000 Hz, measured range ± 70 m/s^2, temperature -54 – +100 °C, weight 4.8 gram, sensitivity 100 mV/ms^{-2}, output CCLD, resonance frequency 25 kHz.

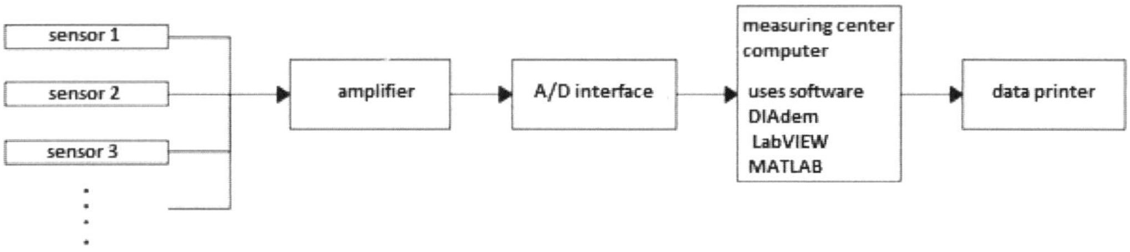

Fig. 2: Measuring line.

Fig. 3: Accelerometer Deltatron 4508 B.

M68D1 amplifier was used, Fig. 4. The amplifier is intended for connection of piezoelectric acceleration sensors. It enables functions such as amplification, low-pass or high-pass filtering, signal integration for speed or displacement measurement. It can be used in laboratory conditions as well as in the field.

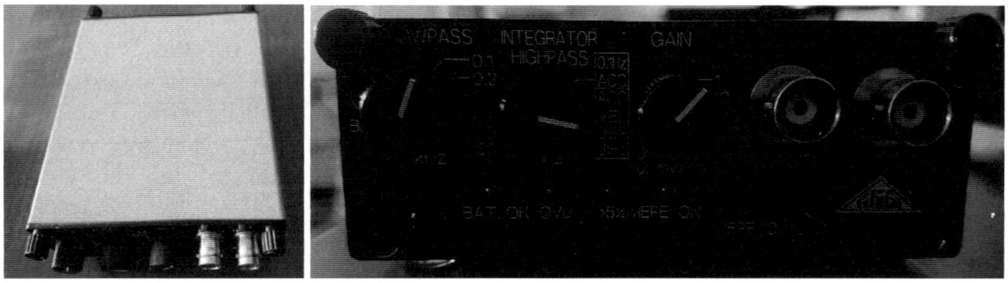

Fig. 4: Amplifier M68D1.

A/D converter NI 9215 was used, Fig. 5. The module contains four simultaneously sampled analog input channels, a sequential approximation register SAR and a 16-bit analog-to-digital converter ADCs.

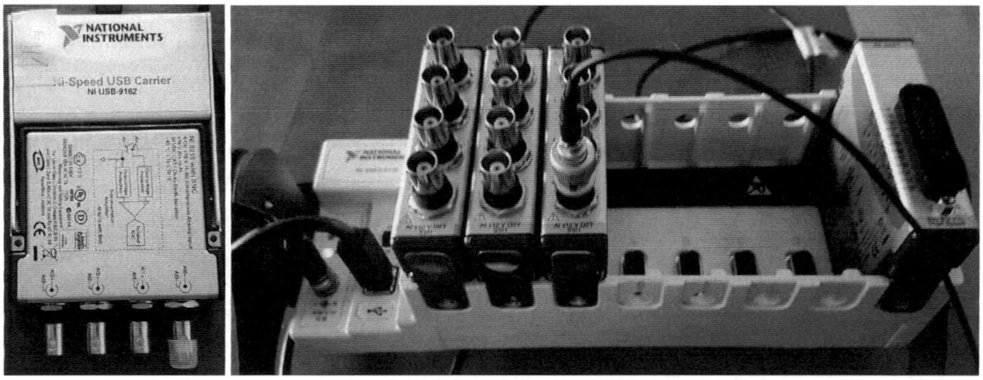

Fig. 5: A/D converter NI 9215.

The response of the vehicle was monitored on a selected section of road 100 m long. The vehicle moved at different speeds in the range from 15.18 to 52.95 km/h. The total weight of the vehicle was 21,500 kg. The sampling frequency of 200 Hz was used for the measurement. The measurement was performed in a LabVIEW environment using the DIAdem system.

3 Processing of measured signals

Signal processing was done in MATLAB. Examples of possible analyzes are documented in the following text. In Fig. 6 is an example of time courses of vertical accelerations at test No. 4, vehicle speed $V = 33.84$ km/h, from sensors A1 – gravity center and A2 – front axle.

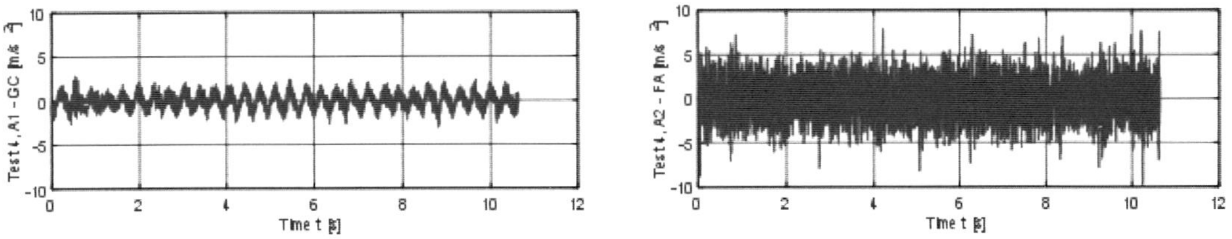

Fig. 6: Time courses of accelerations A1-GC, A2-FA, test No. 4, vehicle speed $V = 33.84$ km/h.

It is possible to make a basic statistical analysis of each record. For example the basic statistical characteristics of records A1 – gravity center, A2 – front axle and A3 – rear axle at test 4, vehicle speed $V = 33.84$ km/h are in Tab. 1.

Tab. 1: Basic statistical characteristics of records, test 4, speed $V = 33.84$ km/h.

Test 4, $V = 33.84$ km/h	A1-GC	A2-FA	A3-RA
Mean value $\bar{a}$ [m/s^2]	0.012876	0.009878	0.030961
Arithmetic mean deviation R_a	0.710678	1.494229	1.658621
Root mean square average deviation R_q	0.857459	1.885612	2.083721
Dispersion σ^2	0.735236	3.555535	4.341895
Effective value RMS	0.857555	1.885638	2.083951
Asymmetry coefficient R_{sk}	-0.105586	0.011212	0.024257
Kurtosis R_{ku}	2.456701	3.144322	3.033628
The greatest depth of record	-3.114865	-8.202233	-8.036912
The largest height of record	2.431183	7.859604	8.499235
Overall height of the record	5.546049	16.061838	16.536148

All statistical characteristics depend on the speed of the vehicle movement. Such a theoretical dependence has a large number of local maxima and spikes. It is a characteristic property for each quantity expressed on the speed of vehicle movement when solving the effects of moving loads on structures. The dependence of the RMS values of acceleration A-GC on the vehicle speed is shown in Fig. 7.

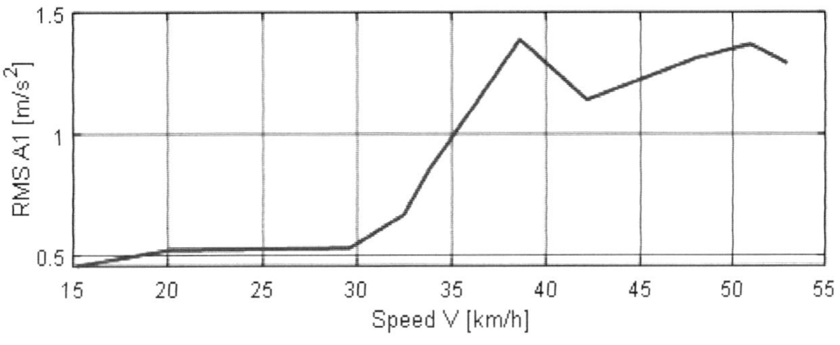

Fig. 7: RMS values of acceleration A1-GC on the vehicle speed.

The probability distribution histograms of acceleration from sensors A2-FA and A3-RA satisfy very well with Gaussian distribution law (red line), Fig. 8. The probability distribution histograms of acceleration from sensor A1-GC have in some cases the character of a bimodal distribution, Fig. 9.

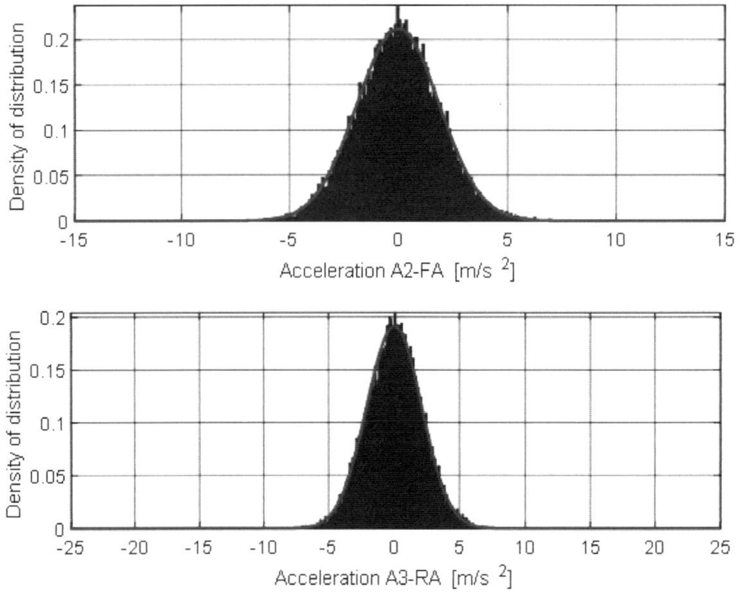

Fig. 8: Histograms of acceleration from sensors A2-FA and A3-RA, test 4, speed V = 33.84 km/h.

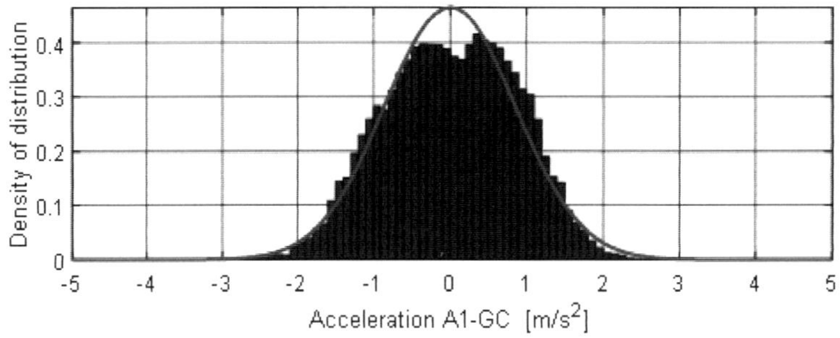

Fig. 9: Histogram of acceleration from sensor A1-GC, test 4, speed V = 33.84 km/h.

Auto-correlation functions (ACF) and cross-correlation functions (CCF) can be quantified from acceleration records. It is suitable to work with standardized correlation functions, where the value of the correlation function at the zero point (lag = 0) is equal to zero. The maximum value $C_{\max}$ of CCF oscillates in a wide interval. Under certain circumstances and under certain mutual lag, the CCF value can be more than 40 times higher than when the records are not lagged relative to each other. For example, test 12, V = 52.95 km/h (14.7091 m/s) the lag is +0.2375 s, what is 3.493 m and $C_{\max}$ = 40.4969. The theoretical distance between the front axle wheel and the front wheel of the rear axle is 3.55 m. The rear axle wheel follows the same road profile as the front axle wheel, but with a lad of 0.2375 s, which results in a high $C_{\max}$ value, Fig. 10.

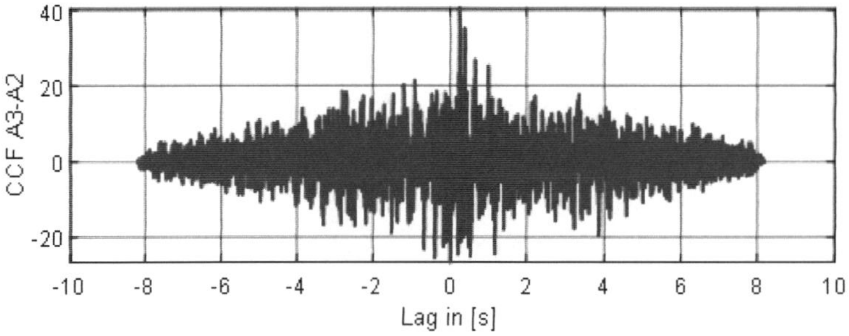

Fig. 10: Cross-correlation function A3-A2, test 12, speed V = 52.95 km/h.

To solve dynamic problems in the frequency domain, it is necessary to know the frequency spectra. A fast Fourier transform (FFT) is used to calculate the frequency spectra. The number of 2^n samples is used. The spectra can be amplitude or phase. It is worked with different types of amplitude spectra. The difference between them is only in scale. Power spectral densities (PSD) are often used to provide information on the power of individual frequency components. PSD of accelerations from sensors A1, A2, A3 for test 12, speed V = 52.95 km/h in the frequency range 0-50 Hz are shown in Fig. 11–13.

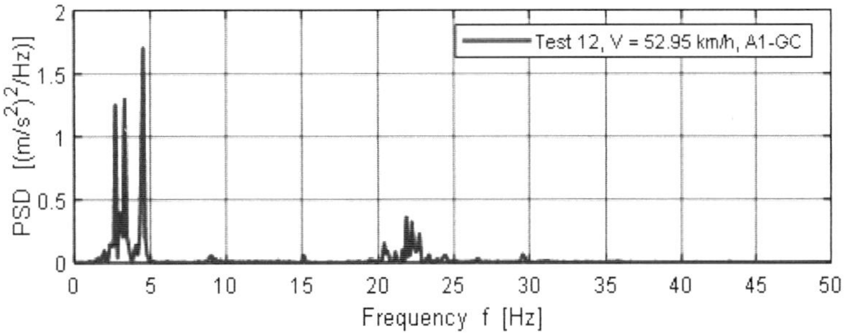

Fig. 11: PSD of accelerations, sensor A1-GC, test 12, speed V = 52.95 km/h.

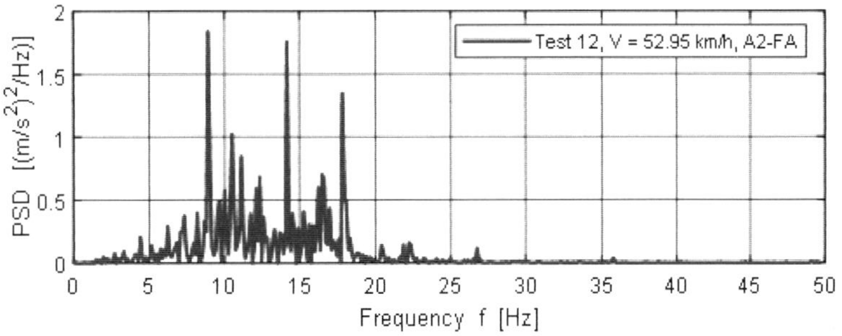

Fig. 12: PSD of accelerations, sensor A2-FA, test 12, speed V = 52.95 km/h.

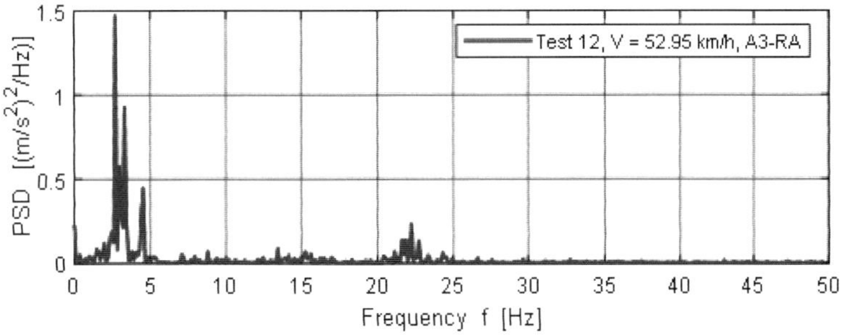

Fig. 13: PSD of accelerations, sensor A3-RA, test 12, speed V = 52.95 km/h.

4 Conclusion

The experiment is the only tool for finding objective reality. It is advantageous to measure accelerations when monitoring the vehicle response when driving along the road. There are many suitable sensors for this purpose. The signal from the sensor to the measuring center can be transmitted via cables or wireless transmission can be used. There are also many tools for time and frequency domain signal analysis, such as DIAdem or MATLAB. The quality of the results when using these tools is high. Numerical methods are currently making great progress. Current computing and numerical simulation methods make it possible to simulate real operating states in real time. However, we need to be sure that the calculation model is in line with reality. Therefore, these type of experiments will need to be carried out in the future. Once the computational model is in line with reality, many analyses will be able to be done numerically. This is the progress of these procedures. It saves time, energy, finances and moves knowledge one step further.

Acknowledgement

This work was supported by the Grant National Agency VEGA of the Slovak Republic, G1/0006/20.

References

[1] P. Hejtmánek, O. Blaťák, P. Kučera, P. Porteš, J. Vančura, Measuring the yaw mass moment of inertia of vehicle, MECCA - Journal of Middle European Costruction and Design of Cars XI, (2013), 16–22.

[2] P. Porteš, F. Bauer, J. Čupera, Laboratory-experimental verification of calculation of force effects in tractor's three point hitch acting on driving wheels, Soil & Tillage Research 128, (2013), 81–90.

[3] Y. Lu, S. Yang, S. Li, L. Chen, Numerical and experimental investigation on stochastic dynamic load of a heavy duty vehicle, Applied Mathematical Modelling 34, (2010), 2698–2710.

[4] P. Baranowski, J. Malachowski, Numerical and experimental testing of vehicle tyre under impulse loading conditions, International Journal of Mechanical Sciences 106, (2016), 346–356.

[5] G.J. Heydinger, J.G. Howe, Analysis of vehicle response data measured during severe maneuvers, SAE transactions, Section 6: Journal of passenger cars: Mechanical systems journal 109, (2000), 2154–2167.

[6] J. Melcer, Experimental verification of modal characteristics of vehicle computing model, Proceedings of the 54th International Conference on Experimental Stress Analysis EAN 2016, 2016, Srní, Czech Republic, University of West Bohemia, 1–6, code 123122.

[7] J. Melcer, Wireless signal transmission at testing vehicle response to kinematic excitation, Book of Full Papers, 58th Conference on Experimental Stress Analysis EAN 2020, 2020, on line, Czech Republic, Editing Center, VŠB Technical University of Ostrava, Faculty of Mechanical Engineering, 317–320.

Strain Measurements on Samples of an 3D-printing Maraging Steel by a High-resolution Neutron Diffraction

P. Mikula[1,*], V. Ryukhtin[1], A. Michalcová[2]

[1] *Nuclear Physics Institute CAS, 250 68 Řež, Czech Republic*
[2] *University of Chemistry and Technology, 166 28 Prague 6, Czech Republic*
* *mikula@ujf.cas.cz*

Abstract: In our contribution we present strain measurements on several samples related to the 3D printing material with an unconventional high-resolution neutron diffraction. This method is based on the three-axis diffractometer setting when a conventional two axis neutron diffraction scanner does not provide a relevant resolution.

Keywords: strains; neutron diffraction; high resolution; 3D printing material.

1 Introduction

The measurement of strain/stress fields within polycrystalline components which are formed by inhomogeneous deformation or heat treatment, has been a challenge for engineers for many years. A prime concern of them are usually macrostresses which vary over distances much larger in comparison with the grain size. However, microstresses which vary on a spacial scale comparable to the grain size can also provide information related to the mechanical properties of materials. For scanning the macrostresses specialized dedicated diffractometers (strain scanners) have been developed at almost all medium- and high-flux neutron sources, e.g. [1].

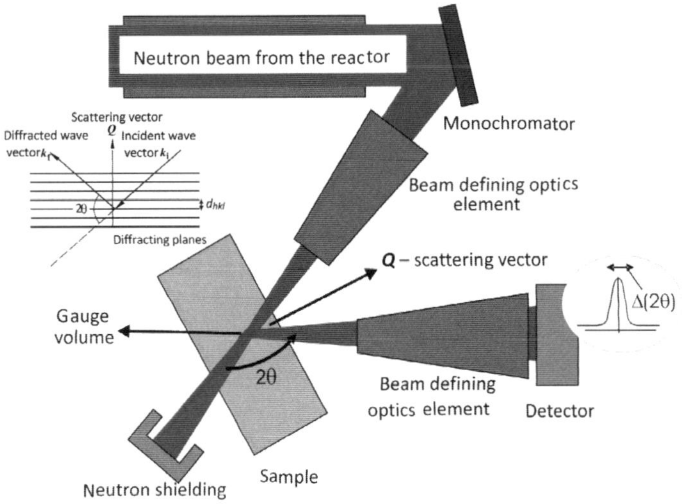

Fig. 1: Schematic illustration of a conventional diffractometer for strain measurements.

The principle of the method is quite simple and based on the measurement of the angular shift (of the sample diffraction profile) brought about by elastic strains (macrostrains) present in the polycrystalline material [2] (see Fig. 1). In neutron and X-ray diffraction the angular positions of the diffraction maxima are related to the values of the lattice spacing through the Bragg equation $2d_{hkl} \cdot \sin\theta_{hkl} = \lambda$ (d_{hkl} – lattice spacing, θ_{hkl} – Bragg angle, λ – the neutron wavelength) which offers a unique non-destructive technique for investigation of stress fields. When a specimen is elastically strained, the lattice spacing changes. Then, when defining the strain ε as $\varepsilon = \Delta d/d_{0,hkl}$, it is related to a relative change in the value of the lattice spacing. As the strain in the material is the tensor, a component parallel to the scattering vector Q perpendicular to the reflecting set of planes is determined from one diffraction measurement. Therefore, the knowledge of the $d_{0,hkl}$ value ($d_{0,hkl}$ lattice spacing of the strain-free material) is a crucial task [1,2]. Then, by differentiation of the Bragg condition we arrive at $\varepsilon = -\cot\theta_{hkl} \cdot \Delta\theta_{hkl}$. The relation for the strain ε indicates that it gives rise to a change in the scattering angle $2\theta_{hkl}$ resulting in an angular shift $\Delta(2\theta_{hkl})$ of the peak position for particular reflecting planes illuminated by a fixed wavelength. In such a way, the shift in the Bragg angle (relative to that of the stress-free material) permits the determination of the average lattice macrostrain over the irradiated gauge volume. The conversion of strains to stresses is carried out by means of the relation [1,2].

$$\sigma_x = \frac{E_{hkl}}{(1-2\nu_{hkl})(1+\nu_{hkl})}\left[(1-\nu_{hkl})\,\varepsilon_x^{hkl} + \nu_{hkl}\left(\varepsilon_y^{hkl} + \varepsilon_z^{hkl}\right)\right], \tag{1}$$

where $\varepsilon_{x,y,z}^{hkl}$ is the x, y, z – component of the lattice strain measured at the hkl crystal lattice planes and E_{hkl} and ν_{hkl} are the diffraction elastic Young modulus and diffraction Poisson ratio.

2 High-resolution three axis diffraction setting

The resolution of the conventional scanners expressed by *FWHM* of the diffraction profile is not sufficient for studies of microstresses which affect just *FWHM* as well as the form of the profile if the resolution is sufficiently high. As can be seen from Fig. 2 conventional two axis neutron scanners usually use, for strain determination, beam optics elements, namely: Focusing and optimally bent monochromator; a system of slits before and after the sample creating a gauge volume element (irradiated by the beam coming from the monochromator) and a position sensitive detector (PSD) imaging the diffraction profile coming from the irradiated gauge volume [3–5]. The resolution of the conventional scanner is thus determined mainly by the thickness and curvature of the monochromator, the widths of the slits (usually in the order of 1–3 millimetres), the divergence of the beam diffracted by the gauge volume and the spatial resolution of PSD. The combination of all these uncertainties results in an uncertainty (*FWHM* of the gaussian image in PSD) of about roughly 5×10^{-3} rad which appears sufficient for the measurement of $\Delta(2\theta_{hkl})$ angular shifts brought about by strains. However, the mentioned resolution is not sufficient for measurements of changes of diffraction line profiles imaged by PSD which would permit to carry out microstrain and grain size studies for plastically deformed polycrystalline samples.

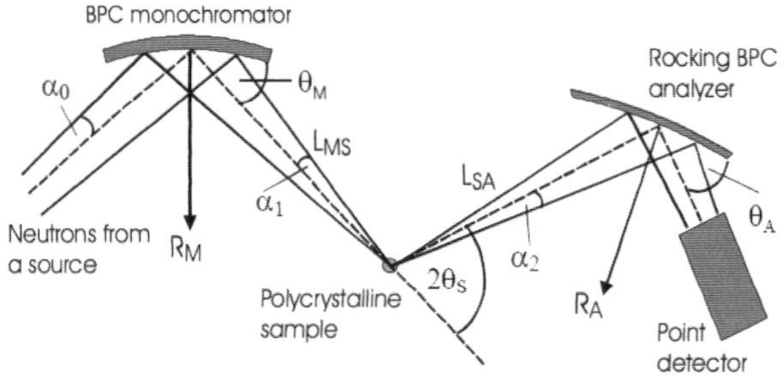

Fig. 2: Three axis diffractometer set-up employing bent perfect crystal monochromator and analyzer as used in the experimental strain studies (R_M, R_A – radii of curvature, θ_M, θ_A – Bragg angles).

As can be seen from Fig. 2, for the investigation of the strain/stress effects of bulk samples (e.g. 10 mm width or more) the dimensions of the irradiated gauge volume substantially blur the resolution of the conventional two axis instrument. Therefore, in such a case the high resolution three axis diffractometer setting equipped with the focusing BPC monochromator as well as focusing BPC analyzer and the sample situated between them was tested as an alternative, where the beam diffracted by a polycrystalline sample is analyzed by rocking the crystal analyzer (see Fig. 2). The scanning of the diffraction profile consists in the step-by-step analysis by rocking the analyzer and the analyzed neutrons are detected by a point detector [6,7]. Following the sketch displayed in Fig. 2 (for small widths of the samples), a maximum resolution from this arrangement can be achieved for minimal dispersion of the whole system. When treating it in momentum space, this means that the orientation of the Δk domains of the wave vector related to the monochromator and analyzer are matched to that of the sample. For $L_{MS}/(R_M \cdot \sin\theta_M) \neq 1$ and $L_{SA}/(R_A \cdot \sin\theta_A) \neq 1$, a general form for focusing in momentum space (not dependent on α_1 and α_2 which minimizes the dispersion between all elements can be derived as [6].

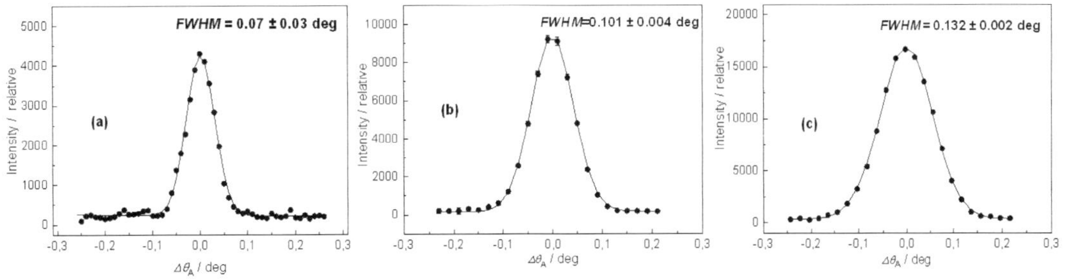

Fig. 3: Analyzer rocking curves for the optimum analyzer curvature as obtained on two vertically situated standard samples of different diameters; (a) ⌀=2 mm, (b) ⌀=5 mm and (c) ⌀=8 mm.

$$2\tan\theta_S = \tan\theta_M / \left(1 - L_{MS}/(R_M \cdot \sin\theta_M)\right) + \tan\theta_A / \left(1 - L_{SA}/(R_A \cdot \sin\theta_A)\right). \qquad (2)$$

When fulfilling the condition (Eq. 2) a maximum peak intensity and a minimum *FWHM* of the analyzer rocking curve can be expected. Fig. 3 shows the result of the optimization procedure on 3 virgin samples of different diameters providing a minimum *FWHM* of the analyzer rocking curves. Of course, this high-resolution performance can be used for macro-strain scanning, but due to the step-by-step analysis with the analyzer, the measurement would be impractical.

However, due to its high resolution, it can be used for plastic strain studies on the basis of diffraction profile analysis as e.g. in deformation studies on samples under the external thermomechanical load [8,9]. Recently, it has been found out that the setting provides a sufficiently high resolution, though slightly relaxed, when wider slits (e.g. of the width of about 10 mm) or wider samples can be used. The experimental tests were carried out on the 3-axis neutron diffractometer installed at the Řež research reactor LVR-15. The diffractometer employs the Si(111)-monochromator and Si(311)-analyzer single crystals which have the dimensions of 200×40×4 mm^3 and 20×40×1.3 mm^3 (length × width × thickness), respectively. The monochromator Si(111) provides the neutron wavelength of 0.162 nm and has a fixed curvature with a radius R_M of about 12 m. The radius of curvature of the analyzer was changeable in the range from 3.6 m to 36 m. It permitted us to determine an optimum analyzer curvature minimizing the dispersion and thus optimizing the resolution of the setting.

3 3D-printing material

Following the recent contributions on 3D printing material at EAN 2020 conference e.g. [10], we have decided to look at a possibility of studies of the microstructure of several related samples by the high-resolution neutron diffraction. The samples were in the form of small plates of the dimensions of 3×5×15–25 mm^3 (thickness × width × length) and installed on the sample place in the vertical position. The cross-section of the incident beam from the monochromator was 20×20 mm^2. For the standard powder sample a V-container of the diameter of 5 mm was used.

The following samples of X3NiCoMoTi 18-9-5 maraging steel were investigated (see Figs. 4–6): P – standard powder, C – as printed, B – annealed one at 490°/ 6 h, D – annealed one at 820°/ 1 h, then air cooled and annealed again at 490°/ 6 h. The chemical composition of the material is shown in the next table.

Tab. 1: Chemical composition of X3NiCoMoTi 18-9-5 maraging steel.

Element	Ni	Co	Mo	Ti	C	Al	Cr	Mn	Si	Fe
Wt/%	19	9.3	5	0.64	≤ 0.03	0.06	0.08	0.04	0.07	Bal.

Similarly to the case of the conventional strain scanner, it is necessary to show how to evaluate the strain $\varepsilon = \Delta d_S/d_0$ from the obtained data which provide the peak shifts $\Delta\theta_A$. In this case the peak shift $\Delta\theta_A$ corresponds

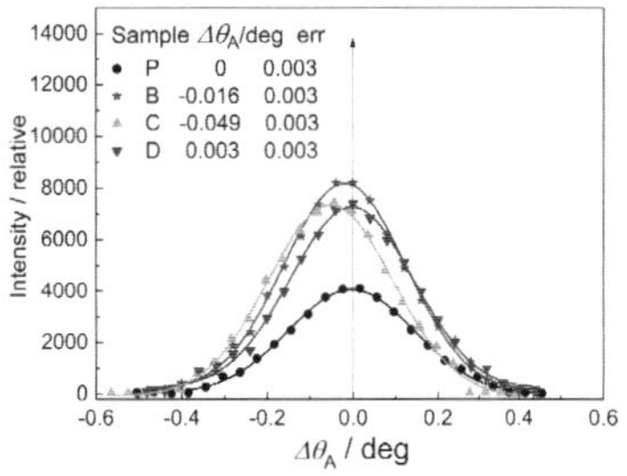

Fig. 4: Analyzer rocking curves for individual samples.

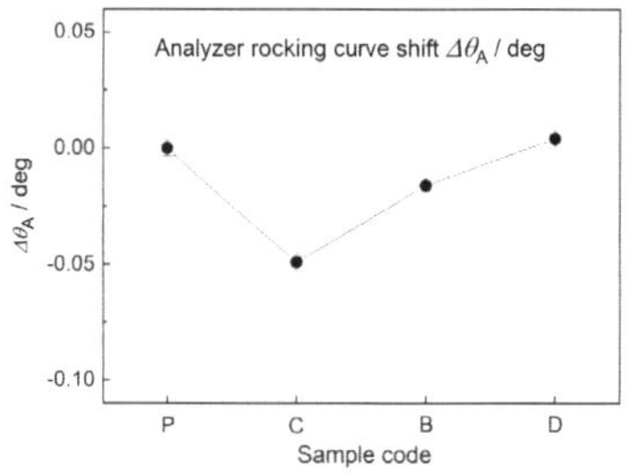

Fig. 5: $\Delta\theta_A$ shifts for individual samples.

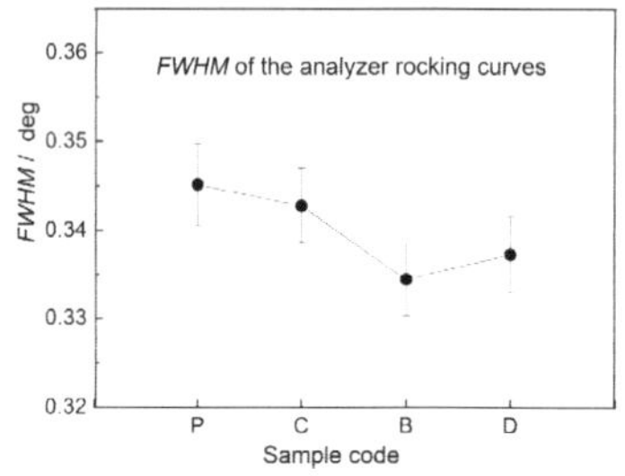

Fig. 6: *FWHM* of the rocking curves shown in Fig. 2.

to the change in the scattering angle $\Delta(2\theta_S)$ as $\Delta\theta_A = -\Delta(2\theta_S)$. Assuming the powder sample P to be a strain free one, calculation gave the strains for the samples B, C and D: $\Delta d_B/d_P = -(3.20 \pm 0.54) \times 10^{-4}$, $\Delta d_C/d_P = -(9.80 \pm 0.60) \times 10^{-4}$, $\Delta d_D/d_P = -(1.00 \pm 1.00) \times 10^{-4}$. As to the *FWHM* of the rocking curves, it is evident that due to plastic deformation, *FWHM* related to the powder and maraging samples is much larger than the values of the virgin samples (see Fig. 3). It proves the expectation that the high-resolution 3-axis diffractometer setting is fully feasible for substructure studies of plastically deformed polycrystalline samples. For example, from a simple comparison of the rocking curves B and D when $\varepsilon = \Delta d_D/d_P$ can be considered zero within the experimental error, we can conclude that the annealing of the sample at 820°/ 1 h probably could be omitted.

4 Conclusion

The methodological experiment on the 3-axis diffractometer instrument was carried out with the aim of verifying its feasibility for studies substructure studies of plastically deformed 3D printing maraging steel material. The instrumental resolution function is characterized by diffraction profiles of representative standard samples. The sensitivity of the instrument to the lattice strain is given by the numerical error of determination of the peak center after using the fitting procedure. Very narrow *FWHM* of the diffraction curves from the standard samples indicate that the main application field of this diffractometer setting lies in plastic deformation studies using shape analysis of diffraction profiles. It can be stated that the presented neutron diffraction set-ups can offer an additional support to complement the information achieved by using the other conventional characterization methodologies.

Acknowledgement

Measurements were carried out at the CANAM infrastructure of NPI CAS Řež. P. Mikula and J. Šaroun acknowledge support from ESS participation of the Czech Republic – OP (CZ.02.1.01/0.0/0.0/16_013/0001794) and from the project ESS Scandinavia-CZ II (LM2018111), respectively. Furthermore V. Ryukhtin acknowledge support from the Czech Academy of Sciences in the frame of the program "Strategie AV21, No. 23". The authors thank B. Michalcová from NPI CAS for significant help with measurements and data elaborations.

References

[1] I.C. Noyan, J.B. Cohen, In Residual stress: Measurement by diffraction and interpretation, 1st edition New York, Springer-Verlag, 1987.

[2] M.T. Hutchings, A. D. Krawitz (Eds.), Measurement of residual and applied stress using neutron diffraction, NATO ASI Series, Applied Sciences 26, Kluwer Academic Publisher, 1992.

[3] P. Mikula, M. Vrána, P. Lukáš, J. Šaroun, P. Strunz, H.J. Ullrich, V. Wagner, Neutron diffractometer exploiting Bragg diffraction optics - A high resolution strain scanner. In Proc. of ICRS-5, June 16-18, 1997, Linköping, T. Ericsson, M. Odén and A. Andersson (Eds.), **2** (1997) 721–725.

[4] R.C. Wimpory, P. Mikula, J. Šaroun, T. Poeste, Junghong Li, M Hoffmann and R. Schneider, Efficiency boost of the materials science diffractometer E3 at BENSC: One order of magnitude due to a double focusing monochromator, Neutron News, **19** (2008) 16–19.

[5] B.S. Seong, V. Em, P. Mikula, J. Šaroun, M.H. Kang, Materials Sc. Forum, **681** (2011) 426–430.

[6] M. Vrána, P. Lukáš, P. Mikula and J. Kulda, Bragg diffraction optics in high-resolution strain measurements, Nucl. Instrum. Methods in Phys. Research, **A 338** (1994) 125–131.

[7] P. Mikula, J. Šaroun, V. Ryukhtin, J. Stammers, An alternative neutron diffractometer perfor-mance for strain/stress measurements, Powder diffraction, **35**, Issue 3, (2020) 185–189.

[8] R. Delhez, T.H. de Keijser, E.J. Mittemeijer, 1982, Determination of crystallite size and lattice distortions through X-ray diffraction line profile analysis, Fresenius' Zeitschrift für analytische Chemie, **312**(1) (1982) 1–16.

[9] V. Davydov, P. Lukáš, P. Strunz, R. Kužel, Single-line diffraction profile analysis method used for evaluation of microstructural parameters in the plain ferritic steel upon tensile straining, Materials Sc. Forum, **571-572** (2008) 181–188.

[10] R. Halama, P. Gál, M. Pagáč, B. Govindaraj, On Cyclic Hardening/Softening Behaviour of Conventional and 3D Printed SS316L, in proc.: 58th Conference on Experimental Stress Analysis (2020), Sobotín, eds. M. Fusek et al., Czech Society for Mechanics, 98–106. ISBN: 9781713827832.

Stress and Strain Analysis of Prefabricated Unfired Earth Wall Made of Commercial Earth Mixture

B. Muzikova[1,*], T. Placek Otcovska[1], P. Padevet[1], J. Ruzicka[1], J. Divis[1], F. Havlik[1]

[1] *Czech Technical University in Prague, Faculty of Civil Engineering, Thakurova 7, 166 29 Prague 6, Czech Republic*

* *barbora.muzikova@fsv.cvut.cz*

Abstract: The paper is focused of comparison an experiment and numerical model of unfired rammed earth made of prefabricated panels. The numerical model was made in four versions and loaded according to the experiment. The deformations were evaluated and compared. Then the verification of the model to the experiment was made. The position of the load was searched to give the same displacements as the data from the experiment. It was found that a quite big eccentricity in the load is needed to obtain the same deformation. The determination of maximal force and displacement for different thicknesses of walls was made. The determination of maximal force and displacement for walls with thicknesses of 200 mm, 300 mm, 400 mm and 500 mm were searched.

Keywords: rammed earth; prefabricated wall; stress and strain analysis.

1 Introduction

The building with unfired earth is nowadays getting back thanks to its positive environmental properties of natural product. But the mechanical properties and the behaviour of earthen constructions are complicated and not well known, because rammed earth was replaced by fired bricks. Material properties, behaviour of the earthen constructions have to be observed for spreading the building of this material. These data can help engineers and architects to use unfired earth as a material [2–4].

An experiment with load bearing earthen wall was made by J. Ruzicka, F. Havlik (the experiment was a part of Havlik's PhD thesis [1]), J. Divis and their team at CTU in Prague in 2013. The wall was made of ten prefabricated rammed earth panels bound with earth mortar. The wall was loaded until it collapsed. The force of the load and deformations were measured and recorded. The aim of the paper is to create a numerical model of the testing wall and

- to made a model of the experiment and made stress and strain analysis to compare the model and the experiment,
- to determine maximal loading forces and deformations for walls of different width.

The first part was divided into several steps

- to define the geometry of model,
- to determine the boundary conditions,
- to set necessary material characteristics that were missing (especially Poisson's ratio),
- to set the position of the load.

2 Prefabricated rammed earth wall experiment

The experiment was conducted on the wall of 1.672 m width, 2.944 m height and 0.200 m depth built from ten prefabricated panels. The whole experiment is elaborately described in [1]. Mortar was used both in vertical and horizontal joints, it was from the same material as the panels. The concentric load was made by the load cell with a hydraulic jack. Eight gauges recording local strain on the surface and two recording buckling were installed. Schmidt hammer testing was realized before performing the full scale load test on the finished wall.

The stress-strain curve was recorded. The maximal force was 525 kN and the maximal horizontal displacement was 42.94 mm. Then the wall collapsed. The wall is shown in the Fig. 1.

The problem of the experiment is that it was done just once. So there is a question if it is a representative sample. To prove that the experiment was a representative one - the experiment can be repeated at least twice to get some statistics data or a numerical model can be created. The goal of the paper is to prove if the experiment was a representative one or there was some kind of deviation.

Fig. 1: Precast earth wall [1].

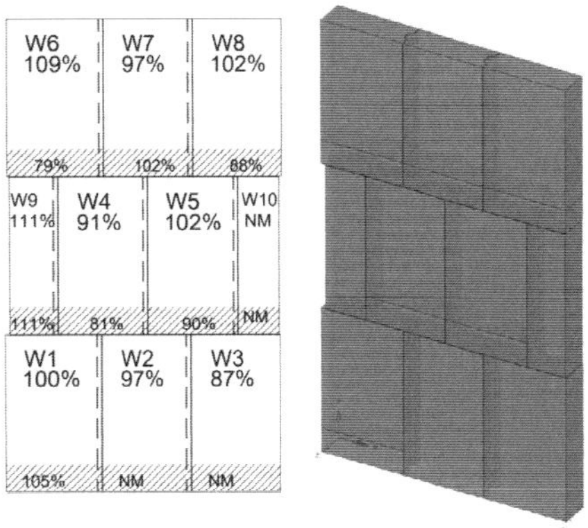

Fig. 2: Schema of the wall and panels.

3 Process of creating the model

The model of the rammed earth was modelled in Ansys 2019 program with a student licence. The model was made in 3D. The issues that need to be solved were the geometry, material characteristics, boundary conditions and position of the load. The results of the model were displacements and values of the loading forces.

3.1 Geometry

The model was made in four versions of geometry. The first one was a simply monolithic wall with no regard to panels and mortar (ver. A). The second one was a wall consisting of panels but without mortar (ver. B). The third one was a wall consisting of panels and mortar (ver. C) and the fourth one was a wall of panels, mortar and different materials in the bottom part of the panel according to the measurement of the Schmidt hammer (ver. D). Non-destructive testing of strength using the Schmidt hammer has shown that the panel bottoms used in this test were weaker than the rest of the panels. The difference is shown in the Fig. 2. The panel W1 was considered as the basic one with the comprehensive strength of value 100 %. Strength measured with the Schmidt hammer on other panels are related to the W1 value.

3.2 Material characteristics

The characteristics needed for the model were volume density, modulus of elasticity, comprehensive strength, tensile strength in bending and Poisson's ratio. Poisson's ratio was the missing value and it was necessary to make new specimens. They were made of the same mixture as the panels. Poisson's ratio was determined from the comprehensive test of three specimens. The value of Poisson's ratio is the negative of the ratio of transverse strain to axial strain. The used material characteristics for panel W1 are described in Tab. 2. Properties of other parts of the wall were modified according to the results of the Schmidt hammer.

3.3 Boundary conditions

The earth wall was built on the ground of the laboratory, it was supported against horizontal displacement in the top and a steel beam was settled on the top of it. The steel beam provided the force distribution to the wall during the test. Schema is shown in the Fig. 4. The simplified 2D model was supported by a clamped

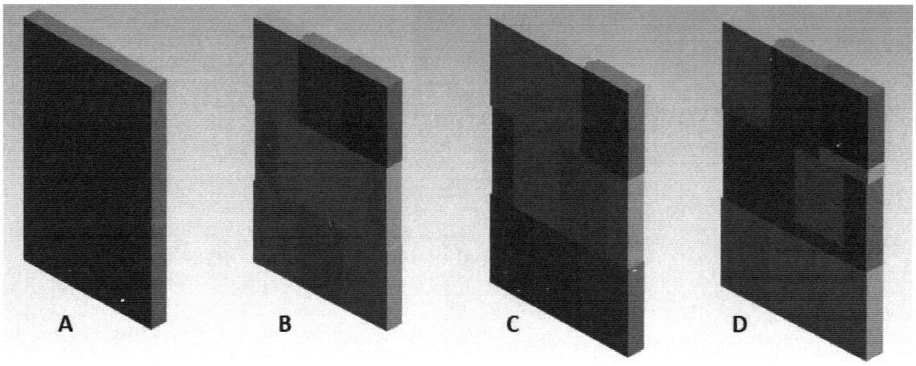

Fig. 3: Four versions of geometry – A) a monolithic wall, B) panels without mortar, C) panels with mortar and D) panels with mortar and different material in the bottom.

Tab. 1: Material characteristics used in the model.

	volume density [kg/m^3]	modulus of elasticity [MPa]	com. strength [Pa]	ten. strength in bending [Pa]	Poisson's ratio [-]
W1	2,079	337	3,350	400	0.2
mortar	1,859	187	3,400	340	0.2

edge at the bottom and by a hinge on top. The wall with these boundary conditions had the maximal horizontal displacement in the top third of the wall. It corresponds with the theoretical deformation. The horizontal deformation for these conditions is the same as was measured during the test. The maximal 40 mm was reached around top third of the wall and 30 mm in the half. These deformations were measured by force 500 kN. The clamped edge in the simplified 2D is caused by the mass of the wall and the relatively big supporting area at the bottom.

The 3D model was supported by setting displacements. The displacements in direction x, y, z (the system is shown in Fig. 5) were settled 0, 0, 0 in the bottom area of the wall and 0, free, 0 in the top area of the wall.

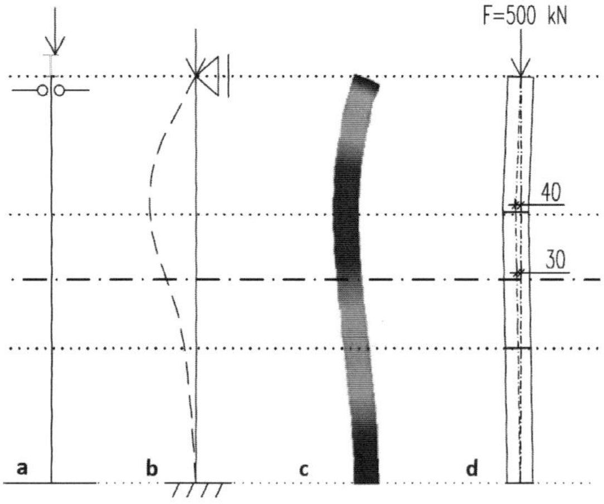

Fig. 4: Boundary conditions – a) the schema of experiment, b)the shape of theoretical deformation, c) the shape of deformation of the model and d) the shape of real deformation of the wall.

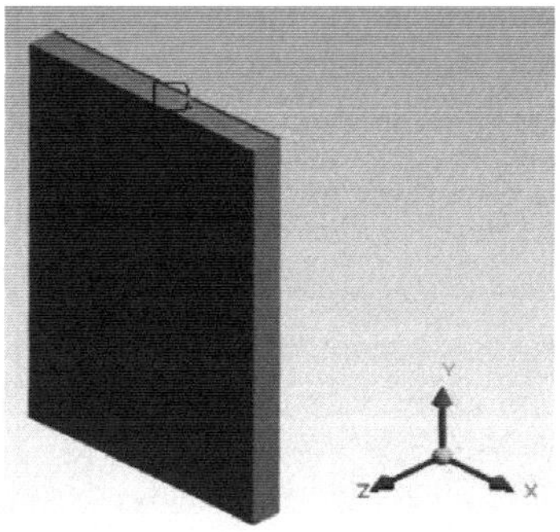

Fig. 5: Coordinate system of the model.

Tab. 2: Deformations from the model and experiment.

model version	deformation (z axis) $e = \frac{1}{6}t$ [mm]	deformation (z axis) (Fig. 5) e according deformation [mm]
A monolithic wall	15.72	39.29
B panels withou mortar	16.23	40.85
C panels with mortar	15.36	38.36
D panels with mortar, weaker bottom	15.74	39.37
experiment	42.94	42.94

3.4 Load

The eccentricity of the load was defined as

$$e = \frac{1}{6}t = \frac{1}{6}200 \sim 35 \text{ mm}. \quad (1)$$

The force was applied as an area load with an eccentric axis 35 mm from the wall axis.

The value of the force was settled according to the experiment. The maximal test force was 519.9 kN. The weight of the steel beam was 475 kg and the maximum test force was 525 kN. The model was also loaded with mass weight caused by earth gravity.

4 Results from the model

The model was made as described above. The wall was loaded with the force of value 525 kN. The results were deformations of the model of the wall that were compared with the deformations of the real wall.

4.1 Eccentricity of load e = 35mm

The deformations of load with eccentricity according to Eq. 1 were observed. The values are shown in Tab. 2. For version A, the deformation is 15.72 mm, version B is 16.32 mm, version C is 15.36 mm, version D is 15.74 mm and the value of the experiment is 42.94 mm.

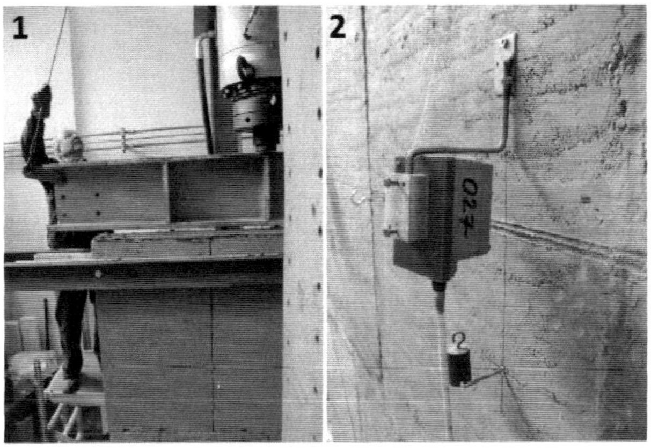

Fig. 6: 1) The setting of the steel beam and position of the load cell with the hydraulic jack and 2) the gauge for recording local strain [1].

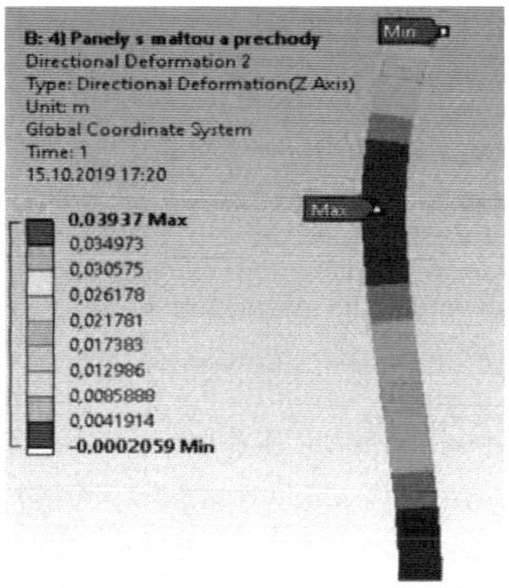

Fig. 7: Deformation of the version D with eccentric load adjusted according to the experiment.

The deformations of all versions are very similar and there is no big difference in it. The differences between versions are very small (Tab. 2) because the comprehensive strength and other material characteristics

Tab. 3: The walls of different thickness and their loading forces and displacements.

thickness of wall [mm]	maximal force [kN]	max. deformation (z axis) (Fig. 5) [mm]
200	525	15.72
300	920	11.51
400	1,300	9.09
500	2,000	9.03

are approximately the same. But the value of the deformation in the real experiment was more 2.6× higher than in the model. Four possible causes of this can be found.

The first one is the imperfection in the transfer of the load through the beam to the wall. A steel beam was placed on the top of the wall, the top of the wall was grinded and leveled with a cement screed and the steel beam was placed on this layer. The position of the beam was probably not exactly in the middle of the wall (the axis of the beam did not lay on the axis of the wall). Because of that, the first inaccuracy was injected in the experiment. The second one is the position of the load cell with the hydraulic jack on the beam (the axis of the jack did not lay on the axis of the beam resp. wall). The stated cube size was 200 mm, it can be seen in the Fig. 6. And the third possibility could be the structural unevenness of the wall. The precision of the vertical plane was controlled with two meter long spirit level lath. Verticality and aligning of panels were adjusted with clips that were left until mortar strengthened and then the next row was built. But the verticality precision was not controlled and measured. There was space for some structural eccentricity. And the last problem could be the process of transferring the load through the beam to the wall. It could have the impact of a solitary force instead of an even pressure on the area.

4.2 Calculated eccentricity of the load

Next, the verification of the model to the experiment was made. The position of the load was searched to give the same displacements as the data from the experiment. For the load strip of width 25 mm from the edge of the wall were found similar deformation as the result of the experiment. In this case, the eccentricity is $e = 87.5$ mm. The calculated eccentricity includes all of the problems described above and simulated them. These displacements are shown in the Tab. 2 in the third column. For version A the deformation is 39.29 mm, version B is 40.85 mm, version C is 38.36 mm, version D is 39.37 mm and the value of the experiment is 42.94 mm. These data showed that there was probably a quite big eccentricity in the load.

5 Walls with different thicknesses

Another target was to determine the maximal force and displacement for different thicknesses of walls. The method was as follows. The value of the normal stress in the place of the maximal deformation was found out. This was done for the wall of 200 mm, loading force of 525 kN and the eccentricity of the loading force was $e = 35$ mm (reference wall). The model of the wall was considered as a monolithic wall. The walls of different thicknesses were made and loaded until the value of the normal stress was at the same value of the reference wall. The displacement in the axis z (Fig. 5) was taken at that moment and considered as the maximal one.

The reference wall of the thickness 200 mm had the loading force of 525 kN and displacement 15.72 mm. The wall of thickness 300 mm had loading force of 920 kN and displacement 11.51 mm, 400 mm wall had the loading force of 1,300 kN and displacement 9.09 mm and 500 mm wall had the loading force of 2,000 kN and displacement 9.03 mm.

The Fig. 8 shows the dependence of the thickness of walls to maximum loading force and shows that the trend of the dependence is linear. The Fig. 9 shows the dependence of the thickness of walls to displacement in z axis. The trend of the dependence seems to be gently exponential.

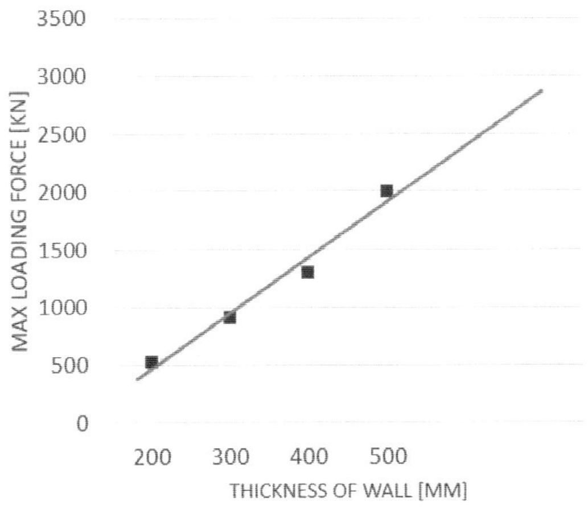

Fig. 8: Dependence of thickness of walls to max loading force.

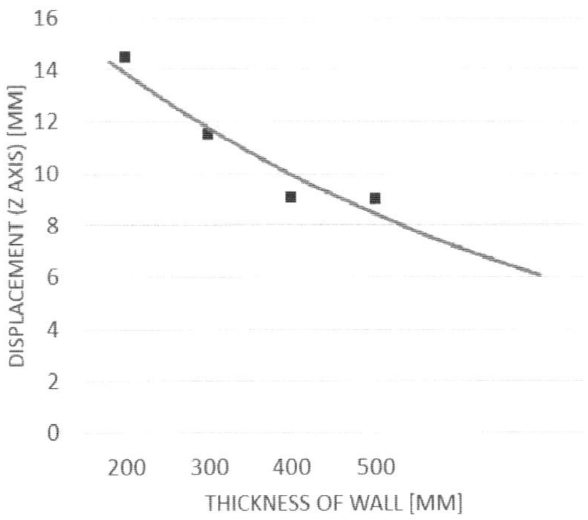

Fig. 9: Dependence of thickness of walls to displacement in z axis.

6 Conclusion

The paper focused on the comparison of experiment and numerical model of unfired rammed earth made of prefabricated panels. The numerical model was made in four versions and loaded according to the experiment, then the deformations were evaluated.

The values of the deformations in the real experiment were more than 2.6× higher than in the model. Four possible causes of this can be found:

- position of the steel beam,
- position of the load cell with hydraulic jack,
- structural eccentricity in vertical precision,
- process of transferring the load through the beam to the wall.

Then the verification of the model to the experiment was made. The position of the load was searched to give the same displacements as the data from the experiment. It was found that a quite big eccentricity in the load is needed to obtain the same deformation.

Finally, the determination of maximal force and displacement for different thicknesses of walls was made. The values were searched for thicknesses of 200 mm, 300 mm, 400 mm and 500 mm.

Acknowledgement

The research was financially supported by Faculty of Civil Engineering Czech Technical University in Prague (SGS project No.SGS19/148/OHK1/3T/11 and No.SGS18/106/OHK1/2T/11).

References

[1] F. Havlik, Development and Experimental Verification of Mechanical-physical Properties of Pre-formed Rammed Earth Wall Panel Title of Book, doctoral thesis, Prague, 2017.

[2] M. Rauch. Refined earth construction and design with rammed earth, vol. 1. Detail, Munich, 2015.

[3] G. Minke. Building with Earth - Design and Technology of Sustainable Architecture, vol. 2. DeGruyter, Berlin, 2009.

[4] I. Zabickova. Hlinene stavby. ERA 21, Brno, 2002.

Effects of Pre-strain, Strain Amplitude and Frequency on Dynamic Compressive Properties of Magnetorheological Elastomeric Composite

T.H. Nam[1,*], I. Petríková[1], B. Marvalová[1]

[1] *Faculty of Mechanical Engineering, Technical University of Liberec, Studentská 1402/2, 461 17 Liberec 1, Czech Republic*

* *nam.tran.huu@tul.cz*

Abstract: Dynamic compressive properties of isotropic magnetorheological elastomeric composite (MEC) under cyclic compression loading have been studied in this paper. Effects of pre-strains from 5 to 20%, strain amplitudes from 0.01 to 0.04, and excitation frequencies from 10 to 50 Hz on the dynamic properties of isotropic MEC under harmonic cyclic compression loading were examined. Results showed that the storage and loss moduli of the isotropic MEC decreased with increasing the strain amplitude, but they enhanced with the rise of the pre-strain and frequency. Although the storage modulus of the MEC changed slightly over time, the loss modulus reduced somewhat with increasing time. The reduction in the loss modulus of the MEC under cyclic compression loading is attributed to increasing material temperature during the testing. Besides, despite the loss angle tended to reduce over time, it changed irregularly with raising the pre-strain, strain amplitude, and frequency as well. In general, the dynamic compressive properties of the MEC strongly depended on the pre-strain, strain amplitude, and frequency of cyclic loading.

Keywords: magnetorheological elastomer; composite materials; dynamic properties; cyclic compression test.

1 Introduction

Magnetorheological elastomeric composites (MECs) have been produced by introducing ferromagnetic particles into a non-magnetic elastomeric matrix. The MECs belong to a group of smart materials, because their rheological and mechanical properties can be controlled by applying an external magnetic field [1]. MECs exhibit a change in their stiffness and damping properties under the magnetic field. With their changing and controllable properties, MECs have been used in a broad range of engineering field applications [2]. MECs are considered to be viscoelastic rubber-like materials, thus they inherit predominant properties of the rubber matrix such as large deformations, stress softening effect, amplitude and frequency dependency, and time-dependent features. Therefore, studies of the dynamic mechanical properties of MECs under various loading conditions are essential for engineering applications.

The viscoelastic mechanical properties of MECs have been extensively studied during the past decade. The measurement of viscoelastic mechanical properties of MECs has been carried out by using either transient or dynamic mode [3–5]. The dynamic mechanical properties of MECs depend on constituent material properties and loading conditions such as preloads, frequency, and amplitude. Besides, the dynamic mechanical properties of MECs can be studied by the equations typically for viscoelastic materials. In this paper, the isotropic MEC was produced from a silicone matrix and micro-sized carbonyl iron powders (CIPs). The main objective of this paper is to study the dynamic properties of the isotropic MEC cylindrical samples subjected to cyclic compression loading. Effects of the pre-strain, strain amplitude, and excitation frequency on the dynamic properties of the isotropic MEC under harmonic cyclic compression loading were examined.

2 Experimental

2.1 Materials

Micro-sized CIPs, silicone rubber ZA13 and its catalyst were used to fabricate the isotropic MEC. The micro-sized CIPs (type: 44890) were supplied by Sigma-Aldrich (USA). The micro-sized CIPs have overall spherical shapes with 2–5 μm ($\geq$ 99.5%) in diameter. The silicone rubber ZA13 and its catalyst were created by Zhermack S.P.A (Italy) and were provided by Havel Composites Ltd. (Czech Republic).

2.2 Processing of isotropic MEC

The isotropic MEC samples were produced as follows. To begin with, the silicone rubber ZA13 mixed with its catalyst in the weight ratio of 1:1. Subsequently, the CIPs with a volume fraction of 27% were added to the mixture. The mixture was then throughout stirred in a glass cup and was put in a vacuum chamber for about 15 min to eliminate air bubbles trapped during the mixing process. Next, the mixture was poured into a plastic mold and was placed in the vacuum chamber for about 10 min to remove thoroughly air bubbles trapped inside the mixture. Lastly, the mixture in the mold was cured for 24 h at room temperature (RT) to create isotropic MEC samples. Microstructural morphology of the MEC can be found in our earlier reports [3–5].

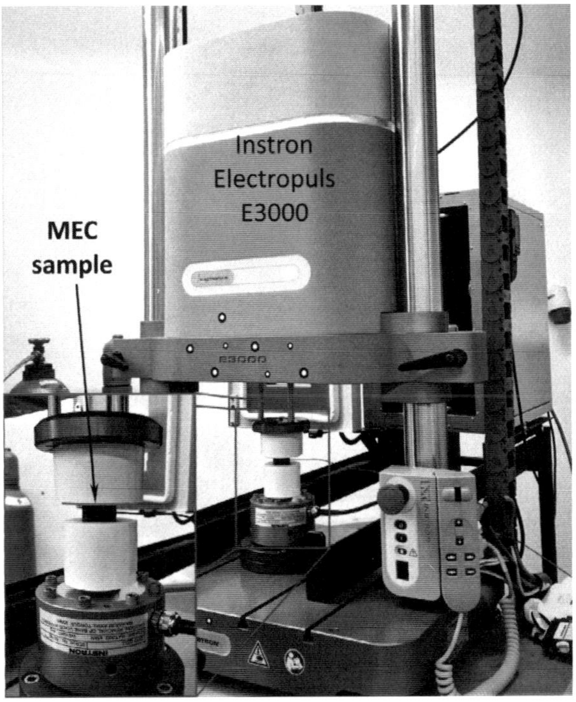

Fig. 1: The experimental setup of cyclic compression test for the MEC samples.

2.3 Cyclic compression test

The cyclic compression test for the MEC samples was performed under harmonic sinusoidal loading in the displacement control mode using Instron Electropuls E3000 testing system (Instron Corp., MA, US). The experimental setup of the cyclic compression test for the MEC samples is depicted in Fig. 1. The dynamic properties of the MEC were measured through cyclic compression tests for 120 min with various pre-strains from 5 to 20%, and strain amplitudes from 0.01 to 0.04, and frequencies from 10 to 50 Hz. Twenty cycles of force and displacement were recorded every 120 sec for dynamic mechanical analysis of the MEC.

2.4 Characterization of the MEC dynamic properties

The cyclic compression test under harmonic sinusoidal loading is conducted to characterize the dynamic properties of the isotropic MEC specimens. Therefore, the compressive stress and strain of the isotropic MEC samples will vary sinusoidally with the same angular frequency ω in the steady condition after initially several cycles. However, the compressive stress and strain will lag by a phase angle δ (loss angle).

The isotropic MEC specimens were loaded in the cyclic compression with a sinusoidal displacement-controlled process $u(t)$ of the following form:

$$u(t) = u_0 + \Delta u \sin(2\pi f t), \tag{1}$$

where u_0 is the static pre-displacement, Δu is the displacement amplitude, and f is the excitation frequency.

To calculate stresses and strains the pre-deformed specimen was chosen as the reference geometry. The sinusoidal strain in the reference geometry was described as follows [6, 7]:

$$\varepsilon(t) = \frac{u(t)}{L_0} = \varepsilon_0 + \Delta\varepsilon \sin(2\pi f t), \tag{2}$$

where $L_0 = L_u - u_0$ is the pre-deformed sample length, L_u is the un-deformed length, $\varepsilon_0 = u_0/L_0$ is the pre-strain, and $\Delta\varepsilon = \Delta u/L_0$is the dynamic strain amplitude.

The force response $F(t)$ can be written as a harmonic sinusoidal function in Eq. 3.

$$F(t) = F_0 + \Delta F \sin(2\pi f t + \delta), \tag{3}$$

where F_0 is the static force depending only on the pre-strain. The force amplitude ΔF and the phase angle δ depend on the pre-strain, frequency, and strain amplitude [6].

The isotropic MEC is assumed to be an incompressible material, i.e. $A_u L_u = A_0 L_0$ in which A_u and A_0 are cross-sectional areas of un-deformed and pre-deformed specimens. The dynamic stress response $\sigma(t)$ can calculate by dividing the force by the cross-sectional area of the pre-deformed specimen:

$$\sigma(t) = \frac{F(t)}{A_0} = \sigma_0 + \Delta\sigma \left[\cos\delta \sin(2\pi f t) + \sin\delta \cos(2\pi f t)\right], \tag{4}$$

where σ_0 is the static stress and $\Delta\sigma$ is the stress amplitude.

The dynamic stress response $\sigma(t)$ normalized by the strain amplitude $\Delta\varepsilon$ can be written as [6, 7]:

$$\sigma(t) = \sigma_0 + \Delta\varepsilon \left[E'(\varepsilon_0, f, \Delta\varepsilon) \sin(2\pi f t) + E''(\varepsilon_0, f, \Delta\varepsilon) \cos(2\pi f t)\right], \tag{5}$$

where E' is the storage modulus and E'' is the loss modulus.

Storage and loss moduli are determined according to the standard ISO-4664 as:

$$E'(\varepsilon_0, f, \Delta\varepsilon) = \frac{\Delta\sigma}{\Delta\varepsilon} \cos\delta\ , \tag{6}$$

$$E''(\varepsilon_0, f, \Delta\varepsilon) = \frac{\Delta\sigma}{\Delta\varepsilon} \sin\delta\ . \tag{7}$$

The loss tangent ($\tan\delta$) is the ratio of the loss modulus E'' to the storage modulus E'. The loss angle δ was determined using the discrete Fourier transform (DFT) at the main excitation frequency [3, 4]. The amplitudes of the compressive stress and strain were determined from experimentally recorded signals.

The storage modulus shows the MEC capability to store the strain energy, which contributes to the MEC stiffness. The loss modulus indicates the ability of the MEC to dissipate the strain energy. Therefore, the loss modulus can be estimated from the dissipated energy [3, 4]. The dissipated energy D in a loading cycle is expressed as:

$$D = \int_0^T \sigma(t)\, \dot{\varepsilon}(t)\, dt = \pi \Delta\varepsilon^2 E^* \sin\delta = \pi \Delta\varepsilon^2 E''. \tag{8}$$

3 Results and discussion

3.1 Effects of pre-strains on the MEC dynamic compressive properties

The dynamic compressive properties of the MEC specimen under the cyclic loading at the frequency of 50 Hz, the strain amplitude of 0.03, and different pre-strains from 5 to 20% were measured for 120 min corresponding to 360,000 cycles. Typical stress-strain hysteresis loops of the isotropic MEC sample at various pre-strains were presented in Fig. 2a. The storage and loss moduli and the loss angle versus time at various pre-strains were described in Figs. 2b-2d. As Fig. 2a shows, the area and slope of the MEC hysteresis loops increase with increasing the pre-strain. It is clearly observed the beak of hysteresis loop for 5% pre-strain. The beak is a sign to indicate a gap between the grip and the MEC specimen when the grip is moved upwards, thus the compressing force is equal to zero at the end of the unloading cycle. This shows that the sample does not instantaneously return to its original position during unloading.

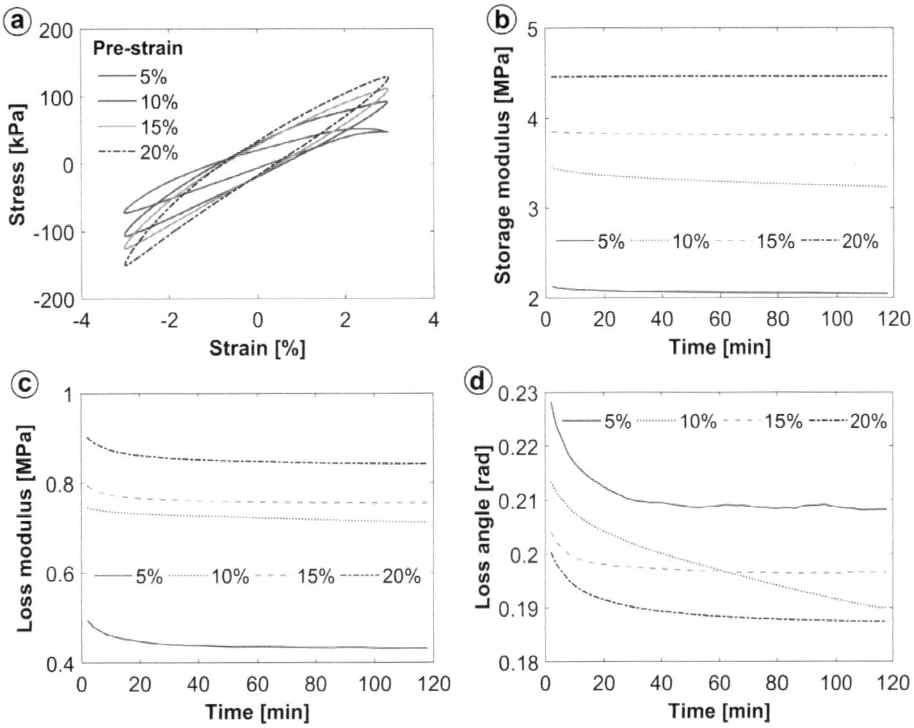

Fig. 2: Typical stress-strain hysteresis loops; (b-d) storage and loss moduli, and loss angle of isotropic MEC under cyclic loading with the frequency of 50 Hz, the strain amplitude of 3%, and various pre-strains.

As seen in Fig. 2, the storage and loss moduli of the MEC enhance with increasing the pre-strain, but the loss angle tends to decrease. Besides, although the storage modulus reduces slightly over time at the pre-strains of 5 and 10%, it almost does not change at the pre-strains of 15 and 20%. The stability of storage modulus appears as a long plateau during 120 min cyclic testing, especially at high pre-strains. The loss modulus declines slightly over time, especially at the beginning period. The loss angle decreases with increasing time, but the change of the loss angle at 10% pre-strain is different from the others. A comparison of loss moduli determined from recorded signals according to Eq. 7 and from the dissipated energy by Eq. 8 was shown in Fig. 3. The values of loss modulus calculated from the energy dissipation are in very good agreement with those determined from recorded signals. The maximal relative error between such loss moduli is less than 0.2%.

3.2 Effects of strain amplitudes on the MEC dynamic compressive properties

The dynamic properties of the isotropic MEC sample under cyclic compression loading at the frequency of 50 Hz, the pre-strain of 20%, and strain amplitudes from 0.01 to 0.04 were measured for 120 min. Typical stress-strain hysteresis loops, storage and loss moduli, and loss angle of the isotropic MEC specimen under

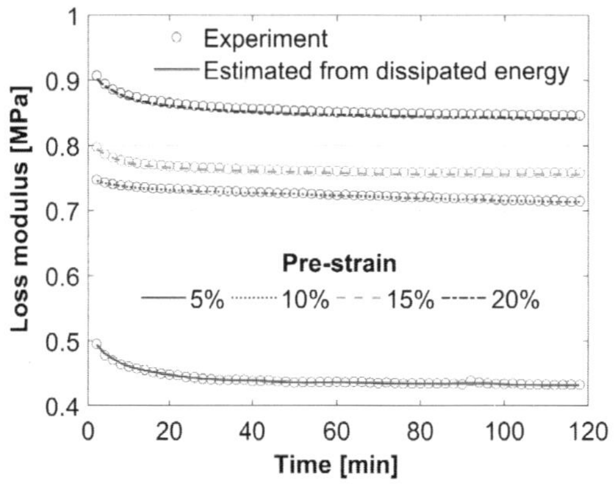

Fig. 3: Loss moduli of the MEC determined from recorded signals according to Eq. 7 and estimated from the dissipated energy by Eq. 8 as functions of time and pre-strain.

cyclic compression loading at strain amplitudes were depicted in Fig. 4. As observed in Fig. 4a, the stress-strain hysteresis loop at low strain amplitudes is nearly elliptic, but it transforms to a non-elliptical shape at high strain amplitudes. This result indicates that the MEC behaves as a nonlinear viscoelastic material at high strain amplitudes. The increase in the strain amplitude results in larger and wider hysteresis loops of the MEC. Besides, the slope of the MEC hysteresis loops slightly decreases with the rise of the strain amplitude, leading to the reduction of the loss modulus (Fig. 4c).

As observed in Figs. 4b-4c, the storage and loss moduli of the MEC reduce with increasing the strain amplitude. The decrease in the storage and loss moduli of the MEC with the rise of the strain amplitude is attributed to increasing the energy of dissipation under cyclic loading, as presented by Tong et al. [8]. Interestingly, the reduction in storage and loss moduli declines with raising the strain amplitude. The loss angle varies irregularly with increasing the strain amplitude (Fig. 4d). Besides, although the storage modulus of the MEC changes slightly during 120 min testing, the loss modulus reduces somewhat over time, especially at the initial period. The change of the loss angle over time is similar to that of the loss modulus. Moreover, the loss moduli calculated from recorded signals are in very good agreement with those determined from the energy dissipation (Fig. 5). The maximal relative error between such loss moduli is less than 0.4%.

3.3 Effects of frequencies on the MEC dynamic compressive properties

The dynamic properties of the isotropic MEC subjected to cyclic compression loading with the pre-strain of 20%, the strain amplitude of 0.03, and different frequencies were measured for 120 min. The stress-strain hysteresis loops, storage and loss moduli, and loss angle of the MEC sample at various frequencies were presented in Fig. 6. Results showed that the increase in the loading frequency results in smaller hysteresis loops. The slope of hysteresis loops enhances slightly with the rise of the frequency. Although storage and loss moduli increase with increasing frequency, the loss angle decreases slightly. Storage and loss moduli grow rapidly with raising the frequency to 30 Hz, and then gain moderately at higher frequencies.

The variation of storage and loss moduli and loss angle of the MEC specimen over time were presented in Fig. 6. Despite the storage modulus changed slightly over time, the loss modulus reduced strongly with increasing the time at the beginning of the test and followed by a slight reduction. The loss moduli determined from recorded signals are in good agreement with those calculated from the energy dissipation (Fig. 7). The maximal relative error between such loss moduli is less than 0.7%. The variation of loss angle over time is similar to that of loss modulus. In general, the dynamic proper-ties of the MEC varied markedly with raising the pre-strain, strain amplitude, and frequency of cyclic loading. Although the storage modulus changed slightly over time, the loss modulus and loss angle decreased with increasing time.

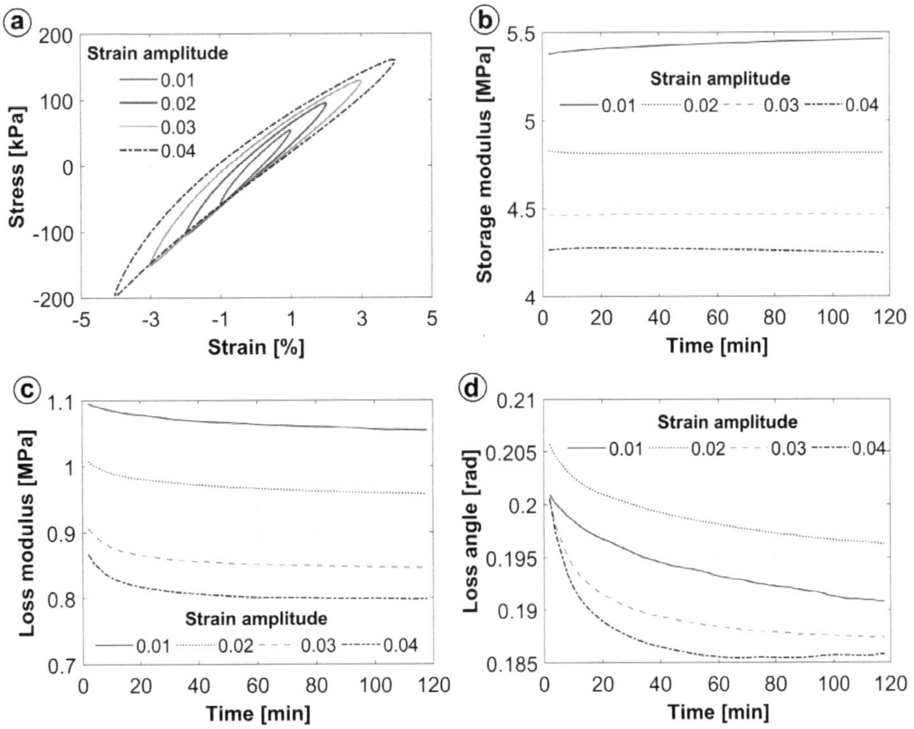

Fig. 4: (a) Typical stress-strain hysteresis loops; (b-d) storage and loss moduli, and loss angle of isotropic MEC under cyclic compression loading with the frequency of 50 Hz, 20% pre-strain, and different strain amplitudes.

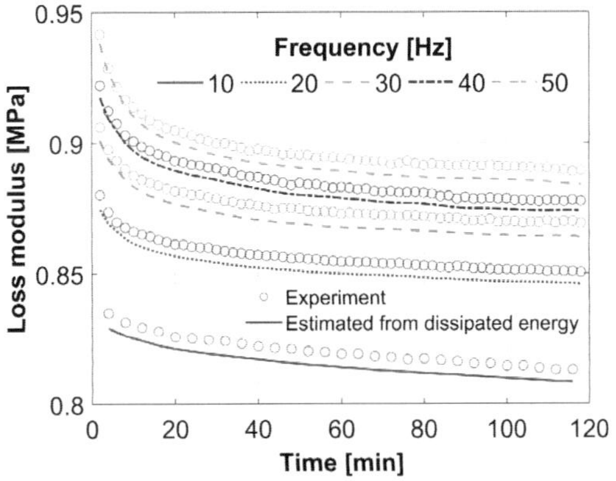

Fig. 5: Loss moduli of the MEC determined from recorded signals according to Eq. 7 and estimated from the dissipated energy by Eq. 8 as functions of time and strain amplitude.

4 Conclusion

The experimental research on the dynamical properties of the isotropic MEC made of the silicone matrix filled with the CIPs under cyclic compression loading has been conducted in this paper. Effects of different pre-strains, strain amplitudes, and loading frequencies on the dynamic properties of the isotropic MEC were investigated. The storage and loss moduli of the isotropic MEC decrease with the rise of the strain amplitude, but they enhance with increasing the pre-strain and frequency as well. Although the storage modulus of the isotropic MEC varies slightly over time, the loss modulus reduces with raising the time. The decrease in the loss modulus of the isotropic MEC under cyclic compression loading is ascribed to increasing material temperature during the testing. In short, the dynamic properties of the MEC under cyclic compression loading strongly depended on the pre-strain, strain amplitude, and excitation frequency.

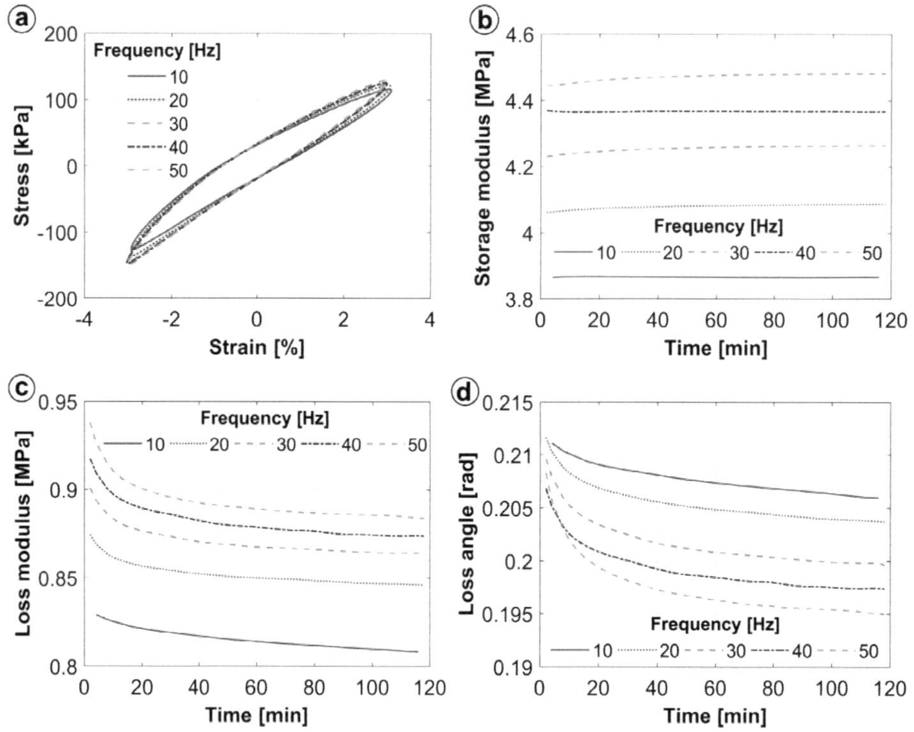

Fig. 6: (a) Typical stress-strain hysteresis loops; (b-d) storage and loss moduli, and loss angle of isotropic MEC under cyclic loading with the pre-strain of 20%, the strain amplitude of 0.03, and different frequencies.

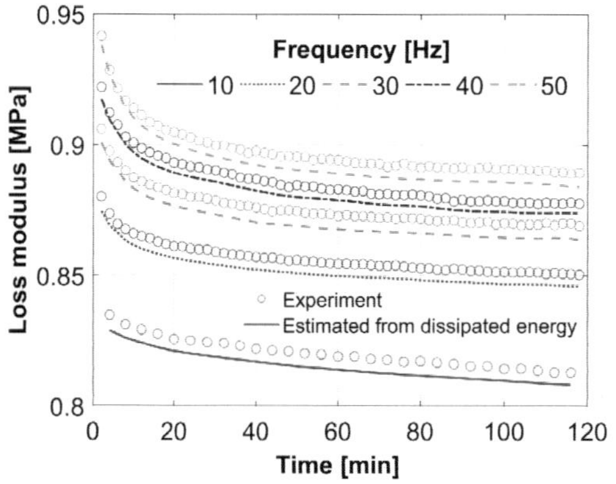

Fig. 7: Loss moduli of the MEC determined from recorded signals according to Eq. 7 and estimated from the dissipated energy by Eq. 8 as functions of time and frequency.

Acknowledgement

This work was supported by the Ministry of Education, Youth and Sports of the Czech Republic and the European Union – European Structural and Investment Funds in the frames of Operational Program Research, Development and Education – project Hybrid Materials for Hierarchical Structures (HyHi, Reg. No. CZ.02.1.01/0.0/0.0/16_019/0000843).

References

[1] A.K. Bastola, et al., Recent Progress of Magnetorheological Elastomers: a review, Smart Mater. Struct. 29 (2020) 123002, doi: 10.1088/1361-665X/abbc77.

[2] T. Liu, Y. Xu, Magnetorheological Elastomers: Materials and Applications, in Smart and Functional Soft Materials, 147–180, IntechOpen, London, UK, 2019.

[3] T.H. Nam, I. Petríková, B. Marvalová, Experimental and Numerical Research of Dynamical Mechanical Properties of Magneto-sensitive Elastomeric Composites,in proc.: Constitutive Models for Rubber XI (2019), eds. B. Huneau et al., Nantes, France, 138–143.

[4] T.H. Nam, I. Petríková, B. Marvalová, Experimental characterization and viscoelastic modeling of isotropic and anisotropic magnetorheological elastomers, Polym. Test. 81 (2020) 106272, doi: 10.1016/j.polymertesting.2019.106272.

[5] T.H. Nam, I. Petríková, B. Marvalová, Experimental and Numerical Research of Stress Relaxation Behavior of Magnetorheological Elastomer, Polym. Test. 93 (2021), 106886, doi: 10.1016/j.polymertesting.2020.106886.

[6] B. Marvalová, I. Petríková, T.H. Nam, Fatigue of Filled Rubber due to Hysteretic Heating, in proc.: Constitutive Models for Rubber XI (2019), eds. B. Huneau et al., Nantes, France, 374–378.

[7] B. Marvalová, I. Petríková, J.Novak, Self-heating of Filled Rubber during Cyclic Loading. Proceedings of 54th International Conference on Experimental Stress Analysis (2016), eds. F. Plánička, J. Krystek, Srní, Czech Republic.

[8] X. Tong, et al, Excitation of Thermal Dissipation of Solid Propellants during the Fatigue Process. Materials & Design 128 (2017), 47–55, doi: 10.1016/j.matdes.2017.04.088.

Influence of the Circular Reinforcement on the Young's Modulus of Elasticity

P. Padevět[1,*], P. Bittnar[1], Z. Bittnar[1]

[1] *Czech Technical University in Prague, Faculty of Civil Engineering, Department of Mechanics, Thákurova 7, 166 29 Prague 6, Czech Republic*

[*] *pavel.padevet@fsv.cvut.cz*

Abstract: The article is focused on the comparison of material characteristics - modulus of elasticity for two different materials. The first material is plain concrete, the second is concrete clamped in metal rings with different thickness and distribution of clamping. The rings into which the cylindrical body is concreted are made of aluminum alloy EN AW 6060 T66. The strength of concrete has a decreasing trend with decreasing number of encircling rings, as well as decreasing ring thickness.

Keywords: Young's modulus of elasticity; concrete; circular reinforcement; compression test; strain measurement.

1 Introduction

Efforts to minimize the energy intensity of the construction industry lead not only to changes in the material properties of building materials, but also to the optimization of load-bearing structures. With the correct design of the supporting structure, it is possible to realize a slimmer, and thus less energy-intensive structure. Concrete is one of the most energy-intensive building materials. Efforts to slim concrete structures lead to the use of the maximum capacity of material properties of concrete. One possibility is to use annular reinforcement to catch the transverse bars in the pressed structures [1]. The external reinforcement of the concrete structure will affect the material properties of the concrete. Attention in this article is paid to mechanical properties. Such properties include compressive strength and Young's modulus, in particular [2].

Prepapation of mixture was realized by cement CEM I 42.5 R [3] as is possible see in Tab. 1.The amount of mixing components in the table is based on 70 kg of mixture. This amount was sufficient for the preparation of bodies within two days, ie. for 18 samples. The next day after concreting, the specimens were demolded and stored in water boxes, but were not immersed in water. Concreting was realized during two days. One day before testing, the surface of the upper pressure surface was treated by grinding. The testing took place at the age of 27 and 28 days, respectively. The homogeneity of the manufactured bodies was checked by determining the bulk density for a set of bodies tested in 1 to 4 days. The volume weight variance was 100 kg/m^3 for sets 1 to 4 as is possible see in Fig 7.

Tab. 1: Mixture of concrete.

Parts	kg
Cement	11.480
Water	4.760
Sand	22.310
Aggregate 4–8	12.600
Aggregate 8–16	14.800
Superplasticizer Stachement F	0.114

Tab. 2: Measured material properties of sets.

Rings/thickness	12/3	8/3	12/2	8/2
Volume weight [kg/m^3]	2581	2509	2471	2466
Compressiion strength [MPa]	77.31	67.35	70.03	63.88
Modulus of Elasticity [GPa]	35.79	35.18	38.47	36.42

2 Preparation of tests

The comparison was performed on concrete cylinders 200 mm high and 100 mm in diameter. The production and concreting of the specimens was carried out in two steps. Firstly was concreted specimens with 3 mm thicknes of rings, and secondly with rhickness of aluminum rings 2 mm. Both mixtures were accompanied by the production of specimens without rings on the surface. This compared the preservation of the same properties of both mixtures. 18 bodies with rings th. 3 mm was made on the first day and 18 bodies with rings thickness 2 mm was made on the third day. Of the 18 bodies, 9 bodies were designated for testing the first set and 9 bodies for testing the second set. The first set contained 4 ordinary cylinders and 5 cylinders with 12 rings. The second set contained 4 cylinders without rings and 5 cylinders with 8 rings. Each set was tested for one day.

The test of each sample was controlled by increasing the force to a load level of 200 kN. Further loading proceeded under the control of the deformation increment. The experiments were carried out until the specimens were completely broken.

3 Description of instrumentation of tested specimens

The specimen testing procedure was performed in accordance with the requirements of the standard for testing concrete in compression [4]. Each specimen was fitted with two extensometers 100 mm long to measure axial deformation. Some specimens with rings were equipped with strain gauges for measuring horizontal deformations on aluminum rings, see Fig. 1 in the region of half the height of the body. The working diagrams were measured up to the failure of the bodies. Breakage of the aluminum rings was noted in the working diagrams, as well. This is very clearly viewable in Fig. 2.

Fig. 1: The cylinder specimen after compression test. The specime was tightened with 12 rings thickness 2 mm.

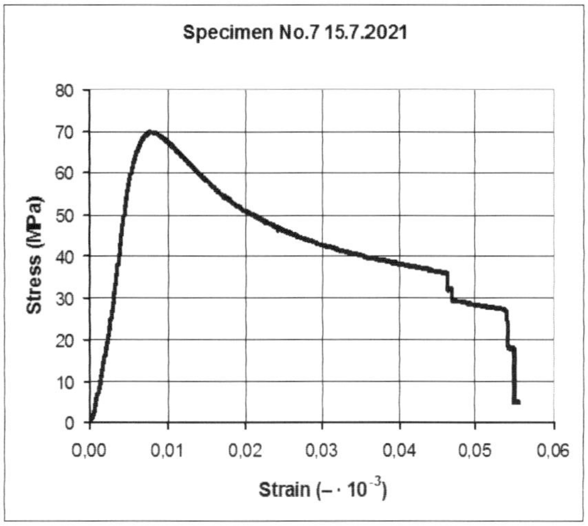

Fig. 2: Stress-strain graph of specimen with 12 rings, thickness 2 mm.

The specimens with rings were sprayed with paint to measure deformations using digital image correlation, as can also be seen in Fig 1. But our focus is on comparing compressive strength as well as modulus of elasticity on bodies with different numbers of rings and different ring thicknesses.

4 Comparison of measured parameters

The compared parameters were the average compressive strength of the specimens and the average value of the modulus of elasticity. The strength was determined as the ratio of the highest achieved force and the loaded area. The modulus of elasticity was determined as the shear value between the initial state and the value of stress and relative strain in 1/3 of the compressive strength of the specimen. The correct method of cracking the concrete cylinders in the compression test was checked so that it corresponded to the correct shape of the test specimens defined by the concrete pressure test [5].

The average compressive strength of ordinary cylinders was 45 MPa and the Young's modulus was 37 GPa. These values were achieved for both sets. The selection of test specimens for the test in one day was random from the number of specimens made from one mix. Therefore, it can be stated that even for the third and fourth tested sets, the average value of compressive strength corresponds to the strength of the first and second tested sets. The properties of the first and second mixtures are equivalent, see Fig 3. Fig. 3 and 5 show the achieved values of compressive strength and Modulus of elasticity of specimens without circular reinforcement.

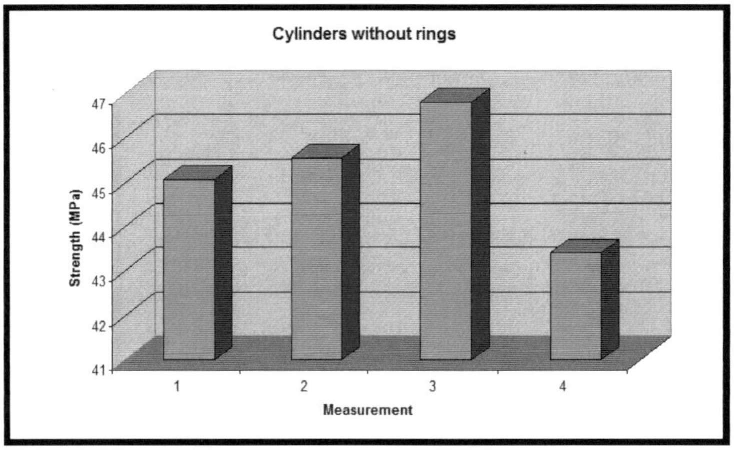

Fig. 3: Compression strength of cylinders without rings .

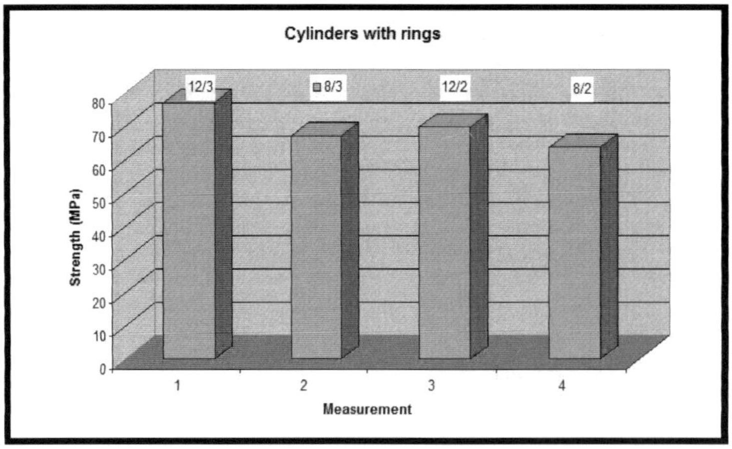

Fig. 4: Compression strength of cylinders with rings .

In contrast, Fig. 4 and 6 show the compressive strengths and the modulus of elasticity of the rings reinforced specimens. The number of rings and the thickness are indicated in the graphs by a fraction in which the first value means the number of rings and the second value is the thickness of the rings. Sets 1 to 4 correspond to Figs. 4 and 6.

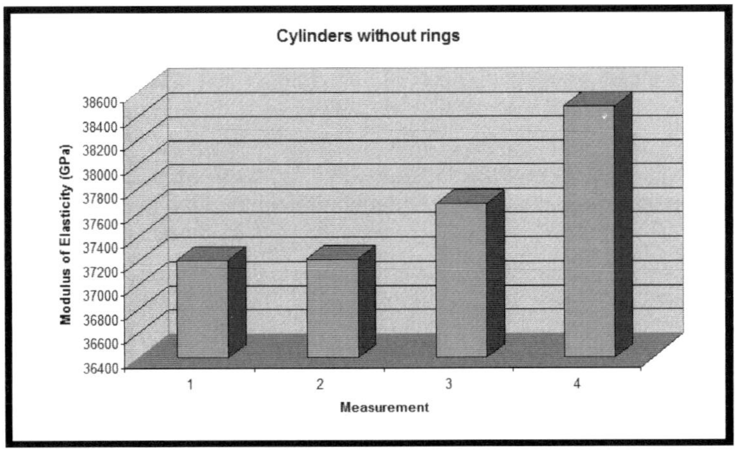

Fig. 5: Modulus of elasticity measured on specimens without rings.

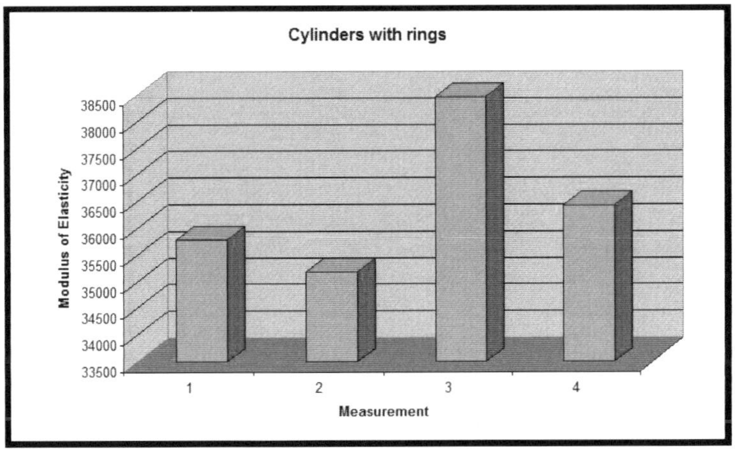

Fig. 6: Modulus of elasticity measured on cylinders with rings.

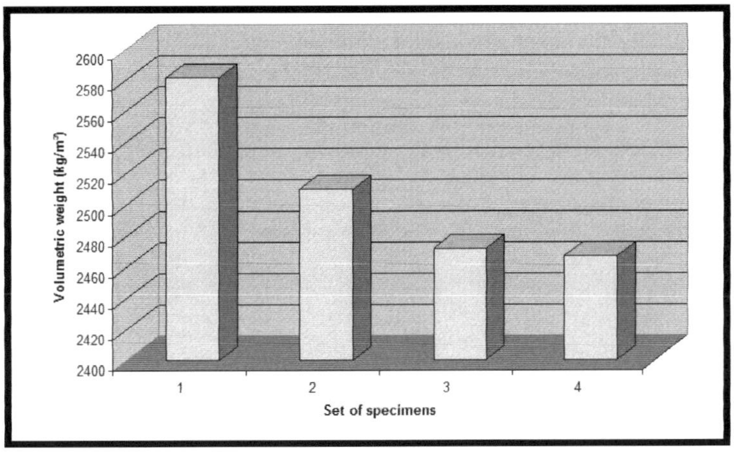

Fig. 7: Volumetric weight sets of speciems used for testing frrm 1st to 4th day.

5 Conclusion

The compressive strength has a decreasing trend, which depends on the thickness of the rings on the bodies, as well as on their number, as is possible see in Tab. 2. This table also shows that the volume density was affected by the amount and thickness of the aluminum rings. The number of rings affected the value of compressive strength. A higher number of rings increased the strength of the concrete. Compared to the basic value of compressive strength, the strength of bodies with 12 rings and 3 mm thick was 72 % higher.

The modulus of elasticity shows the opposite trend. The presence of aluminum rings on the specimens reduced the modulus of elasticity. The lowest value of Youn's modulus of elasticity was reached by specimens with 12 rings 3 mm thick. The reduction in the value of the modulus of elasticity is up to 4 %. Significantly ductile behavior of the strength part of the working diagram can be seen in Fig 2.

As can be seen in the working diagram on Fig. 2, the shape of the descending branch depends on achieving the bearing capacity of the enclosed reinforcement. By supplementing the concrete elements with reinforcement by means of external fastening, there is a significant potential in the use of strength properties, or in the possibility of reducing the amount of concrete in the structure. The failure shape of concrete bodies corresponds to the failure of specimens with circular reinforcement [6].

Acknowledgement

This research has been supported by project Grant Agency of Czech Republic GACR No.19-253125 "Compression tests with confinement for advanced modeling of concrete columns".

References

[1] P. Havlásek, et al., Robust and Inexpensive Test on Passively Confined Concrete for Efficient Design of Concrete Structures, Special Concrete and Composits 2020, AIP Conference Proceedings, Vol. 2322, 2020, doi: 10.1063/5.0041622.

[2] ČSN ISO 1920-10 Testing of concrete – Part 1. Determination of static modulus of elasticity in compression.

[3] CEM I 42.5 R [online] Available on: https://www.heidelbergcement.cz/cs/cement/volne-lozeny-cement/cemi425r

[4] EN ČSN 12390-3 Testing hardened concrete - Part 3: Compressive strength of test specimens.

[5] ASTM C39/C39M-21, Standard Test Methods, for Compressive Strength of Cylindrical Concrete Specimens, 2021.

[6] G. M. van Mier, Fracture Processes of Concrete, Assesment of Material Parameters for Fracture Models, CRC Press London, 1997, 1st edition, ISBN 9780849391231.

Fatigue/lifetime Testing of T and Y Composite Profiles

Z. Padovec[1,*], R. Sedláček[1], T. Zámečníková[1]

[1] *Czech Technical University in Prague, Faculty of Mechanical Engineering, Department of Mechanics, Biomechanics and Mechatronics, Technická 4, 160 00 Prague 6, Czech Republic*

* *zdenek.padovec@fs.cvut.cz*

Abstract: Presented paper describes fatigue/lifetime testing of two composite profiles designed primarily for the joining of aircraft structures. The selected loading levels were derived from the limit load. Once the sample was subjected to 1 000 000 cycles without failure, the experiment was stopped, and the sample was inspected by means of the ultrasonic testing method. The sample was then put aside for subsequent static testing aimed at proving whether a change had occurred in terms of the strength and stiffness compared to the "virgin" samples that had been statically tested previously.

Keywords: fatigue; lifetime; thermoplastic composite.

1 Introduction

The T profile, which is frequently used in the structural assembly of aircraft in the form of clips, brackets, and stringers [1], is manufactured via the application of various methods and using various types of composite materials. However, whichever method is applied, it is necessary to address the problem of the structure and the production of the intersection between the flange and the web [2–4]. This problem was eliminated in our case because of novel shapes of profiles. In previous works [5,6] two novel composite profiles (T and Y), manufactured by thermoforming process from C/PPS 5H satin fabric with $[[(0,90)/(\pm 45)]_2]_s$ lay-up, were tested statically in pull of direction and these experiments were simulated by FEM. The main aim of presented work was to determine the fatigue/lifetime limits of these joining components and compare them with prescribed load.

2 Materials and methods

The fatigue/lifetime tests were conducted in the laboratory on three T profiles and three Y profiles. The R value (stress ratio – minimum peak stress divided by the maximum peak stress) was 0.1 and the frequency of the load was 5 Hz except for sample Y_03 with a frequency of 4 Hz. Sinusoidal type of signal and load control was used. The experiment was performed using the Instron 40 kN hydraulic testing system (the assembly of the fatigue test is shown in Fig. 1). From the prescribed limit load (5 910 N) the selected loading levels were derived – 100% meant loading up to the limit load level, 153% loading up to the ultimate load level, 200% loading up to twice the limit load level and 306% loading up to twice the ultimate load level. Once the sample was subjected to 1 000 000 cycles without failure, the experiment was stopped, and the sample was inspected by means of the ultrasonic testing method.

The sample was then put aside for subsequent static testing aimed at proving whether a change had occurred in terms of the strength compared to the "virgin" samples that had been statically tested previously (see [1,2]). The static experiment was conducted using a TIRA 2300 universal testing machine with a crosshead speed of 2 mm/min. Displacement was measured directly from the displacement of the machine's crosshead and the load cell used during the testing has 100 kN capacity. Tensile loading was applied through the screws in the web of the profile jointed with the jaws of the machine. Fig. 2 illustrates the fixing of the sample in the machine.

Fig. 1: Assembly of the fatigue test.

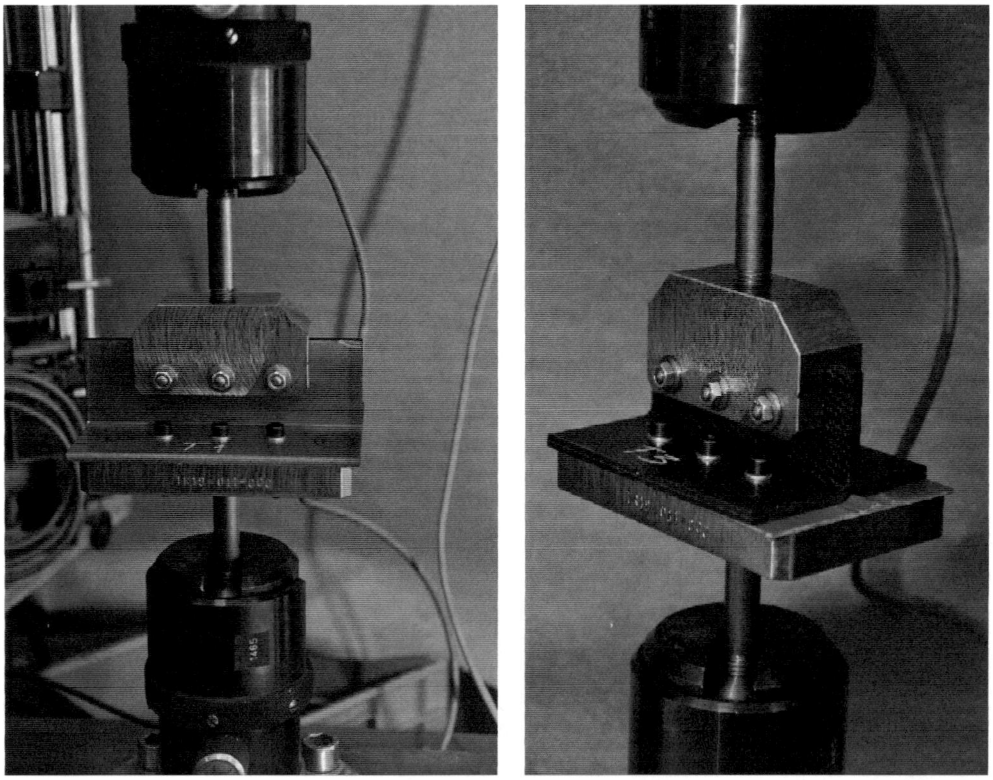

Fig. 2: T profile in the testing machine.

3 Results and discussions

The results and a description of the fatigue/lifetime test are provided in Tab. 1. Sample Y_03 was destroyed after 122 396 cycles following the destruction of the screws (after 83 407 cycles and 108 876 cycles) – see Fig. 3, which were replaced both times. Moreover, the final failure was also accompanied by the destruction of some of the screws. The T-shaped samples and samples Y_01 and Y_02 survived the fatigue testing process (the limit and ultimate upper loads for T-profiles and the ultimate load and twice the limit load for Y-profiles), thus presenting the opportunity to perform static testing aimed at proving whether there had been a change in their behavior. The comparison of tested samples after 1 000 000 cycles and "virgin" samples (see Fig. 4 and 5) shows that stiffness and strenght were not significantly affected by the fatigue effect.

Tab. 1: Lifetime results for the tested profiles.

Sample	Load level	F [N]	Number of cycles N [-]	Result
T_01	100%	5 910	1 000 000	no failure
T_02	100%	5 910	1 000 000	no failure
T_03	153%	9 042	1 000 000	no failure
Y_01	153%	9 042	1 000 000	no failure
Y_02	200%	11 820	1 000 000	no failure
Y_03	306%	18 085	122 396	failure

Fig. 3: Destroyed sample Y_03 after 122 396 cycles.

Acknowledgement

This study received support from project FV30033 of the Ministry of Industry and Trade of the Czech Republic and by the Grant Agency of the Czech Technical University in Prague, under grant No. SGS18/175/OHK2/3T/12.

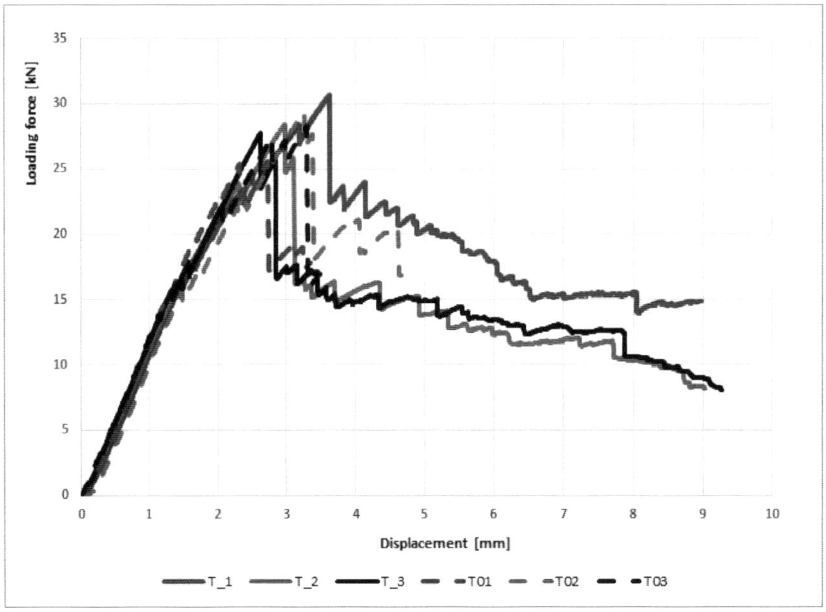

Fig. 4: Relationship between the loading force and the displacement for the tested T profiles ("virgin" and after 1 000 000 cycles).

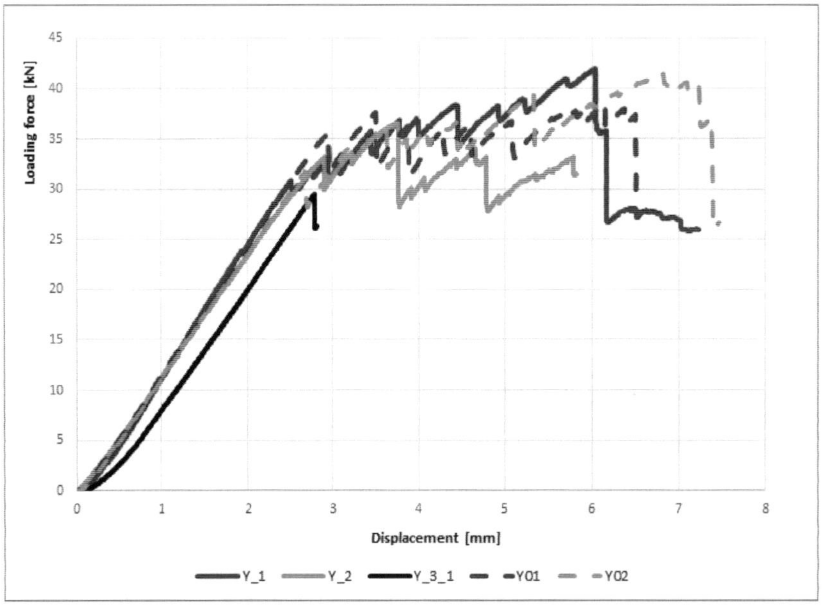

Fig. 5: Relationship between the loading force and the displacement for the tested Y profiles ("virgin" and after 1 000 000 cycles).

References

[1] V. Bühler, A. Wöginger and S. Geh, Thermoplastic Profiles as Part of the Thermoplastic Composite Toolbox by SGL Carbon, in ITHEC 2018, 4th International Conference & Exhibition on Thermoplastic Composites, Bremen, Germany, (2018).

[2] A. Offringa, Thermoplastics in Aerospace, New Products Through Innovative Technology, SAMPE Journal, 41, No. 6, 19–27 (2005).

[3] A. Offringa, The TAPAS Project, in SAMPE Europe Technical Conference, Leiden, Netherlands, (2011).

[4] Ch. Kassapoglou, Design and Analysis of Composite Structures with Applications to Aerospace Structures - Second Edition, John Wiley & Sons, Chichester (2013).

[5] Z. Padovec, R. Sedláček, T. Zámečníková, Testing of C/PPS Element for Aircraft, in: EAN 2019 57th conference on experimental stress analysis CONFERENCE PROCEEDINGS (2019), Czech Society for Mechanics, Luhačovice, 361–364.

[6] Z. Padovec, R. Sedláček, T. Zámečníková, Numerical and Experimental Analysis of C/PPS Y-Shaped Profile, in: EAN 2020 58th conference on experimental stress analysis Book of Full Papers (2020), Czech Society for Mechanics, Ostrava, 340–343.

Determination of Material Model of Hyperelastic Material

P. Palička[1,*], R. Huňady[1], P. Lengvarský[1]

[1] *Department of Applied Mechanics and Mechanical Engineering, Faculty of Mechanical Engineering, Technical University of Košice, Letná 9/B, 042 00 Košice, Slovak Republic*

[*] *peter.palicka@tuke.sk*

Abstract: Hyperelastic materials are characterized by the fact that they do not have a linear region in the stress-strain diagram, where there is a linear relationship between stress and strain. Thus, their response on mechanical loading cannot be described by material constants such as Young's modulus and Poisson's ratio. For that purpose strain energy potential is usually used. The paper deals with the evaluation of the material model of hyperelastic material in the Abaqus/CAE using the material data obtained from experimental measurements. The tested samples were made from a vehicle water channel seal. This material can be referred to as closed cell sponge rubber. It is a porous material made of expanded rubber with an air-filled matrix structure. For the purposes of evaluating the material model, tensile tests were performed at two different loading speeds, where digital image correlation system was used to measure deformation. Marlow, Ogden and reduced polynomial forms of strain energy potential were used to fit the hyperelastic material model on the test data.

Keywords: hyperelasticity; hyperelastic material model; tensile test; sponge rubber; DIC.

1 Introduction

Isotropic, linear elasticity theory is characterized by two important physical constants: Young's modulus and Poisson's ratio. In isotropic, nonlinear elasticity theory, the traditional material constants play a less important role because the material response to mechanical loading is not generally described by constants [1]. Hyperelastic materials are often used in engineering practice for the production of tires, seals, washers, mechanical energy absorbers, etc. Hyperelasticity is a property of elastomers that can extend by hundreds of percent of their original length. The mechanical response of hyperelastic materials is sensitive to variety of factors including strain rate, stress state, temperature, etc. Due to the unique cellular structures in foam materials, the Poisson's ratio has been observed to be positive, zero, or even negative, depending on cellular structures, densities, or matrix material. When the foam materials are subjected to large deformations, the material response becomes nonlinear, usually in the form of a plateau followed by a drastic hardening behavior after densification. Such a nonlinear response results in the Poisson's ratio no longer being a constant, but rather varying with strain [2].

Hyperelastic materials are described in terms of a "strain energy potential $U(\varepsilon)$" that defines the strain energy stored in the material per unit of reference volume (volume in the initial configuration) as a function of the strain at that point in the material. There are several forms of strain energy potentials available in Abaqus/CAE to model approximately incompressible isotropic elastomers. In this paper the Marlow, Ogden and the Reduced polynomial are described. Generally, when data from multiple experimental tests are available (typically, this requires at least uniaxial and equibiaxial test data), the Ogden form is more accurate in fitting experimental results. If limited test data are available for calibration, Reduced polynomial form provide reasonable behavior. When only one set of test data (uniaxial, equibiaxial, or planar test data) is available, the Marlow form is recommended. In this case a strain energy potential is constructed that will reproduce the test data exactly and that will have reasonable behavior in other deformation modes [3].

2 Hyperelastic material models

Marlow form

The form of the Marlow strain energy potential is

$$U = U_{dev}(\overline{I}_1) + U_{vol}(J_{el}), \tag{1}$$

where U is the strain energy per unit of reference volume, with U_{dev} as its deviatoric part and U_{vol} as its volumetric part; $\overline{I}_1 = \overline{\lambda}_1^2 + \overline{\lambda}_2^2 + \overline{\lambda}_3^2$ is the first deviatoric strain invariant, where the deviatoric stretches $\overline{\lambda}_i = J^{-\frac{1}{3}}\lambda_i$, J is the total volume ratio, J_{el} is the elastic volume ratio and λ_i are the principal stretches. The deviatoric part of the potential is defined by providing either uniaxial, equibiaxial, or planar test data; while the volumetric part is defined by providing the volumetric test data, defining the Poisson's ratio, or specifying the lateral strains together with the uniaxial, equibiaxial, or planar test data [3].

Ogden form

The form of the Ogden strain energy potential is

$$U = \sum_{i=1}^{N} \frac{2\mu_i}{a_i^2}(\overline{\lambda}_1^{a_i} + \overline{\lambda}_2^{a_i} + \overline{\lambda}_3^{a_i} - 3) + \sum_{i=1}^{N} \frac{1}{D_i}(J^{el} - 1)^{2i}, \tag{2}$$

where N is a material parameter; and μ_i, a_i and D_i are temperature-dependent material parameters. The initial shear modulus and bulk modulus for the Ogden form are given by $\mu_0 = \sum_{i=1}^{N} \mu_i$, $K_0 = \frac{2}{D_1}$ [3].

Reduced polynomial form

The form of the reduced polynomial strain energy potential is

$$U = \sum C_{i0}(\overline{I}_1 - 3)^i + \sum \frac{1}{D_i}(J^{el} - 1)^{2i}, \tag{3}$$

where C_{i0} and D_i are temperature-dependent material parameters. The initial shear modulus and bulk modulus are given by $\mu_0 = 2C_{10}$, $K_0 = \frac{2}{D_1}$ [3].

Hyperelastic material stability

It is common for the hyperelastic material model determined from the test data to be unstable at certain strain magnitudes. Abaqus performs a stability check to determine the strain magnitudes where unstable behavior will occur. Once the strain energy potential is determined, the behavior of the hyperelastic model in Abaqus is established. However, the quality of this behavior must be assessed: the prediction of material behavior under different deformation modes must be compared against the experimental data. An important consideration in judging the quality of the fit to experimental data is the concept of material or Drucker stability. The Drucker stability condition for an incompressible material requires that the change in the stress, $d\sigma$, following from any infinitesimal change in the logarithmic strain, $d\varepsilon$, satisfies the inequality

$$d\sigma : d\varepsilon > 0. \tag{4}$$

Using

$$d\sigma = D : d\varepsilon, \tag{5}$$

where D is the tangent material stiffness, the inequality becomes

$$d\varepsilon : D : d\varepsilon > 0, \tag{6}$$

thus requiring the tangential material stiffness to be positive-definite [3].

3 Experimental measurements

The sample dimensions and tensile test methodology were performed according to STN ISO 37 [4]. The samples were made from a vehicle water channel seal, from the upper part that can be referred to as closed cell sponge rubber (Fig. 1). It is a porous material made of expanded rubber with an air-filled matrix structure.

Tensile tests were performed on a FPZ 100/1 testing machine. Digital image correlation system Q-450 with one high-speed camera was used to measure deformation of the samples (Fig. 2a). Digital image correlation (DIC) is an optical method that allows to measure displacements and surface strains of an object [5]. Deformation fields are determined by correlation of corresponding sub-images called facets before and after loading. Each pattern must contain a characteristic part of the contrast stochastic speckle pattern that needs to be applied on the surface of the sample.

Fig. 1: Rubber seal.

In this work, the pattern was created by dotting with a white marker. This method proved to be the most suitable for creating a pattern on the surface of the rubber sample. The use of pre-printed adhesive vinyl foil was not an option due to its relatively high stiffness and the threat of peeling under large deformations. Creating a pattern using white spray paint also proved inappropriate. The paint had a tendency to fall off after drying. The pattern formed using white marker was stable until the sample was torn. The size of the speckles was optimal with respect to the dimensions of the sample (Fig. 2c). The resulting pattern ensured good image correlation. The deformation field remained compact until the last step, as shown in Fig. 3a. The obtained surface contour of the sample is given by a virtual grid of points. The size of the facets was set to 9 px and the grid size to 5 px. Tensile tests were performed at two speeds 84 mm/min and 600 mm/min. The sampling frequency of the camera was set to 50 fps in the first case and 100 fps in the second case.

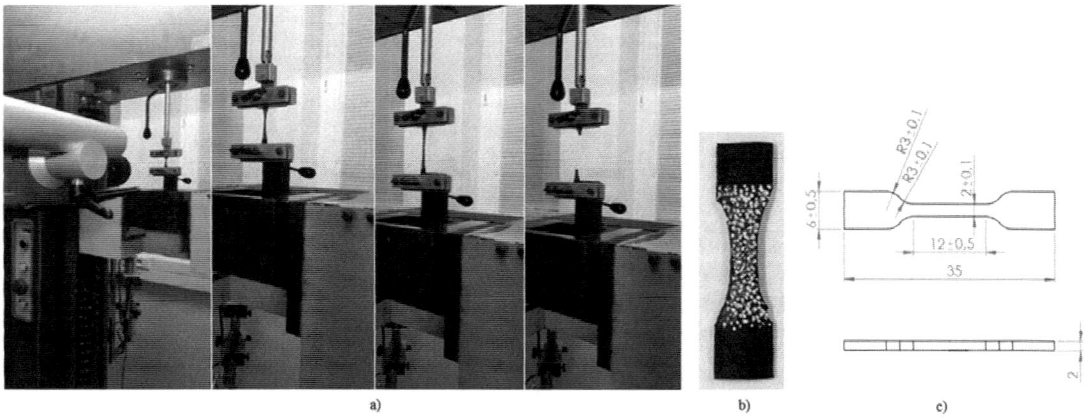

Fig. 2: a) Tensile test; b) Sample with stochastic pattern; c) Sample dimensions.

The tensile force was synchronously measured using a force sensor integrated into the testing machine. The measurements were evaluated in Istra 4D software (Fig. 3a). Nominal stress and nominal strain were calculated according to the procedure defined in the standard [4]. For that purpose, the virtual line gauges were used (Fig. 3b). The nominal strain was determined based on the relative distance of these gauges.

From strain field (see Fig. 3a), the values needed to determine the Poisson ratio were obtained. The Poisson's ratio μ determines the ratio of relative transverse shortening to its relative elongation, i.e.

$$\mu = \frac{-\varepsilon_x}{\varepsilon_y}, \tag{7}$$

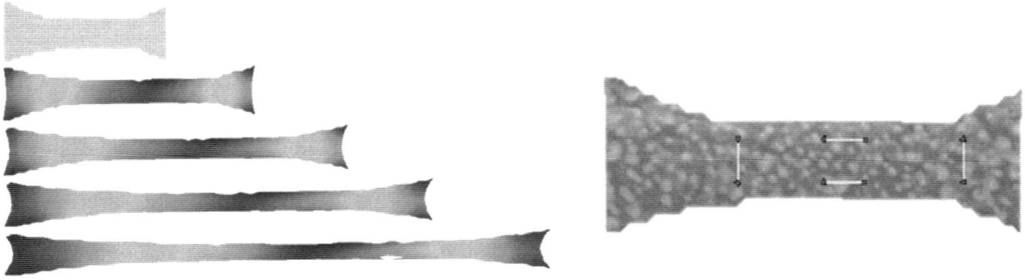

Fig. 3: a) Longitudinal strain field; b) Virtual gauges.

where ε_x and ε_y represent strain in transverse and axial direction, respectively. Their values were obtained from Istra 4D using a virtual line gauges (see Fig. 3b).

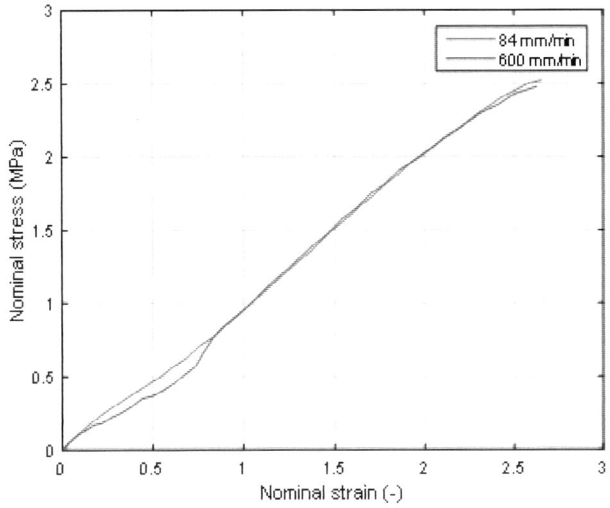

Fig. 4: Stress-strain diagram of tested rubber sponge samples.

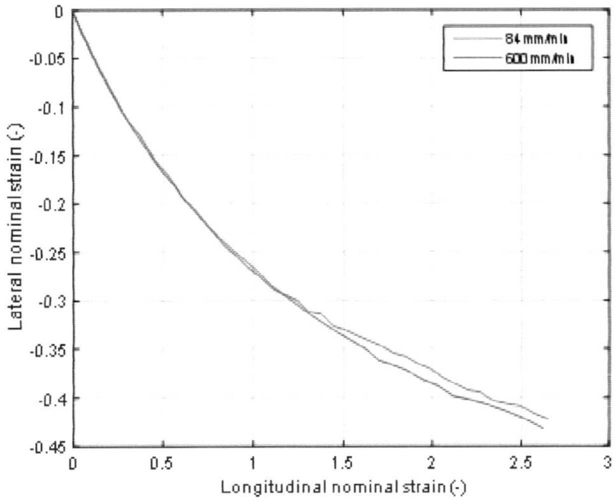

Fig. 5: Lateral vs longitudinal strain of tested rubber sponge samples.

The results of the measurements are shown in the following figures. Fig. 4 shows the stress-strain curve of tested rubber sponge samples for two different loading speed. The dependence of the lateral nominal strain on the elongation is shown in Fig. 5. Since the strains were evaluated in two perpendicular directions, it was possible to show how the Poisson's ratio decreases with strain (see Fig. 6).

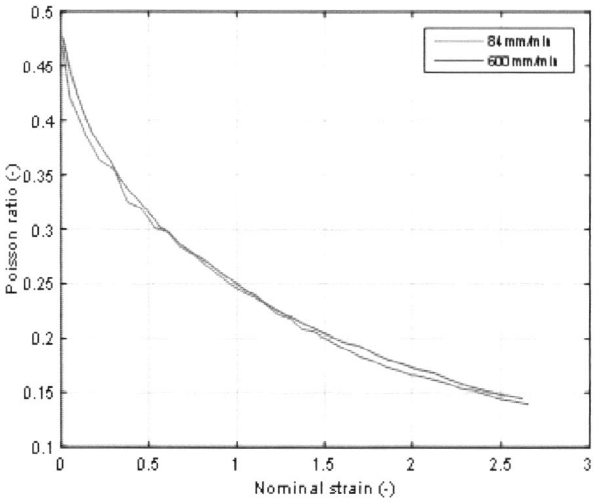

Fig. 6: Poisson's ratio vs longitudinal strain of tested rubber sponge samples.

4 Calibrating a material model

The meausred material data were imported into the Abaqus/CAE to calibrate the hyperelastic material using the "Evaluate" function. This feature allows to predict the behavior of a material model. Calibration process is based on fitting the hyperelastic model to experiment test data. The best fit model must meet two criteria: stability and the least possible fitting errors. Fig. 7 and Fig. 8 show the material characteristics obtained by the fitting process. In addition to the Marlow form, Ogden and Reduced polynomial forms of the higher order were used. The fulfillment of the criteria is given in Tab. 1 and Tab. 2.

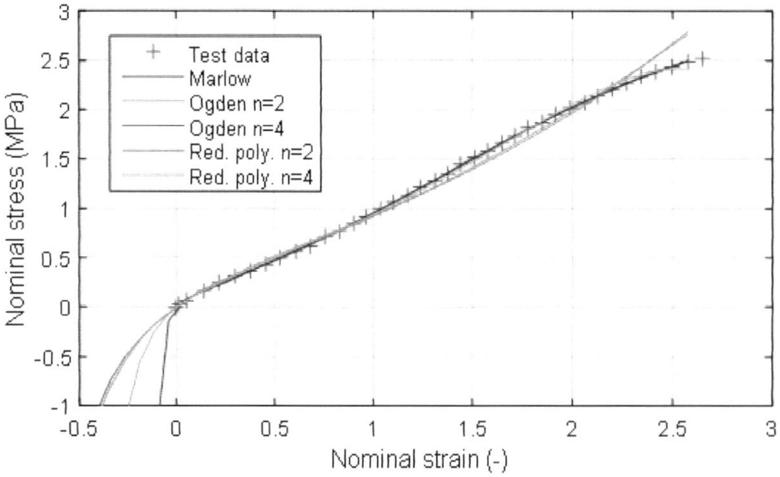

Fig. 7: Fitted stress-strain curves of the sample loaded at a speed of 84 mm/min.

Based on the evaluation of the calibration results of the material models, it was concluded that the Marlow, Ogden 2nd order and Reduced Polynomial 2nd order models are the most suitable with respect to stability for specimens loaded at 84 mm/min and Marlow and Reduced Polynomial 3rd order models for specimens loaded at 600 mm/min. The coefficients of the selected models are given in Table 3.

Tab. 1: Stability and fitting error of the material model at a tensile speed of 84 mm/min.

Stability	Forms				
	Marlow	Ogden n=2	Ogden n=4	Reduced polynom. n=2	Reduced polynom. n=4
Uniaxial tension	Stable	Stable	Stable	Stable	< 2.6800
Uniaxial compression	Stable	Stable	> -0.0930	Stable	> -0.8565
Planar tension	Stable	Stable	< 0.0900	Stable	< 2.600
Planar compression	Stable	Stable	> -0.0826	Stable	> -0.7222
Volumetric tension	Stable	Stable	Stable	Stable	Stable
Volumetric compression	Stable	Stable	Stable	Stable	Stable
Fitting error	-	0.21	0.05	0.31	0.24

Tab. 2: Stability and fitting error of the material model at a tensile speed of 600 mm/min.

Stability	Forms				
	Marlow	Ogden n=2	Ogden n=3	Reduced polynom. n=2	Reduced polynom. n=4
Uniaxial tension	Stable	< 1.7100	< 2.6000	Stable	< 2.7800
Uniaxial compression	Stable	Stable	> -0.9228	Stable	> -0.8628
Planar tension	Stable	Stable	< 2.6000	Stable	< 2.6900
Planar compression	Stable	Stable	> -0.7222	Stable	> -0.7290
Volumetric tension	Stable	Stable	Stable	Stable	Stable
Volumetric compression	Stable	Stable	Stable	Stable	Stable
Fitting error	-	8.99	10.20	13.10	10.18

Tab. 3: Coefficients of Reduced polynomial 2nd order forms.

84 mm/min		600 mm/min
Ogden n=2	Reduced polynomial n=2	Reduced polynomial n=3
$\mu_1 = 0.32738$	$C_{10} = 0.2374$	$C_{10} = -2.4761 \times 10^{-4}$
$\mu_2 = 0.2856$	$C_{20} = -4.8579 \times 10^{-4}$	$C_{20} = 0.11091$
$\alpha_1 = 1.8600$		$C_{30} = -5.5865 \times 10^{-3}$
$\alpha_2 = -3.9997$		
The other coefficients are zero.		

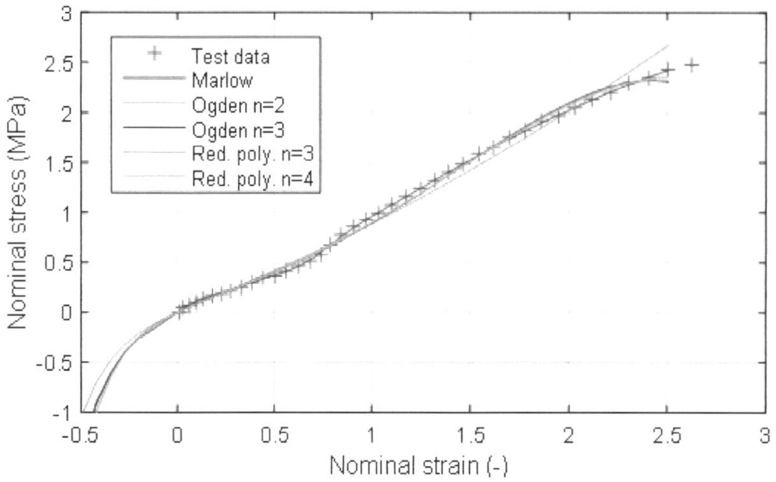

Fig. 8: Fitted stress-strain curves of the sample loaded at a speed of 600 mm/min.

5 Conclusion

The paper presented the process of calibration of hyperelastic material model in the Abaqus/CAE program based on data obtained during tensile tests performed at two load rates. Analyzed material was a porous expanded rubber with an air-filled matrix structure. Marlow, Ogden and Reduced Polynomial forms were used in the fitting process. The selection of the most suitable model depends on the fitting accuracy and on the deformation interval in which the material response is stable. The stability criterion evaluates whether the deformation energy is positive. When a material model is unstable and a component deforms through that instability, the result is non-physical behavior that can cause nonconvergence. Since the stress-strain response of hyperelastic materials is loading mode dependent, there is important to minimize model fit errors in that loading mode that is predominant. For this reason, the visual evaluation of the user is as important as the mentioned criteria. Based on the presented results, it can be stated that the best fitting and stable model was provided by Marlow form.

Acknowledgement

This work was supported by the grant project VEGA 1/0355/18 funded by Vedecká grantová agentúra MŠVVaŠ SR and SAV.

References

[1] M.F. Beatty, D.O. Stalnaker, The Poisson function of finite elasticity, Journal of Applied Mechanics (1986), 807–813, https://www.researchgate.net/publication/245358311_The_Poisson_Function_of_Finite_Elasticity.

[2] B. Sanborn, B. Song, Poisson's Ratio of Hyperelastic Foam Under Quasi-Static and Dynamic Loading, International Journal of Impact Engineering (2019) 48–55, ISSN 0734-743X.

[3] Abaqus documentation, retrieved from: https://abaqus-docs.mit.edu/2017/English/SIMACAEMATRefMap/simamat-c-hyperelastic.htm.

[4] STN ISO 37: Rubber, vulcanized or thermoplastic. Determination of tensile stress-strain properties (in Slovak), 2008.

[5] M. Hagara, R. Huňady, F. Trebuňa, Stress Analysis Performed in the Near Surrounding of Small Hole by a Digital Image Correlation Method, In: Acta Mechanica Slovaca 18 (3-4), 2014, p. 74–81.

Analysis of Thermomechanical Load of GT Turbine

J. Pařez[1,*], T. Vampola[1]

[1] *Department of Mechanics, Biomechanics and Mechatronics, Faculty of Mechanical Engineering, Center of Advanced Aerospace Technology, Czech Technical University in Prague, Technicka 4, 160 00 Prague 6, Czech Republic*

[*] *Jan.Parez@fs.cvut.cz*

Abstract: The published article describes the thermal and mechanical loading analysis of a generator turbine assembly of turboprop aircraft engine. The methodology of determining the critical point with maximum equivalent stress is presented. The input information of the generator turbine stress determination methodology is the loading conditions and the geometry of the assembly. The paper presents a mechanical analysis, where the detailed temperature analysis is already described in the previous article.

Keywords: generator turbine; mechanical analysis turbine loading; turbomachinery.

1 Introduction

The aim of mechanical analysis is to define the inputs and then determine the maximum stress in the component. Since the external time-varying input values are essentially non-stationary, it is necessary to determine a steady response to these inputs. The article therefore deals with the analysis of stress in the steady state of the engine on defined power modes.

The overall load analysis is divided into two steps. In the first step, the response of the system to the thermal load was determined. This response is read as a boundary condition in the stress analysis calculation presented in previous article [1].

For these boundary conditions, the problem of stress field distribution in the GT turbine assembly is solved. The problem assumes the solution of an elastic problem without considering the mechanism of plasticity, which would be calculated using, for example, Neuber's solution presented in [2] or in Nagode research [3].

2 Loadings

The GT turbine assembly is a highly thermally and mechanically stressed part with a whole range of loads, which can be divided into the following categories:

- Temperature loading,
- Rotation (centrifugal loading),
- Aerodynamical forces,
- Contact loading.

2.1 Temperature field

Temperature analysis is used as an input boundary condition for subsequent stress (mechanical) analysis. The load cycle was calculated based on the proposed speed modes theoretically calculated based on the procedures set out in [4].

Steady state is solved by linear temperature analysis in steady state engine operation. This article considers the temperature field, which has already been determined in the article mentioned in [1] and are also described in [5, 6] and [7].

Dividing the task into two steps is for practical reasons. The distribution consists in the type of calculation of the steady response. For each task, it is necessary to set different calculation conditions with regard to networking and other FEM calculation settings.

Furthermore, the possible use of advanced analysis involving the conditions of variable temperature inputs along the blade length, which would be calculated using CFD software and subsequently serve as the input temperature load to the temperature analysis, is also envisaged. The temperature distribution on the components is read from the temperature analysis. In Fig. 1 shows the distribution of the temperature field on a GT turbine assembly.

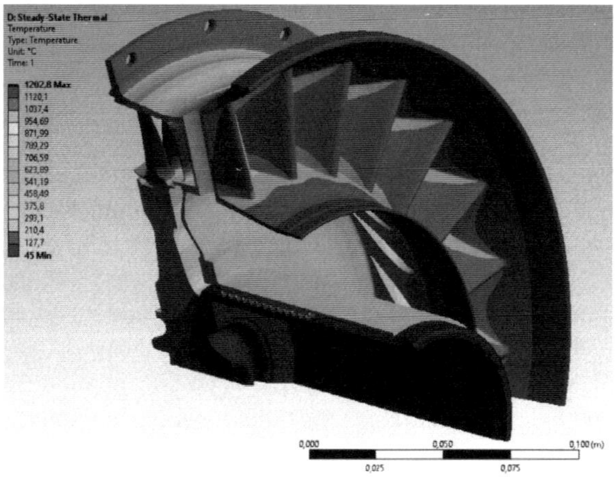

Fig. 1: Temperature field distribution on generator turbine assembly.

2.2 Centrifugal forces

The second significant load is the load from centrifugal forces. The general formula for calculating the centrifugal stresses is given in Eq. 1.

$$\sigma_{od} = r m_r \omega^2, \tag{1}$$

where r is the radius of curvature, m_r is the mass of the centrifugal part as a function of the radius of curvature and ω is the angular velocity for the given flight mode.

The centrifugal force is implemented in the ANSYS calculation program. The input parameters for this calculation are the material density, geometry and considered speed of the GT wheel disc. In Fig. 2 shows the above load. A fixed bearing is the bearing is selected. Axial displacement is ensured by binding D and C. Further, in Fig. 1 shows the speed of the GT disc and shaft.

2.3 Aerodynamical forces

The load from aerodynamic forces is prescribed on the blades of the stator and the rotor. It's based on the assumption set out in [8]. The specific forces are determined based on the relations for the tangential specific load Eq. 2 and axial specific load Eq. 3.

$$q_y \approx \frac{P_\omega}{\omega z} \frac{2}{D_1^2 - D_2^2}, \tag{2}$$

where ω is the power of the turbine stage as a function of speed, z is the number of disc blades and D_1, D_2 are the head and foot diameter of the blade.

$$q_x \approx (p_{1(\omega)} - p_{2(\omega)}) \frac{2\pi D}{z}, \tag{3}$$

where p_1, p_2 are the inlet and outlet pressures of the blade grid stage for the given flight mode, z is the number of disc blades and D is the diameter at which the specific load is calculated.

Fig. 2: Centrifugal forces on rotating parts.

In Fig. 3 shows the above load. Further, in Fig. 3 is shown individual components of aerodynamic forces on the blades from the flue gas stream.

Fig. 3: Aerodynamical forces on stator and rotor blades.

2.4 Contact loading

The generator turbine assembly is simplified for the requirement of mechanical analysis of the system. The shaft, which transmits the torque from the turbine part to the compressor part, is guided by a cut at the bearing location. At this point, the bearing under consideration is then the corresponding bearing in the bearing. Another contact surface is the connection of the stator part in the motor assembly. It is fixed by a structural connection of screws on the outer edge of the stator. Next, a section is made through the stator cover of the shaft.

In the analysis, a system without internal additional stresses is considered. For this reason, there is zero external force on all contact surfaces. The system considers only the internal load of the system. For the analysis of contact surfaces, it is necessary to know all the following structural elements solved within a large system.

3 Boundary conditions

The assumptions for mechanical analysis are based on thermal analysis stated in a previously published article. This article provides a calculation for a steady flight mode at maximum engine speed. The rotor shaft is embedded in the bearing location. The forces were prescribed on the surface according to Fig. 3.

3.1 Contacts

Contact problems are crucial for the choice of bonding conditions. The problem of contacts is their correct definition, which requires sufficient computing power and it is necessary to understand the physics of the problem.

The bonds of the individual parts were created by a combination of contact and kinematic boundary conditions. The contact was selected as the contact between two "Flexible-to-Flexible" bodies. In terms of discretization of contact surfaces, the so-called "Surface-to-Surface" method is chosen for the solution. Due to the time-consuming solution of the contact condition, the contact was defined only on the parts immediately in contact with the target solved part – the turbine wheel.

A so-called "Bonded" bond was chosen for the contact solution, as a small mutual movement of the contact pairs relative to each other is assumed during the solution. The contact is therefore evaluated on the same (small) limited number of nodes, which of course speeds up the calculation itself.

Other contact conditions are "No Separation" and "Frictionless". The difference between these bonds is the normal direction of the force, where a non-friction bond can be caused to settle.

Contact definitions are prescribed in the following sections:

- GT disc - Rotor shaft (No Separation area),
- GT disc – Stator Case (Frictionless surface),
- Disc Seal – Rotor Shaft (Frictionless surface),
- Disc Seal – Stator Case (No Separation).

In the mentioned ANSYS software, it is possible to specify the areas between which a binding condition is created, and the software itself finds the nearest pairs of nodes to which it prescribes identical DOF. The two surfaces are therefore firmly connected, and the contact algorithm is not solved between them.

3.2 FEM model

Finally, the element model follows the thermal analysis for the given assembly. The following model is chosen for the requirement of mechanical analysis. A necessary prerequisite for correct analysis is the creation of finite-element model mesh. The mesh option is created with respect to the assembly geometry. Defining the meshing correctly and its size is a key step in achieving correct results.

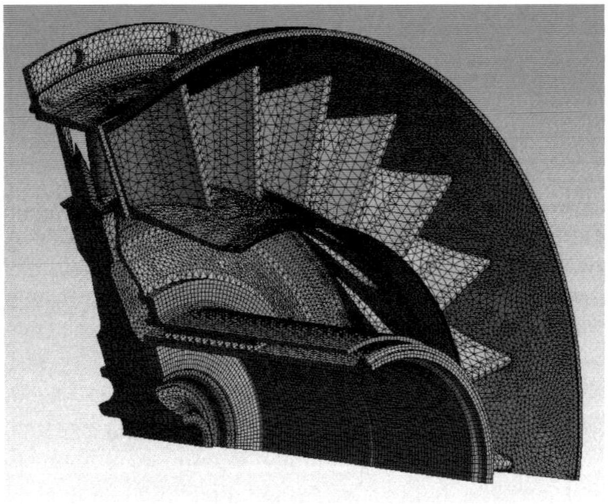

Fig. 4: Mesh of generator turbine assembly.

The mesh was chosen by Automatic elements method due to the rotational strength task in the size according to Fig. 4. As a critical point, a blade corners were assumed, as already mentioned. The mesh has been refined around this section. Excessively refined mesh is also not a suitable solution, so refinement was chosen by the number of elements around the perimeter of the hole. The size of the mesh was different from the mesh size at the edge of the hole due to the computational software that calculates based on the "beam" structure.

3.3 Loading cycle

The course of the load corresponds to steady values and is shown in Fig. 5. It therefore assumes a cycle between zero values at the maximum values of thermal and mechanical load. The resulting load is then a superposition of partial loads.

In Fig. 5 is shown the resulting reduced (Misses) stress distributed in the GT disc assembly. In Fig. 6 represent critical point of the assembly, connection of blade and case.

Fig. 5: Equivalent Misses stress distribution.

In the following Fig. 6, the critical point with the largest accumulated reduced stress is indicated. The calculation corresponds to a steady state at maximum speed and maximum temperature.

Fig. 6: Maximum equivalent stress zone.

3.4 Preparation of experimental measurement

After a detailed numerical analysis of the generator turbine behavior, in order to proceed with the experimental investigation of the stress and strain field, there is preparing the configuration of the test rig. The turbine engine with the analyzed generator turbine section, tested and studied during the research activity of CASR [9], will be first integrated with additional measuring components and forthcoming experiments. Then will be stator generator turbine instrumented with the strain gauges for the strain measuring in the radial direction. In terms of functionality and operational reasons, the location of the strain gauges will be inside the stator cooling blades and will be taken out of the motor's flow path.

Finally, the temperature field will be measured for the clarification of assumptions presented in previous research [1]. Static temperature on stator blades of the generator turbine section during the experimental tests will be measured with NiCr-NiAl thermocouples type K. After completing the installation and the calibration of all the sensors and of the equipment necessary to conduct the experiments, the experimental tests on the test rig will be carried out.

4 Conclusion

The mechanical analysis consisted in calculating the stress response to a steady state field determined by a linear steady state temperature analysis. Stress analysis considers the transfer of heat from the flue gas and the solution of the contact problem between the parts. The result is a linear elastic stress behavior of the GT wheel assembly. The critical point in the figures was determined - the place with the largest accumulation of reduced Misses stress.

Different temperature distribution has different system stiffness due to temperature-dependent parameters (modulus of elasticity is considered depending on temperature). Due to the distribution and size of this temperature field, the stress field varies depending on the speed modes.

Due to the high temperature, at a stable temperature field, determined by linear temperature analysis, there is a greater redistribution of deformation due to lower stiffness in different parts. The course of temperatures and stresses was marked in the figures.

The area of stator blades and the rotor disc area were determined as critical points from the analysis. The accumulation of the maximum reduced stress occurs at the connection of the blades and the case, due mainly to the different thermal stress and expansion of the blades. The connection area of the labyrinth seal and the case stator could not be analyzed, and a more detailed analysis is necessary in case of fatigue behavior.

Acknowledgement

Authors acknowledge support from the ESIF, EU Operational Programme Research, Development and Education, and from the Centre of Advanced Aerospace Technology (CZ.02.1.01/0.0/0.0/16_019/0000826), Faculty of Mechanical Engineering, Czech Technical University in Prague.

References

[1] J. Pařez, T. Vampola, Thermo-mechanical Modeling of Generator Turbine, 28th Workshop of Applied Mechanics. Prague, November 2020. ISBN 978-80-01-06791-8.

[2] J. Pařez, Methodology for the thermomechanical fatigue prediction of generator turbine disc, Praha, 2020, Master thesis, available: https://dspace.cvut.cz/handle/10467/90008.

[3] M. Nagode, M. Hack, M. Fajdiga, Low cycle thermo-mechanical fatigue: damage operator approach. Fatigue & Fracture of Engineering Materials & Structures. 33.3 (2010).

[4] S. Farokho, Aircraft Propulsion. Second edition. The University of Kansas, USA: Wiley, 2014, ISBN 9781118806777, pp. 743–746.

[5] J. Han, Fundamental Gas Turbine Heat Transfer. Journal of Thermal Science and Engineering Applications, ASME, June 2013, Vol5(2), doi: 10.1115/1.4023826.

[6] G.S. Azad, Heat transfer and pressure distributions on a gas turbine blade tip. Journal of turbomachinery 122.4 (2000), pp. 717–724.

[7] M. Kaviany, Principles of Heat Transfer: Wiley-Interscience publication [online]. John Wiley, 2002, ISBN 0471434639.

[8] J. Statečný, F. Sedlář, Z. Doležal, Pevnost a životnost leteckých turbínových motorů. Fakulta strojní: Vydavatelství ČVUT, 1995, ISBN 80-01-00420-01, pp. 17–21.

[9] Center of Aviation and Space Research, ČVUT FS, 2022, available: https://www.fs.cvut.cz/en-ustavy/en-sekce-center-of-aviation-and-space-research/en-center-of-aviation-and-space-research-12203/en-center-12203/.

Analysis of Acoustic Emission Pressure Test of Mortar Specimens with Different Materials

L. Pazdera[1,*], L. Topolar[1], R. Dvorak[1], M. Hodulakova[1]

[1] *Brno University of Technology, Faculty of Civil Engineering, Veveří 95, 602 00 Brno, Czech Republic*
[*] *pazdera.l@fce.vutbr.cz*

Abstract: The acoustic emission method is one of the tools for describing the behavior of a specimen during (mechanical) deformation. The compression test is one of the most conventional destructive methods for testing structural composites. Material engineers strive to find a compound with optimal properties. This paper aims to show the possibilities of a more detailed analysis of acoustic emission signals recorded during the loading of test specimens made from two different materials and two different sizes.

Keywords: accoustic emission; mortar; tensile; analysis; alkali activated.

1 Introduction

The main function of concrete elements in structures is the transmission of strength actions of loading when the concrete is exposed to various stresses. One of the most common building materials today is concrete. Its advantages include reasonable workability, good compressive strengths, durability, fire resistance and high heat accumulation. The traditional binder employed in its production has been Portland cement, whose production is not ideal from the environmental point of view. It is, however, relatively inexpensive. Additions of alkali-activated activators can result in the reduction of cement in the final mixture. The experiments discussed in this paper were conducted in order to compare an alkali-activated slag optimized in regard to chemical resistance with a conventional cement-based mortar. The experiment also studied the influence of specimen size on acoustic emission signals within the testing of prismatic strength [1, 2].

Compressive strength of concrete is a priority property of concrete or mortar for building construction. Compressive strength of concrete is the most important characteristic feature of concrete. The most commonly used is cubic compressive strength $f_{c,cube}$, which is determined on cubes. In our case, we used measurements of prismatic strength $f_{c,prism}$. Prismatic strength $f_{c,prism}$ is therefore measured on prism-shaped specimens. It is determined on prism-shaped specimens with the base to height ratio of 1:4 (1:3 is also allowed). Prismatic strength is lower than cubic strength $f_{c,cub}$:

$$f_{c,prism} = (0.7 \div 0.85) \times f_{c,cube}. \tag{1}$$

Since the aim of this paper is to monitor not only the resulting prismatic compressive strength, but also the behaviour of the specimens during loading, the acoustic emission method was selected to suit both tasks. There are not many methods that allow monitoring of changes in the structure of a material during loading. The acoustic emission method is based on the principle of recording acoustic responses inside a specimen using suitable sensors. Loading of a material results in formations of areas where the mechanical stress accumulates. Once this stress is released, e.g. a crack forms, a part of that energy is emitted in the form of a mechanical wave. The sensor then detects this wave. However, the recorded wave carries information not only about the source, but also about the path of the wave and the material the wave propagated through. Evaluation of the acoustic emission method is therefore far from a simple task [3–5].

2 Experimental set up

The experiment monitored the behaviour of test specimens with dimensions of 40 mm × 40 mm × 160 mm (marked as m) and 100 mm × 100 mm × 400 mm (marked as v) were. The specimens were designed of two

Tab. 1: Composition of the cement-based mixture

Component of mixture	Weight [g]
CEM III/A 32.5 R	1,000
Standardised sands	3,000
Water	450

Tab. 2: Composition of the mixture based on alkali-activated slag

Component of mixture	Weight [g]
Granulated Blast Furnace Slag	1,000
Stabdardised sands	3,000
Water	331.4
50% NaOH	113.8
Sodium water glass (M_s = 1.89)	93.9

different fine-grained mixtures, one based on cement (marked as c) and the other based on alkali-activated slag (marked as o). Loading of the specimens was on bases, i.e. 40 mm $\times$ 40 mm (1.6×10^3 mm^2) for the small specimens and 100 mm $\times$ 100 mm (10^4 mm^2) for the large specimens. The mixtures are in Tab. 1 and Tab. 2.

The specimens in the testing machine were loaded with a constantly increasing stress (0.6 MPa/s) until destruction. A photo of the testing machine is in Fig. 1b. Two acoustic emission sensors were placed on the specimens each time. The distance of the sensors from the end of the specimen was 10 cm. Both sensors were placed on the same surface of the specimen. The locations where the sensors were attached using beeswax are evident from Fig. 1a. Typical acoustic emission parameters, such as acoustic emission RMS, the number of hits over a specified acoustic emission level, etc., were detected every second, i.e. over a 0.6 MPa increase in the mechanical stress. Significant individual events were also recorded during loading [6, 7].

3 Results

In the case of all the measurements, the monitored prismatic compressive strengths reached approximately identical average values, around 50 MPa. The cubic strength according to the Eq. (1) would then reach values in the range of 60 to 70 MPa.

Based on the information included in Tab. 3, it can be concluded that the influence of material composition on the velocity of longitudinal propagation is more evident in specimens from alkali-activated structures. Different propagation velocities can be observed for specimens of different sizes. In the case of cement specimens, the observed velocity is practically identical within the statistical error.

Fig. 2 presents the basic characteristics of the dependence of the acoustic emission Root Mean Square (RMS_{AE}) of the measured acoustic emission signals on the pressure stress (R). Thus, the RMS value was recorded and calculated at regular loading intervals.

Tab. 3: Wave propagation velocity.

Marked	Material	Volume [mm^3]	c [mm/µs]	R [MPa]
o-v	alcali activated	$100 \times 100 \times 400$	3.7	51
c-v	concrete	$100 \times 100 \times 400$	4.3	48
o-m	alcali activated	$40 \times 40 \times 160$	4.1	44
c-m	concrete	$40 \times 40 \times 160$	4.3	50

(a) Specimen, the arrows mark the placement of the acoustic emission sensors

(b) Placement of the specimen in the testing machine

Fig. 1: Specimen and experiment setup.

$$RMS_{AE} = \sqrt{\frac{\sum U_{AE}^2}{N}}, \tag{2}$$

where U_{AE} is the voltage at the acoustic emission sensor and N is the number of specimens during a stress change of 0.6 MPa, as stated in the measurement description. The graph in Fig. 2a incorporates a linear scale for RMS, while the graph in Fig. 2b incorporates a logarithmic resp. decibel scale.

$$RMS_{dB} = 20 \times \log_{10} \frac{U_{AE}}{U_{ref}}, \tag{3}$$

where U_{ref} is equal to 1 μV.

It can be deduced from Fig. 2 that specimens of identical size exhibit similar behaviour. Thus, the specimens with dimensions 100×100×400 mm^3 designated with the symbol v have a lower value of acoustic emission activity than the specimens with smaller dimensions, i.e. 40×40×160 mm^3 designated m. The strengths of all the monitored specimens are approximately the same – see Tab. 3.

The RMS values in Fig. 3. have been divided based on the size of the test specimens. A significant change in the acoustic emission activity in large specimens (see Fig. 3b) occurs from the compressive stress value of 20 MPa. In the case of smaller specimens (see Fig. 3a), the change is less evident but occurs again in the region of the compressive stress of 20 MPa.

In order to make the comparison of the change of the acoustic emission activity for the same specimens but of different dimensions clearer, the graphs in Fig. 4 have been created. The presented graphs clearly show a lower recorded acoustic emission activity for larger specimens.

Fig. 5 shows the number of recorded acoustic emission events (n) at a compressive stress (R), both cumulatively (Fig. 5a) and cumulatively relatively (Fig. 5b). Both graphs in Fig. 5 indicate a change in the activity at the knee regions of the curves, i.e. around 20 MPa. In addition, the recorded acoustic emission activity of the large specimens (designated $c - v$ and $o - v$) is lower than that of the small specimens (designated $c - m$ and $o - m$).

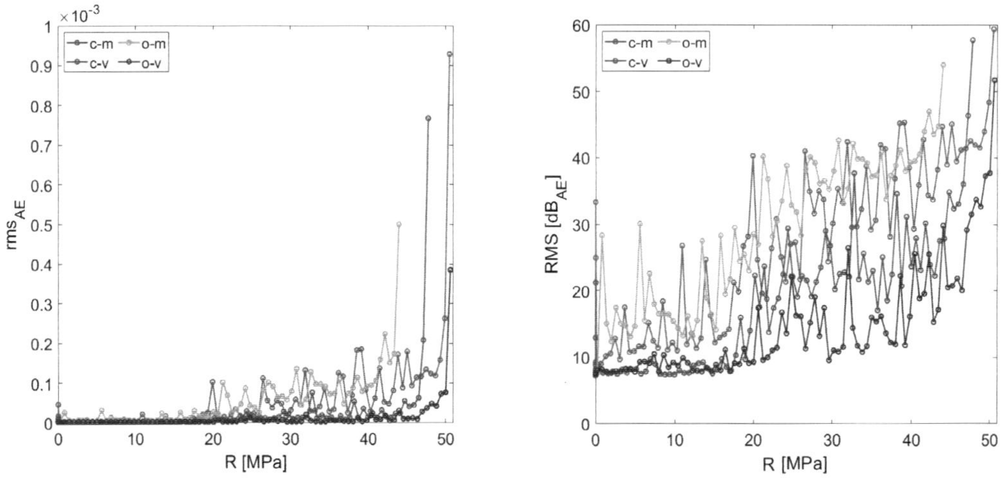

(a) RMS in a linear scale (b) RMS in logarithmic (dB) scale

Fig. 2: The course of the RMS parameter for all types of specimens.

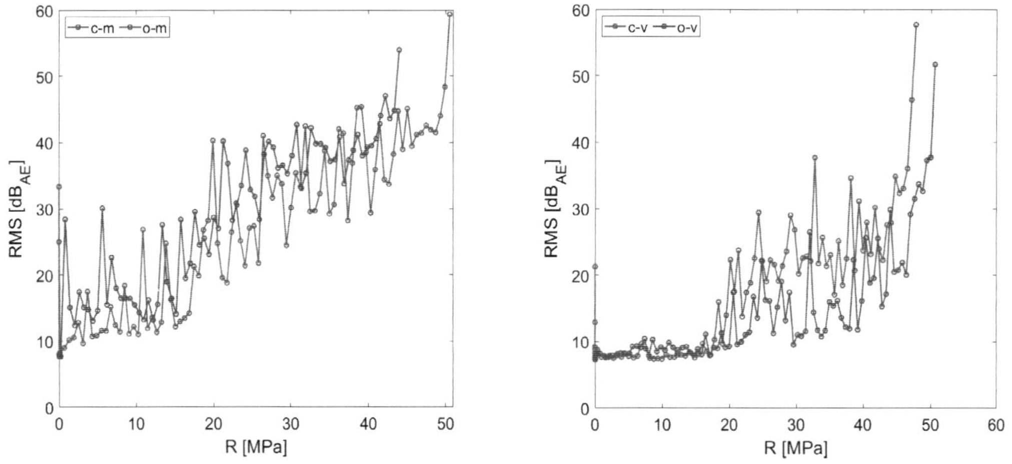

(a) Small specimens (b) Large specimens

Fig. 3: The course of the RMS (in dB) parameter for individual sizes of specimens.

The frequency spectra in Fig. 6 and Fig. 7 presented in the 1/24 octave version indicate similar characteristics of all the measured acoustic emission events. Note that the upper graph shows the order of the recorded event (ns) on the horizontal axis, the frequency (f) on the vertical axis, and the relative spectrum with different colour intensity. The lower graph then shows the development of stress (R) for individual events (ns). It is obvious from the graphs that a significant monitored range is in the frequency range from 10 kHz to 1 MHz. Since the signal was sampled at 10 MHz, it also met the Nyquist's theorem. This means that the sampling frequency was more than twice as high as the frequency contained in the event signals.

There are two basic types of waves in materials: fast longitudinal and slower shear waves. Due to the wave components and their different propagation velocities, the shape of the waveform is also affected. AE signals of a shear origin tend to acquire longer durations since the main energy arrives later, while tensile or compressive signals acquire shorter durations and higher frequencies. The R_a value is defined as [8]

$$R_a = \frac{\text{rise time}}{\text{peak amplitude}} \tag{4}$$

and the average frequency f_r as [9]

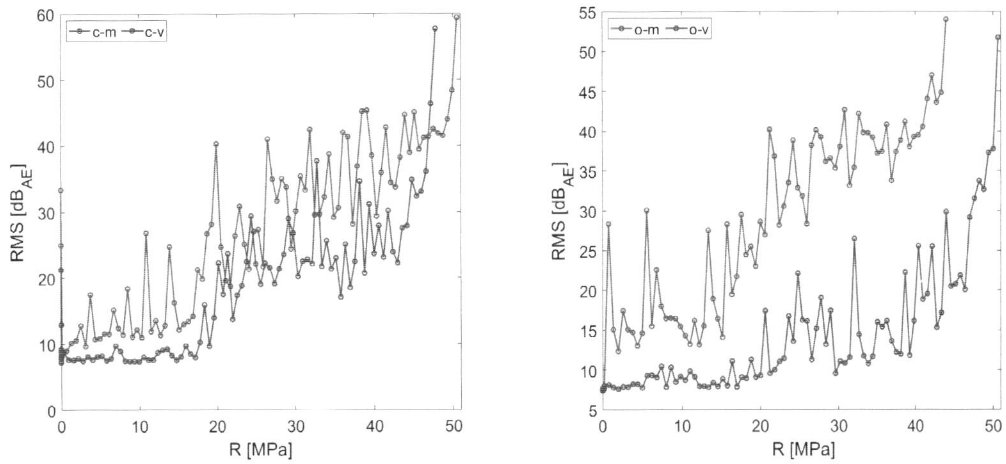

(a) Cement specimens of various sizes

(b) Alkali-activated specimens of various sizes

Fig. 4: The course of the RMS (in dB) parameter for all types of specimens.

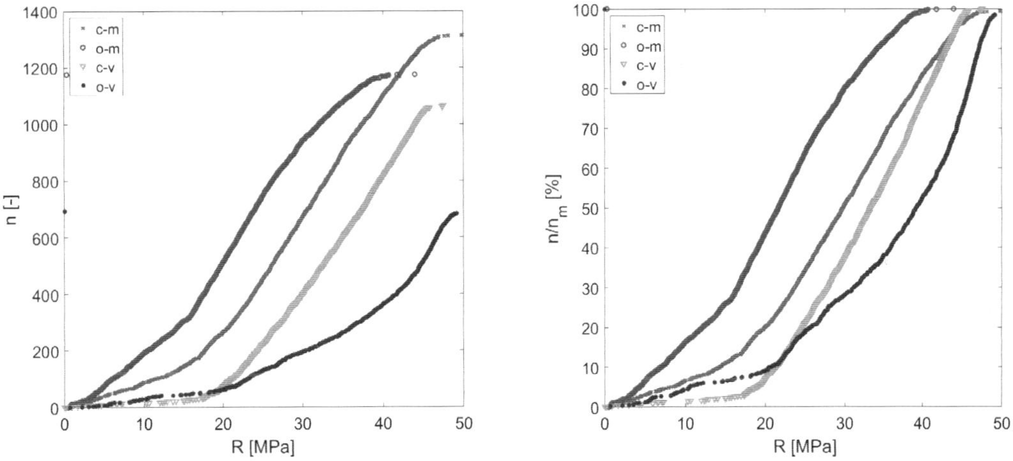

(a) Cumulative curves

(b) Cumulative curves in relative

Fig. 5: Acoustic emission activity in the number of recorded events (n) depending on the pressure stress (R).

$$f_r = \frac{\text{ring down}}{\text{duration}} \,. \tag{5}$$

Fig. 8, Fig. 9, Fig. 10 and Fig. 11 then show the resolution of the compressive and shear stress. The 3D graphs in the figures designated a show the dependence of the R_a and f_r parameters on the development of compressive stress R. The 2D graphs in the figures designated b show a view from the direction of the pressure axis, i.e. the distribution of compressive and shear stresses determined by the acoustic emission method. The blue dots indicate the tensile stress and the red dots indicate the shear stress. The 2D graphs (b) also include a line separating the compressive and shear stresses. The shear stresses are therefore below the line. The presented graphs describing the pressure tests suggest that approximately 1/5 (20%) of the events originate in shear stresses. A comparison of Fig. 8a and, Fig. 9a with Fig. 10 and Fig. 11 clearly shows that a significant activity occurs in the compressive stress region above 20 MPa. However, in the case of smaller specimens, Fig. 8a and, Fig. 9a, the shear stress occurs earlier than in the case of large specimens Fig. 10a and Fig. 11a.

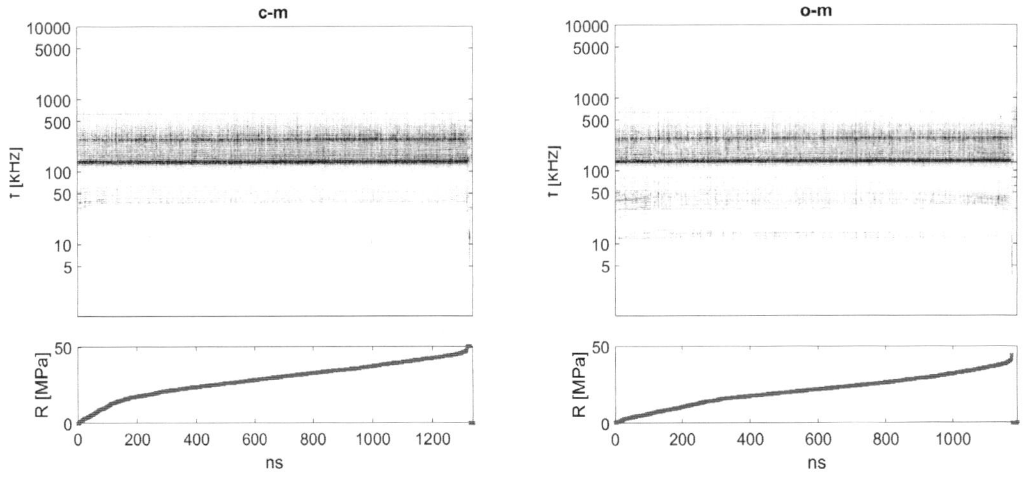

(a) Cement specimen (b) Alkali-activated specimen

Fig. 6: Frequency dependence of acoustic emission events in the case of small specimens.

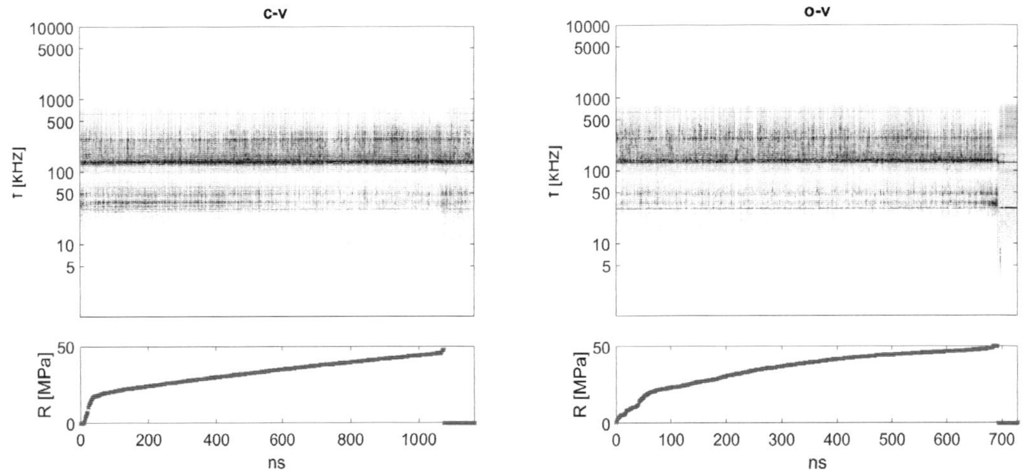

(a) Cement specimen (b) Alkali-activated specimen

Fig. 7: Frequency dependence of acoustic emission events in the case of large specimens.

4 Conclusion

The comparison of the mechanical properties of individual materials and test specimen sizes provided the following conclusions:

- with regard to compressive strength, the specimens produced from common mortar, i.e. cement mortar, reach approximately the same values as the specimens produced with the addition of alkali-activated slag;
- the results of the acoustic emission method follow a similar philosophy, i.e. there is no significant difference between the alkali-activated specimens and the specimens with common cement;
- of interest is the region around 20 MPa, where we start to see increased AE activity in larger samples that were "silent" up to this value.

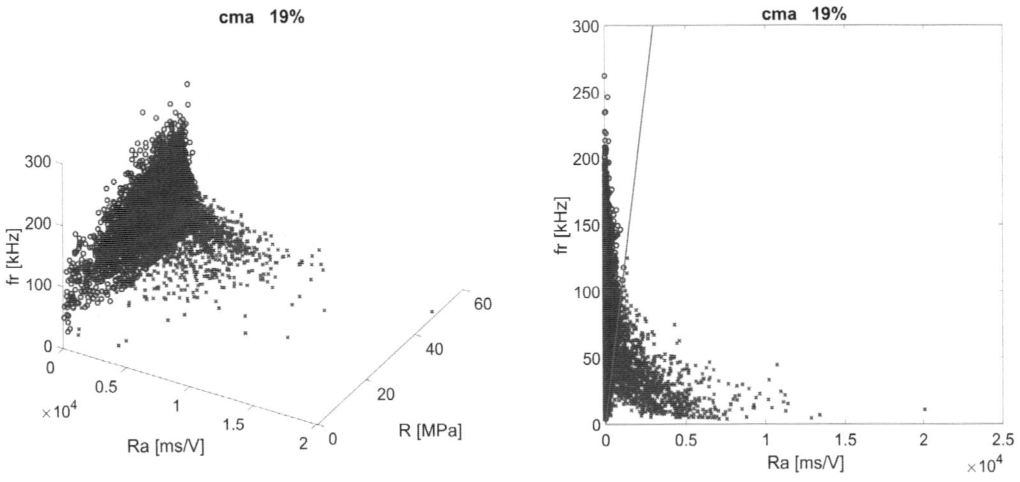

(a) 3D view (b) 2D view

Fig. 8: Probable distribution of compressive and shear stresses in a small cement specimen.

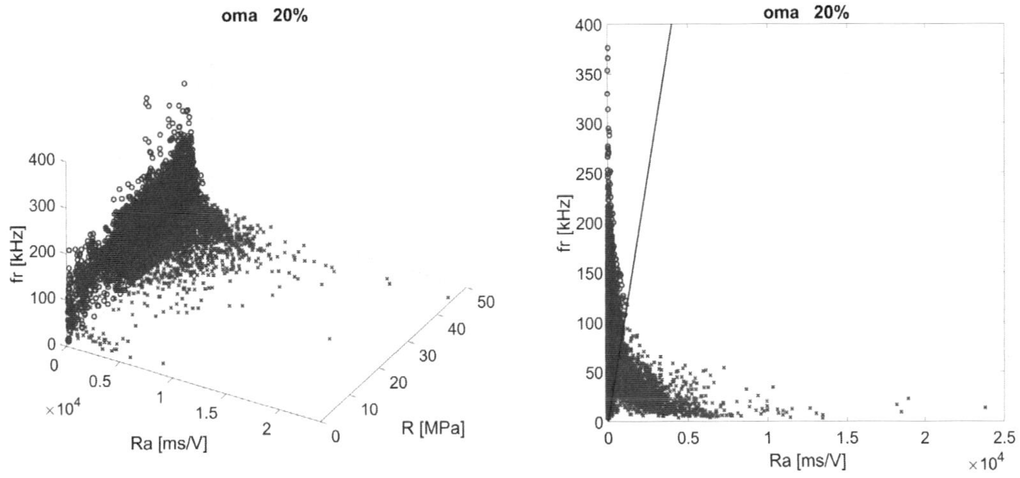

(a) 3D view (b) 2D view

Fig. 9: Probable distribution of compressive and shear stresses in a small alkali-activated specimen.

Acknowledgement

This paper has been worked out under the project Czech Science Foundation GACR No. GA 19-04703S "The use of non-destructive methods for testing of the condition of degraded alkaliactivated concrete" supported by Czech Science Foundation.

References

[1] L. Topolar, et al., Testing the Influence of the Material Bonding System on the Bond Strength of Large-Format Tiles Installed on Concrete Substrate under Mechanical Loading 13(14) (2020), doi: 10.3390/ma13143200.

[2] D. Kocab, et al., The analysis of acoustic emission signals detected during the loading of cement-based materials 99 (2019), 18–25, doi: 10.1016/j.engfailanal.2019.01.002.

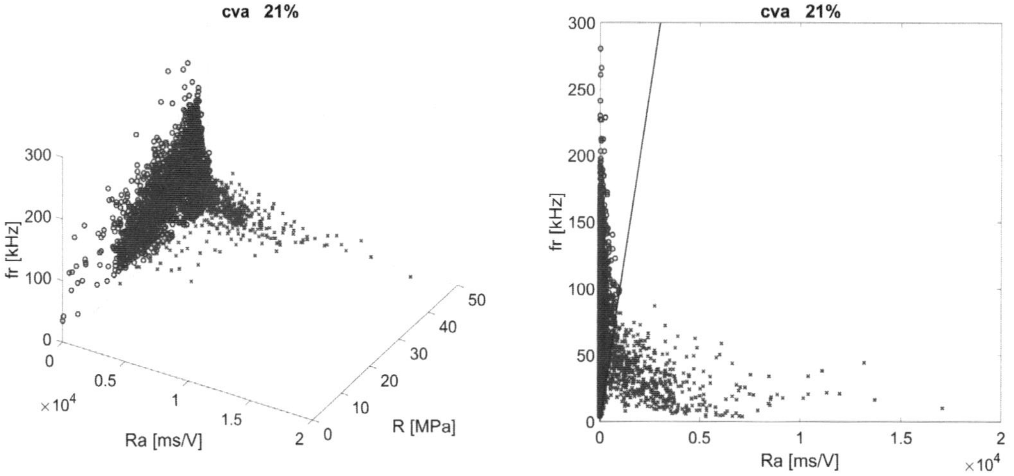

(a) 3D view (b) 2D view

Fig. 10: Probable distribution of compressive and shear stresses in a large cement specimen.

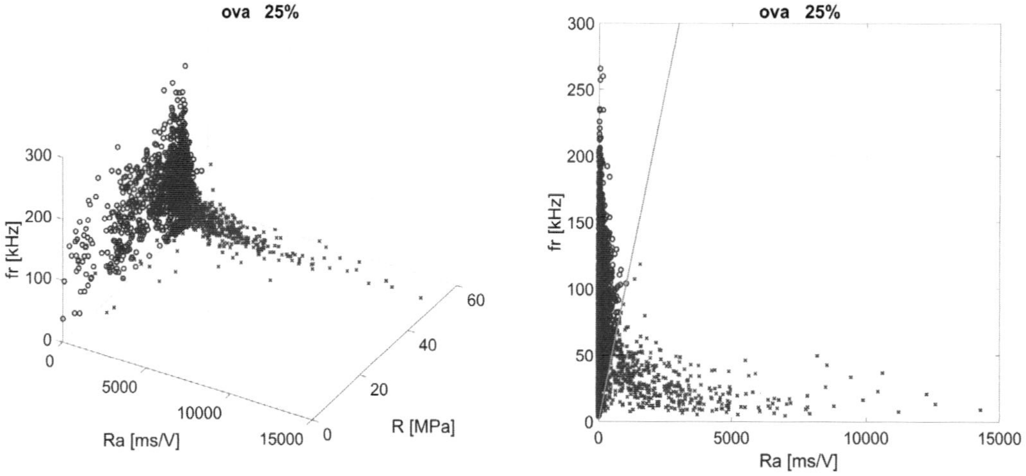

(a) 3D view (b) 2D view

Fig. 11: Probable distribution of compressive and shear stresses in a large alkali-activated specimen.

[3] D. Kocab, et al., Destructive and non-destructive assessment of the frost resistance of concrete with different aggregate, in proc.: 21st International Conference Building Materials, Products and Technologies (2018) Ceskovice Czech rep., doi: 10.1088/1757-899X/379/1/012022.

[4] D. Kocab, et al.: Development of the Elastic Modulus of Concrete under Different Curing Conditions, in proc.: 18th International Conference on Rehabilitation and Reconstruction of Buildings (2017), Brno Czech rep., 96–101, doi: 10.1016/j.proeng.2017.04.529.

[5] L. Topolar, et al.: The Acoustic Emission Parameters Obtained during Three-point Bending Test on Thermal-stressed Concrete Specimens, in proc.: 3rd International Conference on Structural and Physical Aspects of Construction Engineering, (2017) Slovakia, 111–117, doi: 10.1016/j.proeng.2017.05.315.

[6] B. Kucharczykova, et al.: Evaluation of Shrinkage, Mass Changes and Fracture Properties of Fine-aggregate Cement-based Composites during Ageing, in proc.: 3rd International Conference on Structural and Physical Aspects of Construction Engineering (2017) 357–364, doi: 10.1016/j.proeng.2017.05.349.

[7] L. Topolar, et al.: Applying method of acoustic emission for monitoring of different concrete mixture during setting and hardening, Solid State Phenomena (249) (2016) 91–95, doi: 10.4028/www.scientific.net/SSP.249.91.

[8] Y.X. Gan, et al.: Evaluation indexes of granite splitting failure based on RA and AF of AE parameters, Rock and Soil Mechanics (2020) 2324–2332, doi: 10.16285/j.rsm.2019.1460.

[9] Q.H. Han, et al.: Acoustic emission data analyses based on crumb rubber concrete beam bending tests, Engineering fracture Mechanics (210) (2019) 189–202, doi: 10.1016/j.engfracmech.2018.05.016.

Mechanical Response of the Human Aorta to the Delamination at Different Loading Velocities

Z. Petřivý[1], L. Roubalová[1], L. Horný[1,*], J. Kronek[1], H. Chlup[1],

T. Suchý[1,2], A. Blanková[3], T. Adámek[3]

[1] *Czech Technical University in Prague, Faculty of Mechanical Engineering, Technická 4, 160 00 Prague 6, Czech Republic*

[2] *Institute of Rock Structure and Mechanics of The Czech Academy of Sciences, V Holesovickach 94/41, 182 09 Prague 8, Czech Republic*

[3] *Regional Hospital Liberec, Department of Forensic Medicine and Toxicology, Husova 357/10, 460 63 Liberec, Czech Republic*

* *lukas.horny@fs.cvut.cz*

Abstract: Our study deals with the delamination strength of the human aorta. Rectangular strips of the aorta were obtained from three different locations – ascending thoracic aorta, descending thoracic aorta, and abdominal aorta. Peeling experiments were designed to take into account possible effects of arterial anisotropy and a rate-dependent mechanical response. The peeling was carried out at four different loading rates: 0.1, 1, 10, and 50 mms^{-1}. The experiments carried out with samples received from the ascending thoracic aorta and from the abdominal aorta did not show significant effect of the velocity on the delamination strength. On the other hand, the results suggest that in the thoracic region descending aorta delaminates at significantly higher force when teared at 50 mms^{-1} in comparison with the delamination conducted at speed of 0.1 mms^{-1}. This result was found for both circumferential and longitudinal orientation of samples.

Keywords: aorta; crack; dissection; peeling; strength.

1 Introduction

Aortic dissection is a life-threatening disease manifested by a separation of the layers of the aortic wall [1–3]. It occurs most frequently in the thoracic part of the aorta, but it can spread along its entire length. Although aortic dissection is relatively rare disease, the incidence is typically reported as ranging from 3 to 6 cases per 100 000 per year, the lethality of the dissection is rather high. According to [4], 37% of patients who reach the hospital alive die within the next 30 days, and approximately 20% of patients die before they receive medical intervention [5].

Our study seeks to deepen knowledge of the biomechanics of aortic delamination by means of the experimental investigation of the delamination strength of the human aortas. The peeling experiment is used to this end. It is a basic instrument utilized to show how the delamination strength depends on the anatomical site, loading velocity, and crack tip direction.

2 Methods

Segments of human aortas were obtained during regular autopsies conducted in the Department of Forensic Medicine and Toxicology at the Regional Hospital Liberec. The post-mortem use of human tissue was approved by the Ethics Committee of the Regional Hospital Liberec. Rectangular samples, approx. 8×40 mm, were cut from aortic segments. These segments were excised from the ascending as well as descending thoracic aorta, and from the abdominal aorta. Rectangular samples were cut, aligned with both the longitudinal and circumferential direction of the vessel.

A method adopted to characterize the delamination properties of the aorta is the so-called peeling test. This experimental protocol was, in the context of the biomechanics of the aortic dissection, introduced by Sommer et al. in 2008 [1] and has been used in further studies focused on the delamination properties of arteries [6–8]. It resembles the mode I crack opening that is widely used in fracture mechanics, see Fig. 1.

The experiments were carried out with the help of the multipurpose tensile testing machine Zwick/Roell (Messphysik). Both the delamination force F (the force that is necessary to increase a tear length) and the tear length were recorded on a PC. The delamination force was measured by HBM U9C +/- 25N force transducers. The tear length was determined from the movement of the clamps, which was recorded at 1 μm resolution. This data was complemented with the recordings carried out by a built-in video-extensometer, which measured the distance between the marks made on the surface of the samples. To determine whether delamination strength depends on loading rate, the experiments were carried out with clamps' velocity set to 0.1, 1, 10, 50 mms^{-1}.

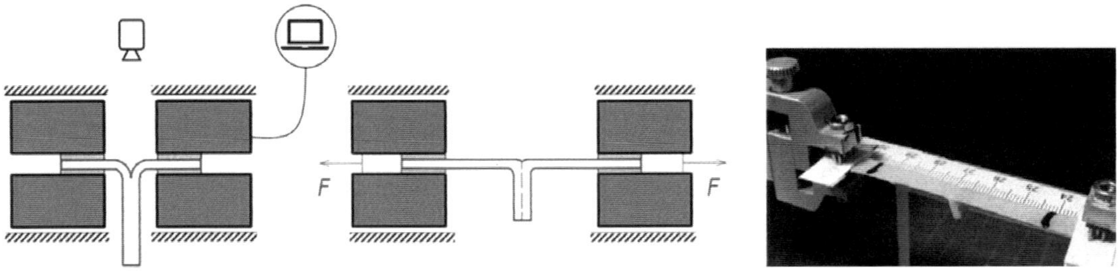

Fig. 1: Left panel illustrates the peeling experiment in which the force necessary to extend the crack that delaminates arterial layers is recorded. Right panel shows delaminated sample.

3 Results and discussion

Total number of conducted peeling experiments was 463. It consists of strips that were cut from 21 cadavers at three anatomic locations (thoracic ascending and descending, and abdominal) with two orientations (longitudinally and circumferentially) for peeling experiments carried out at four different speeds (0.1, 1, 10, and 50 mms^{-1}). This results in average number of 19 samples in each experimental group.

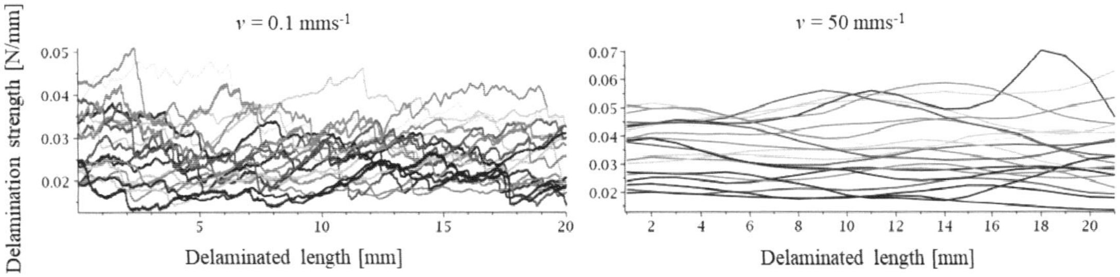

Fig. 2: Left panel depicts force-displacement relationships recorded in the peeling experiments carried out with longitudinal samples of the human descending thoracic aorta at the clamps' speed 0.1 mms^{-1} whereas records obtained at 50 mms^{-1} speed are depicted in the right panel. The sampling rate of 20 Hz was used in both cases which explains why the left panel shows curves exhibiting significant oscillations whereas the right panel shows rather flattened curves. Results are presented for 20 mm of the delamination length (distance through which a crack propagated).

Fig. 2 depicts examples of records obtained in the peeling experiment. Left panel in Fig. 2 shows delamination force as a function of a crack length how it was determined at slow loading velocity (0.1 mms^{-1}), whereas the right panel depicts responses obtained at the highest speed of clamps (50 mms^{-1}). Both records correspond to data obtained in the descending thoracic aorta where delamination strength seems to be depending on loading rate.

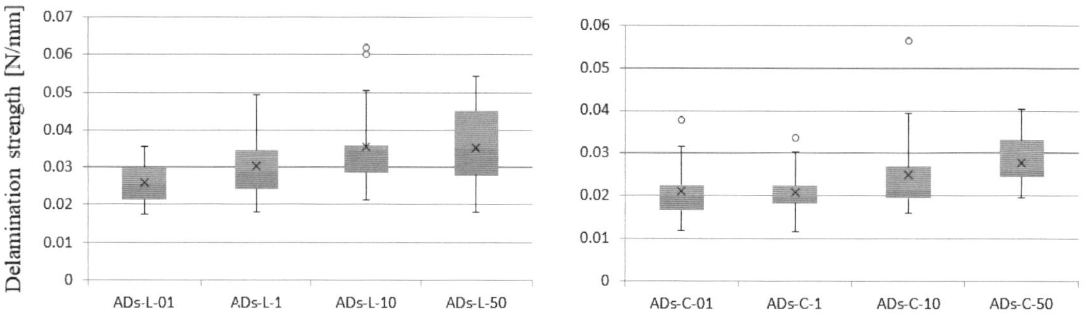

Fig. 3: Box plots of medians of the delamination strength determined with longitudinal samples (left panel) and with circumferential samples (right panel). In both longitudinal and circumferential response, 0.1 ms^{-1} differs significantly from 50 mms^{-1}.

To be more specific, neither ascending thoracic samples nor samples excised form abdominal aorta did show a dependence on the loading rate. On the other hand, results recorded in experiments conducted with samples from descending thoracic aorta show statistically significant differences in average delamination strength. Fig. 3 documents it in the box plots. The left panel shows results for longitudinal samples and the right panel contains results determined with samples oriented circumferentially. Data were evaluated by means of analysis of variance in the form of Kruskal-Wallis test which was followed by Dunn test to determine pairwise differences. Their results confirmed that medians in case of longitudinally oriented samples groups differ significantly (Kruskal-Wallis $p < 0.01$), and that delamination strength obtained at 0.1 mms^{-1} differs from the strength measured at 10 mms^{-1} (Dunn test $p < 0.01$), and at 50 mms^{-1} (Dunn test $p < 0.01$), and that also response at 1 mms^{-1} differs significantly from 10 mms^{-1}. Differences computed for 1 mms^{-1} and 50 mms^{-1} were at the border (p = 0.055). In the circumferential response, significant differences were determined in 0.1 mms^{-1} vs. 50 mms^{-1} (Dunn test $p < 0.01$), 1 mms^{-1} vs. 50 mms^{-1} (Dunn test $p < 0.01$), and in comparison between 10 mms^{-1}and 50 mms^{-1} (Dunn test $p = 0.04$).

4 Conclusions

Present contribution delivers preliminary results of our project that deals with the delamination strength of the human aorta. In this project, our goal is to evaluate anisotropy, inhomogeneity, age-related changes, and loading rate dependence in the peeling properties of the aorta. In this article, an effect of the loading velocity is the main subject. This effort is justified by a fact that general mechanical response of our arteries is viscoelastic thus it cannot be rejected with experimental confirmation that viscous or rate-dependent effects play a role in a propagation of discontinuity.

Our data suggests that delamination strength may depend on the loading velocity. However, this effect seems to be site specific. Neither ascending thoracic samples nor samples excised form abdominal aorta did show a dependence on the loading rate. On the other hand, results recorded in experiments conducted with samples from descending thoracic aorta show statistically significant differences in average delamination strength.

It is not clear at present, why different results at different anatomical sites were obtained. Further research is needed to elucidate it. One possible explanation could arise in microstructural observations. Especially quantitative histological studies could be helpful to this end. This is our plan how we would like to continue in our research.

Acknowledgement

This study has been supported by Czech Science Foundation in the project GA20-11186 entitled "Mechanics of Arterial Delamination and Crack Propagation".

References

[1] G. Sommer, et al., Dissection properties of the human aortic media: An experimental study. Journal Biomechanical Engineering 130 (2008) art. no. 021007, doi: 10.1115/1.2898733.

[2] C.A. Nienaber, R.E. Clough, Management of acute aortic dissection. The Lancet 385 (2015) 800–811, doi: 10.1016/S0140-6736(14)61005-9.

[3] T.-Y. Yeh, et al., Epidemiology and medication utilization pattern of aortic dissection in Taiwan: A population-based study. Medicine 94 (2015) art. no. e1522, doi: 10.1097/MD.0000000000001522.

[4] C. Olsson, et al., Thoracic aortic aneurysm and dissection: Increasing prevalence and improved outcomes reported in a nationwide population-based study of more than 14 000 cases from 1987 to 2002. Circulation 114 (2006) 2611–2618, doi: 10.1161/CIRCULATIONAHA.106.630400.

[5] I. Mészáros, et al., Epidemiology and clinicopathology of aortic dissection: A population-based longitudinal study over 27 years. Chest 117 (2000) 1271–1278, doi: 10.1378/chest.117.5.1271.

[6] J. Tong, et al., Variations of dissection properties and mass fractions with thrombus age in human abdominal aortic aneurysms. Journal of Biomechanics 47 (2014) 14–23, doi: 10.1016/j.jbiomech.2013.10.027.

[7] Y. Wang, et al., Quantitative Measurement of Dissection Resistance in Intimal and Medial Layers of Human Coronary Arteries. Experimental Mechanics 54 (2014) 677–683, doi: 10.1016/j.jmbbm.2016.02.011.

[8] M. Kozuń, Delamination properties of the human thoracic arterial wall with early stage of atherosclerosis lesions. Journal of Theoretical and Applied Mechanics 54 (2016) 229–238, doi: 10.15632/jtam-pl.54.1.229.

Measured and Calculated Static Equilibrium Position of Water-lubricated Tilting-pad Journal Bearing

J. Pokorný[1,*], T. Návrat[1]

[1] *Institute of Solid Mechanics, Mechatronics and Biomechanics, Faculty of Mechanical Engineering, Brno University of Technology, Technická 2896, 616 69 Brno, Czech Republic*

* *jan.pokorny4@vutbr.cz*

Abstract: This paper deals with the measurement and computational modelling of the static equilibrium position of the hydrodynamic water-lubricated tilting-pad journal bearing. The measurement is carried out on a developed test rig with two bearing housings. A computational model is presented based on the generalized Reynolds equation coupled with the 2D energy equation in the bearing midplane. This model includes the lubricant mixing, deformation and tilting of the pads. The steady-state solution is reached using the finite volume method. To find the static equilibrium position of the bearing, the Anderson method with bounds is implemented. Results of the equilibrium position in dependence on the journal speed are given for a particular water-lubricated journal bearing. These results are validated with the measured data.

Keywords: tilting pad journal bearing; water lubricated; measurement; numerical analysis; static equilibrium.

1 Introduction

The knowledge of the static equilibrium position of the hydrodynamic tilting-pad journal bearing is the basis for the determination of its static and dynamic performance. Therefore, it is necessary to establish this equilibrium position. A tilting-pad bearing is in the static equilibrium position if the force balance between the external load and the hydrodynamic pressure resultant and the moment balance of all the pads to their pivots are satisfied at the same time [1]. This position can be determined experimentally or by computational modelling.

The beginning of computational modelling can be dated back to 1886 when O. Reynolds gave the basis for the theory of the thin lubricant film [2]. Since then, many scientists have worked on this problem and the theory has been modified to incorporate some other effects. These effects include the consideration of variable viscosity depending on temperature, which led to the development of the generalized Reynolds equation [3]. Furthermore, the effect of turbulence has been involved using turbulence coefficients [4]. Mixing of hot and cold lubricant in the lubricant grooves has been proposed by Heshmat and Pinkus [5]. The accuracy of the computational models has also increased by considering the deformation and tilting of the pads [6]. However, Reynolds' basis remains.

To calculate the equilibrium position, root-finding methods are used. These methods have been divided into three groups by the author [1]. The first one includes the bisection method, which was used to determine the equilibrium position of lemon-bore bearings by the author [7], or the twofold secant method for tilting-pad bearings published by Zhou et al. [8]. The second group contains optimization methods, for example, the multiple directions method [9]. The third group comprises the Newton-Raphson method used e.g. in [10] but also the very efficient bounded quasi-Newton methods proposed by the author [1].

For the validation of the computational model, a comparison of the computational results with the experiment is preferably performed. The measurement of the equilibrium position has been addressed by many authors, e.g. [11–13].

2 Methods

2.1 Test rig and instrumentation

Fig. 1 shows a test rig for measuring the static performance of the hydrodynamic journal bearings. This device consists of a rigid frame allowing axial displacement of all components and ensuring their correct alignment. The frame is fitted with two bearing housings containing water-lubricated hydrodynamic tilting-pad journal bearings. A rigid rotor is mounted in these bearings, which is driven by a high-speed electric motor up to 18,000 rpm via a flexible coupling. The stiffness of the device is increased by two crossbars, on which four eddy current proximity probes are placed near the bearing housings for measuring rotor displacements relatively to the housings. The rotor speed is measured with an optical tachometer up to 500,000 rpm. For health protection reasons, the device includes a hinged safety guard.

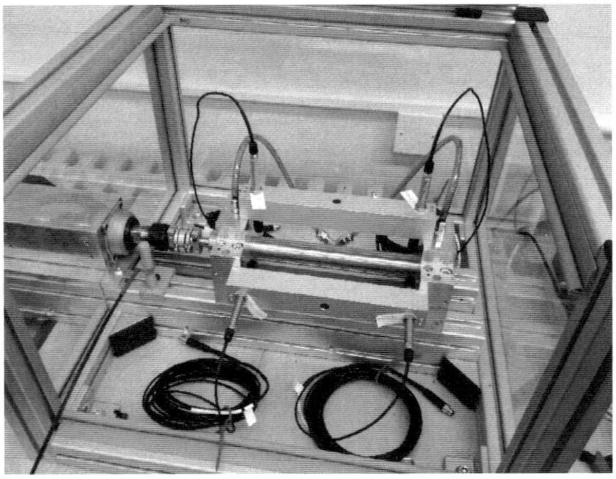

Fig. 1: Test rig for water-lubricated hydrodynamic journal bearings.

2.2 Measuring procedure

The speed maximum of 17,000 rpm was reached with a linear ramp in 120 s. The displacements of the journal were measured during the start-up, thanks to which the instantaneous position of the journal and its trajectory was determined. Fig. 2 presents a scheme of this measurement, including the used coordinate system. To compare measured data of the journal trajectory during the start-up with calculation, a computational model was created.

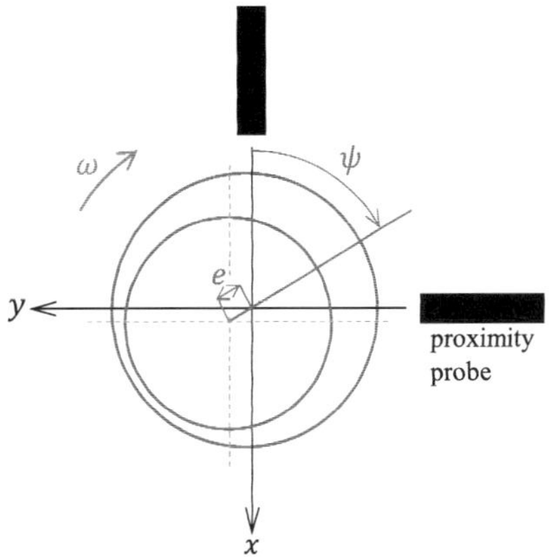

Fig. 2: Hydrodynamic journal bearing scheme with two proximity probes.

2.3 Computational model

The load-carrying capacity of a hydrodynamic bearing is given by the hydrodynamic pressure in the lubricant film. The computational model is therefore based on a numerical solution of the generalized Reynolds equation [14], which describes this pressure:

$$\frac{\partial}{\partial\theta}\left(\frac{H^3}{K_x}\bar{G}\frac{\partial\bar{p}}{\partial\theta}\right)+\frac{\partial}{\partial\bar{y}}\left(\frac{H^3}{K_y}\bar{G}\frac{\partial\bar{p}}{\partial\bar{y}}\right)=\frac{\partial}{\partial\theta}\left(\bar{G}_R H\right), \tag{1}$$

where $H = h/c_b$ is the dimensionless lubricant film thickness, $c_b = R - R_j$ is the radial assembly clearance, R is the bearing radius, R_j is the journal radius, $\bar{p} = pc_b^2/U\eta_0 R$ is the dimensionless pressure, U is the sliding velocity, η_0 is the initial dynamic viscosity, $\theta = x/R$ is the angular coordinate, $\bar{y} = y/R$ is the dimensionless axial coordinate, $\bar{G}$ and $\bar{G}_R$ characterize the variable lubricant viscosity and density, and K_x and K_y models effect of local turbulent flow and Taylor vortices. Detailed information is provided by the author in [1].

Due to the dissipation of energy in the lubricant film, the energy equation is solved simultaneously with the Reynolds equation, based on which the lubricant temperature can be determined. This is because the lubricant characteristics such as viscosity and density are temperature dependent. Dimensionless temperature distribution $\bar{T} = T/T_0$ in the radial and circumferential directions of the lubricant film is expressed by the following energy equation:

$$\frac{\partial}{\partial\theta}\left(\bar{U}\bar{T}\right)+\frac{\partial}{\partial\bar{z}}\left(\bar{W}\bar{T}\right)=\frac{\partial}{\partial\theta}\left(\Lambda_\theta\frac{\partial\bar{T}}{\partial\theta}\right)+\frac{\partial}{\partial\bar{z}}\left(\Lambda_z\frac{\partial\bar{T}}{\partial\bar{z}}\right)+K_T\phi, \tag{2}$$

where T_0 is the water supply temperature, $\bar{U}$ and $\bar{W}$ are the modified dimensionless circumferential and radial velocities, Λ_θ and Λ_z are the modified dimensionless thermal diffusivities, K_T is the dissipation number, ϕ is the dissipation function, and $\bar{z} = z/h$ is the dimensionless radial coordinate (more details in [1]).

Due to the use of water as a lubricant, the effect of temperature is not as significant, unlike oil-lubricated bearings, whose viscosity is approximately two orders of magnitude higher. Nevertheless, the energy equation is given here for the model completeness.

The computational model further incorporates lubricant mixing, reaching a steady-state solution, tilting and deforming the pads with the subsequent establishment of the static equilibrium position of the journal [1]. This equilibrium position occurs once a balance is found between the external load and the hydrodynamic pressure resultant and the moment equilibrium of all segments with respect to the pivot is satisfied at the same time. The multidimensional root finding quasi-Newton method, namely the bounded Anderson method [1], is used to find the static equilibrium position of the bearing. This method and the used computational model are described in detail by the author [1].

3 Results

Tab. 1 summarizes the parameters of the investigated bearing.

Tab. 1: Characteristic data of the test bearing.

Number of tilting pads	3, load on pad
Bearing radius [mm]	15
Bearing length [mm]	18
Radial assembly clearance [μm]	25
Radial machined clearance [μm]	41
Pivot offset [-]	0.5
Rotational speed [rpm]	0–17,000
Bearing load [N]	8.2
Water supply temperature [°C]	25

Fig. 3 shows hydrodynamic pressure distribution in the bearing mid-plane operating at a rotational speed of 2,000 rpm and 16,000 rpm. Maximum pressures are 66.7 kPa and 187.8 kPa. In addition to the pressure, Fig. 3 presents lubricant film thickness which is 50 times magnified. The film thickness at the lower pad increases with rotational speed. Minimum film thickness at 2,000 rpm and 16,000 rpm at the lower pad is 12.6 μm and 19.3 μm, respectively. From the pads position, it is also evident that they are tilted into the converging wedge, which supports the formation of hydrodynamic pressure. Fig. 4 shows this pressure distribution in the circumferential and axial direction of the bearing. In the axial direction from the bearing mid-plane, hydrodynamic pressure decreases to atmospheric pressure, at which point lubricant leaves the bearing.

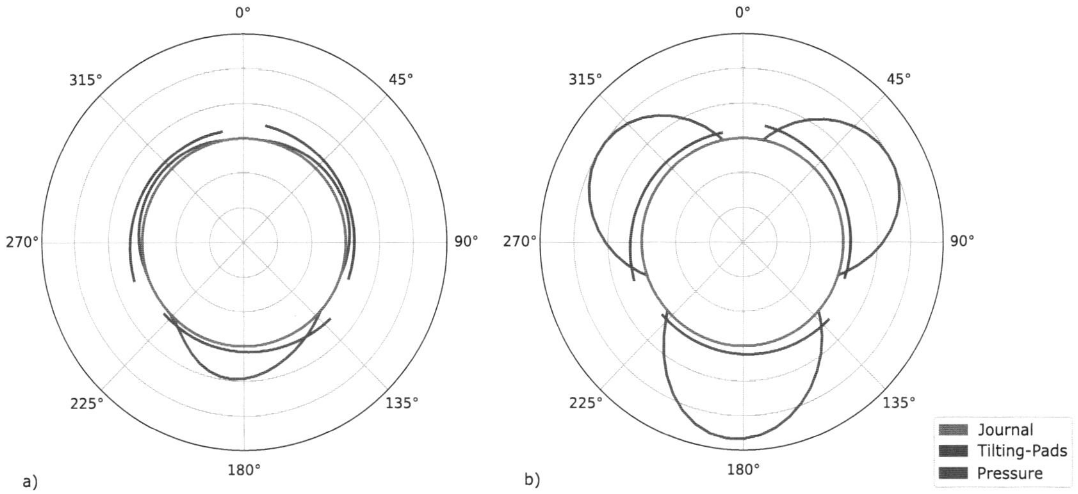

Fig. 3: Lubricant film thickness (50× magnified) and distribution of hydrodynamic pressure, (a) n = 2,000 rpm, (b) n = 16,000 rpm.

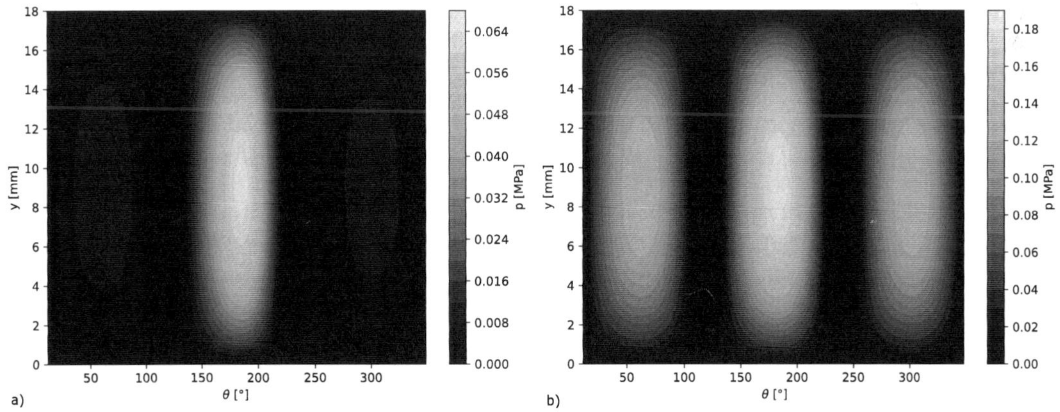

Fig. 4: Distribution of hydrodynamic pressure, (a) n = 2,000 rpm, (b) n = 16,000 rpm.

Fig. 5 presents the temperature change in the lubricant film and pads. Fig. 5 was scaled so that the lubricant film thickness was the same size as the thickness of the pads. The shaft temperature was assumed to be constant. The lowest temperatures are in the area of leading edges of the pads, the highest in the area of the trailing edges. The temperature distribution is almost the same on all the pads, which is due to the low external load on the journal in combination with relatively high rotational speed.

Fig. 6 shows the comparison of the measured and computed position of the journal center during start-up. The computational model works only with a fully developed hydrodynamic lubricant film, therefore the calculation cannot be performed at zero speed. In our case, the minimum rotational speed for the calculation was 80 rpm. At this speed, a very thin lubricant film of minimum thickness 1.8 μm was found, at which the bearing would not be possible to operate. Thus, the lower speeds were not considered only from the point of

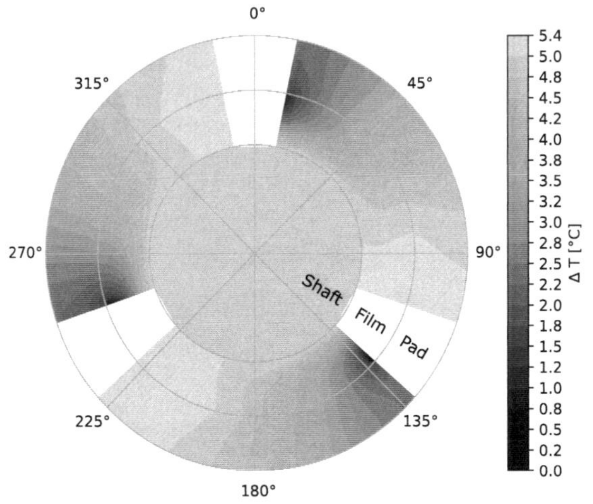

Fig. 5: Temperature change in the lubricant film and pads in the bearing mid-plane, n = 16,000 rpm.

view of the lubricant film thickness but also the point of view of the validity of the lubrication theory based on the Reynolds equation. Conversely, performing the measurement is not a problem without rotating the shaft. The biggest issue with the measurement is the noisy data, from which the noise needs to be removed. For this reason, a lowpass filter was used with a maximum rotational speed threshold. Furthermore, the data were approximated using least squares by a 6th order polynomial with fixed endpoints to gain the journal trajectory.

It can be stated that the difference between the calculated and measured data is not very large. At high speeds, the data almost match. In the region between approximately 2,000 rpm and 200 rpm, the calculated data deviates slightly in the y-direction. It is worth noting that the load on pad configuration does not have a clearly defined rest position of the rotor. Unlike the load between pads configuration where the rotor at rest lies between the pads. In the load on pad configuration, the rotor may lie to the left or right of the lower pad pivot and must first get out of this position when starting up. This corresponds to the initial part of the measured trajectory. The computational model is not able to capture this phenomenon, it determines the equilibrium position of the journal only after this transient process.

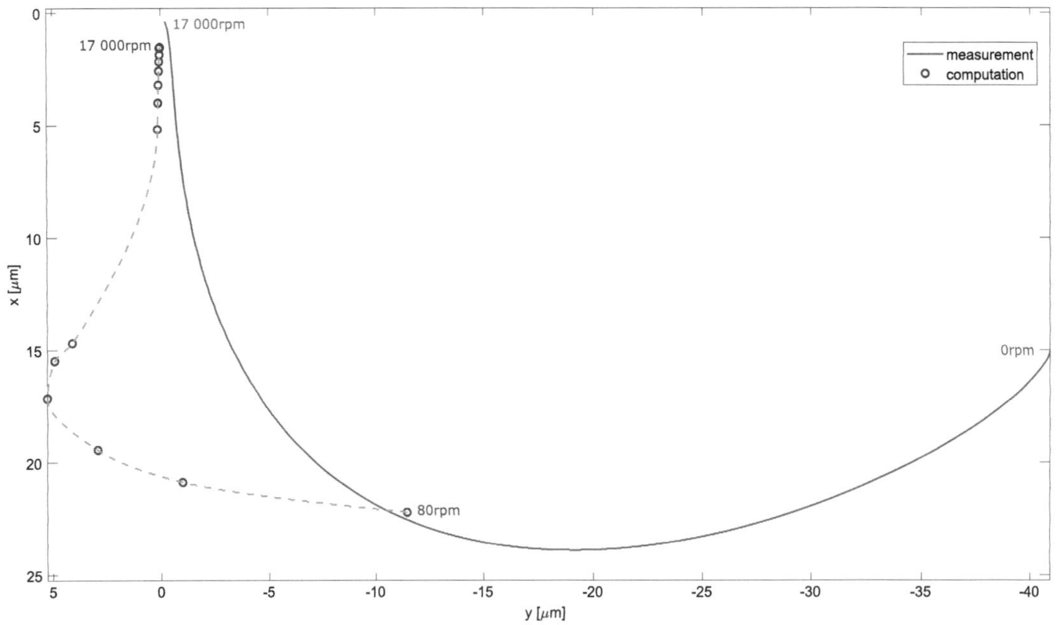

Fig. 6: Dependence of the journal locus on the rotational speed: a comparison of the computed and measured data.

4 Conclusion

The paper compared the measurement and calculation of the static equilibrium position of the water-lubricated tilting-pad journal bearing during rotor start-up. The test rig, its instrumentation and measuring procedure were presented. Furthermore, the computational model was given based on the generalized Reynolds equation solved simultaneously with the energy equation at the bearing mid-plane in the circumferential and radial direction. To find the equilibrium position, the model involved lubricant mixing, reaching the steady-state solution, tilting and deforming the pads.

To conclude, the comparison showed a good agreement between measurement and calculation at high speeds. The calculation could not be performed from the zero speed of the rotor because there is a minimum speed limit given by the full development of the lubricant film. In contrast, the measurement can be made in the rotor rest position. This position is not clearly defined in the load on pad configuration, which probably causes deviations in the results at lower speeds. However, other influences may play a role, such as the effect of the flexible coupling compensating for the misalignment of the rotating parts. Further research might be focused on the spring and damping coefficients of the hydrodynamic tilting-pad journal bearings due to their relationship to the static equilibrium position.

Acknowledgement

This work was partially supported by the project FV 21-23 funded by The Ministry of Education, Youth and Sports of the Czech Republic; and the project "Problems of the strength of materials and dynamics and their engineering applications" (FSI-S-20-6164).

References

[1] J. Pokorný, An efficient method for establishing the static equilibrium position of the hydrodynamic tilting-pad journal bearings, Tribol. Int.153 (2021), 106641, doi: 10.1016/j.triboint.2020.106641.

[2] O. Reynolds, On the theory of lubrication and its application to Mr. Beauchamp tower's experiments, including an experimental determination of the viscosity of olive oil, Philos. Trans. R. Soc. London. 177 (1886) 157–234, doi: 10.1098/rstl.1886.0005.

[3] D. Dowson, A generalized Reynolds equation for fluid-film lubrication, Int. J. Mech. Sci. 4 (1962) 159–170, doi: 10.1016/S0020-7403(62)80038-1.

[4] H.G. Elrod, C.W. Ng, A theory for turbulent fluid films and its application to bearings, J. Tribol. 89 (1967) 346–362, doi: 10.1115/1.3616989.

[5] H. Heshmat, O. Pinkus, Mixing Inlet Temperatures in Hydrodynamic Bearings, J. Tribol. 108 (1986) 231–244, doi: 10.1115/1.3261168.

[6] C.M.M. Ettles, The analysis and performance of pivoted pad journal bearings considering thermal and elastic effects, J. Tribol. 102 (1980) 182–191, doi: 10.1115/1.3251465.

[7] J. Pokorný, Calculation of Static Equilibrium Position of Hydrodynamic Journal Bearings, in: Eng. Mech. (2019), Czech Academy of Sciences, Svratka, 291–294, doi: 10.21495/71-0-291.

[8] W. Zhou, X. Wei, L. Wang, G. Wu, A superlinear iteration method for calculation of finite length journal bearing's static equilibrium position, R. Soc. Open Sci. 4 (2017), doi: 10.1098/rsos.161059.

[9] S. Yu, S. Chen, An approach searching for the steady-state equilibrium position of tilting-pad journal bearing for supporting motorized spindle, J. Comput. Theor. Nanosci. 9 (2012) 1710–1714, doi: 10.1166/jctn.2012.2269.

[10] A. Cerda Varela, B. Bjerregaard Nielsen, I.F. Santos, Steady state characteristics of a tilting pad journal bearing with controllable lubrication: Comparison between theoretical and experimental results, Tribol. Int. 58 (2013) 85–97, doi: 10.1016/j.triboint.2012.10.004.

[11] S. Taniguchi, T. Makino, K. Takeshita, T. Ichimura, A thermohydrbdynamic analysis of large tilting-pad journal bearing in laminar and turbulent flow regimes with mixing, J. Tribol. 112 (1990) 542–550, doi: 10.1115/1.2920291.

[12] T. Hagemann, S. Kukla, H. Schwarze, Measurement and prediction of the static operating conditions of a large turbine tilting-pad bearing under high circumferential speeds and heavy loads, in: Proc. ASME Turbo Expo, 2013, doi: 10.1115/GT2013-95004.

[13] K.D. Wygant, R.D. Flack, L.E. Barrett, Measured Performance of Tilting-Pad Journal Bearings over a Range of Preloads—Part I: Static Operating Conditions, Tribology Transactions. 47 (2004) 576–584, doi: 10.1080/05698190490504154.

[14] N. Mittwollen, J. Glienicke, Operating conditions of multi-lobe journal bearings under high thermal loads, J. Tribol. 112 (1990) 330–340, doi: 10.1115/1.2920261.

Influence of Calcination Level of Recycled Gypsum on the Physical Properties of Hardened Gypsum Slurry

Z. Prošek[1,2*], P. Tesárek[1]

[1] *CTU in Prague, Faculty of Civil Engineering, Thákurova 7, 166 29 Prague 6, Czech Republic*

[2] *CTU in Prague, University Centre for Energy Efficient Buildings, Třinecká 1024, 273 43 Buštěhrad, Czech Republic*

* *zdenek.prosek@fsv.cvut.cz*

Abstract: This article examines the effect of incomplete calcination of wasted recycled hardened gypsum. Waste gypsum originated from the process of recycling plasterboard. As part of the processing, the cardboard is separated from the hardened gypsum and the gypsum is physical and mechanical activated (calcined) by means of high-speed grinding. As part of recycling, there is no complete calcination of hardened gypsum. In the resulting gypsum composite, these uncalcined grains have a fundamental influence on the process of hydration and its resulting physical properties of hardened gypsum slurry. In the research, the percentage of uncalcined grains was simulated by adding non-activated recycled hardened gypsum to the mixture in the required amount from 0 to 1 wt. %. Compressive strength and flexural tensile strength were measured on hardened gypsum slurries measuring 40 × 40 × 160 mm. The possible processing time of gypsum slurry (setting time) was other monitored parameters.

Keywords: recycled gypsum; calcination; compressive strength; flexural strength.

1 Introduction

Currently, the issue of recycling building materials is being dealed worldwide. In connection with this issue, concrete is most often mentioned material for recycling. Concrete is not the only material used in the construction industry. There are other widely used materials that should be recycled and further processed. However, the recycling of some building materials is problematic, as is the case with waste from other industries and is therefore landfilled [1, 2]. Among the materials that are not effectively processed today is plasterboard waste.

In recent years, there has been a growing interest in dry construction and with it interest in plasterboard systems. With the growing demand for plasterboards, the amount of waste generated from plasterboard production, constructions, reconstructions and demolitions is also increasing [3]. At present, the vast majority of plasterboards waste is landfilled and a relatively small part is reused [4].

What is usually forgotten in the concept of solutions for recycling plasterboards and gypsum products is the fact that in terms of recycling and reuse, it is a completely unique material that is 100% recyclable to infinity. And above all, it is unique in that the properties of gypsum do not change during recycling [5]. From the point of view of construction, gypsum is a unique material that has been used for several thousand years (eg in ancient Egypt), it is mainly valued for its good thermal-technical properties, mechanical properties and fire resistance. Other advantages of gypsum include the possibility of its modification by means of additives and admixtures which modify its useful properties. Calcined gypsum ($CaSO_4 \cdot \frac{1}{2}H_2O$) is not found in nature (only in the form of $CaSO_4$ anhydrite or $CaSO_4 \cdot 2H_2O$ calcium sulphate). Calcined gypsum is made from calcium sulphate (which is either of natural or artificial origin) by using high temperatures, so-called calcination. The calcination usually takes place at temperatures above 180 °C in such a way that water is released from the gypsum (calcium sulphate dihydrate) at a temperature above 150 °C [6].

The basic equation of gypsum production is as follows (Eq. (1)):

Tab. 1: Composition of individual samples.

Mixtures	Gypsum binder [g]	Waste gypsum [g]	Water [g]	Setting time [min]	Bulk density [kg/m^3]
ref.	1,000	-	640	6.1	1,165 $\pm$ 16
0.25	997.5	2.5	640	3.2	1,155 $\pm$ 7
0.50	995	5	640	2.4	1,122 $\pm$ 4
0.75	992.5	7.5	640	1.5	1,120 $\pm$ 14
1.00	990	10	640	1.3	1,132 $\pm$ 16

$$CaSO4 \cdot 2H_2O + heat = CaSO_4 \cdot \frac{1}{2}H_2O + 1.5H_2O\,. \tag{1}$$

Rehydration of gypsum (formation of hardened gypsum) occurs after mixing with water according to equation Eq. (2):

$$CaSO_4 \cdot \frac{1}{2}H_2O + 1.5H_2O = CaSO_4 \cdot 2H_2O + heat\,. \tag{2}$$

According to the above equations, it is therefore possible to recycle the gypsum several times at relatively low costs using high temperatures (calcination) and grinding to the required grain size. The properties of the resulting calcined gypsum are directly dependent on the grain size and the level of calcination (amount of uncalcined gypsum grains). Grain size and the amount of uncalcined gypsum grains affect the process of hydration. The smaller grain size increases the specific surface area of the grains and thus leads to a faster reaction during hydration. Uncalcined grains create nucleation centers for the formation of calcium sulphate and thus the whole hydration process is accelerated. Thus, both parameters accelerate hydration time and thus reduce the processability time and setting time [7].

2 Materials and samples

Individual samples were composed of gypsum binder from the company Knauf Praha and uncalcined recycled waste gypsum. The gypsum binder was formed by calcination of energy gypsum (FGD gypsum). FGD gypsum is a by-product of the combustion of brown coal by the flue gas desulphurisation process in a power plant. In our case, it was FGD gypsum from the Počerady power plant, which has a very high purity (on average 96 %) and is used as the main material in the production of plasterboards.

Uncalcined waste recycled gypsum originated from plasterboard waste. Gypsum plasterboard waste originated from the construction process and it was gypsum plasterboard with the designation A for common use according to the ČSN EN 520 + A1 standard [8]. During recycling, cartons were separated from the gypsum slurry and the waste gypsum slurry was ground using a high speed mill. The resulting uncalcined waste gypsum was of high purity and was 95 wt. % composed of gypsum. In the research, the percentage of uncalcined waste gypsum was simulated by adding non-activated recycled hardened gypsum to the mixture.

Tab. 1 shows the composition of the individual mixtures. Each mixture consisted of 6 samples measuring 40 × 40 × 160 mm were cast to the rectangular molds. After casting, these samples were kept in the molds for 1 h at the laboratory with the temperature 22 °C. After demolding, these samples were deposited in the laboratory environment. All mixtures had the same water content of 0.64. The setting time was measured during sample preparation.

3 Experimental methods

The samples were destructively tested after 7 days to determine the flexural strength and compressive strength by using FHF Strassentest device. The flexural strength was determined by a three-point bending test. The testing of flexural strength was displacement controlled at a constant rate of 0.5 mm/min. The distance between supports for three-point bending test was equal to 100 mm. The compressive strength was determined by using uniaxial compressive test. The uniaxial compressive test was performed on the broken halves of the samples after bending test with effective dimensions of 40 × 40 × 40 mm. The testing of compressive strength was displacement controlled at a constant rate of 1 mm/min.

All tests were performed according to ČSN 72 2301 standard [9]. The resulting average values of compressive strength and flexural strength were calculated excluding the highest and lowest values determined during testing.

4 Results and discussion

The results showed that all mixtures had the same bulk density (the differences were up to the size of the standard deviation) and the uncalcined particles had a high effect on the setting time (Tab. 1). The results show that uncalcined gypsum particles form nucleation centers and thus accelerate the entire hydration.

The results of the destructive tests can be seen in Fig. 1 and Fig. 2. Due to the accelerated hydration, more heat is released in a shorter time. As a result, microcracks form in the sample due to volume changes. For example, sample with 1 wt. % of uncalcined gypsum had an average flexural strength of 3.04 ± 0.22 MPa. In contrast, the reference sample had a flexural strength of 4.20 ± 0.16 MPa. Microcracks have a major effect on flexural strength. For this reason, the flexural strength is reduced due to uncalcined gypsum. In the case of compressive strength, a similar effect can be observed. Fig. 2 shows the decrease of compressive strength due to uncalcined gypsum and a mixture with 1 wt. % of uncalcined gypsum has about half the average compressive strength.

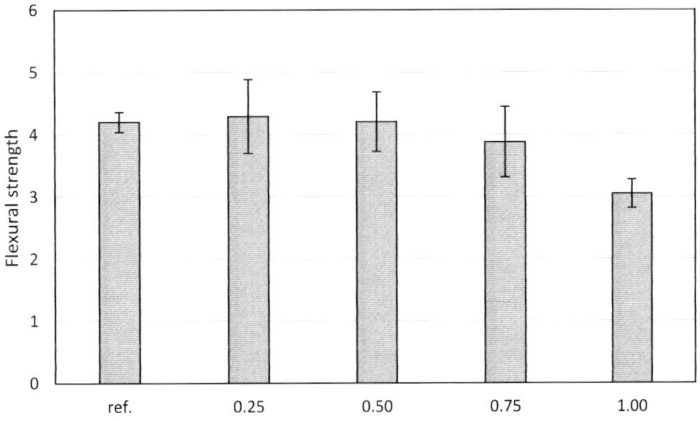

Fig. 1: Comparison of flexural strength (with standard deviations).

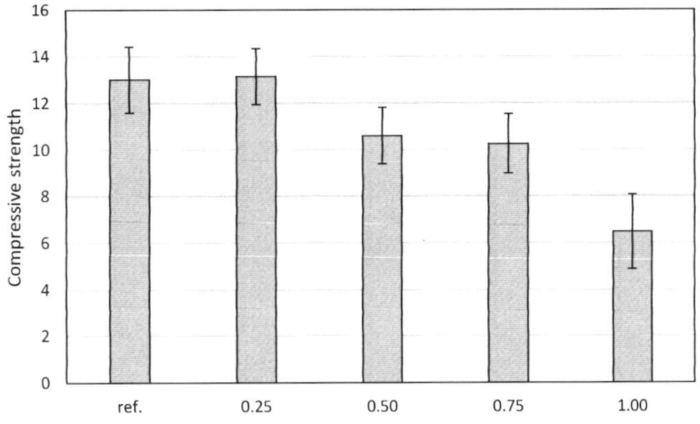

Fig. 2: Comparison of compressive strength (with standard deviations).

5 Conclusion

The results of this work will be used as a basis for the design of a recycling line, which will be used to grind and calcine gypsum-based waste materials. The results show that the calcination level of the resulting recycled material must be 99.5 %. Thus, the recyclate may contain 0.5 wt. % of uncalcined gypsum. In the case of a lower level of calcination, the mechanical properties will be affected. In the future work will deal directly with calcined recycled material from the recycling line and the possibility of using setting retarders.

Acknowledgement

This outcome was supported by Technology Agency of the Czech Republic (TA CR) research project FW03010054, which is being solved in cooperation with LAVARIS, MORAVOSTAV Brno and KNAUF Prague.

References

[1] J. Trejbal et al., Recyklace sádrových výrobků pomocí vysokorychlostního mikromletí, in proc.: RECYCLING 2019 – Recyklace a využití stavebních odpadů jako druhotných surovin (2019), ed. M. Škopán, Vysoké učení technické Brno, Brno, 113–116.

[2] H. Sekavová et al., The influence of the use of recycled gypsum on the properties of gypsum products in labor conditions, Acta Polytechnica CTU Proceedings 26 (2020) 81–85, doi: 10.14311/APP.2020.26.0081.

[3] Z. Prošek et al., Alternativní možnosti využití recyklované sádry ve stavebnictví, in proc.: RECYCLING 2019 – Recyklace a využití stavebních odpadů jako druhotných surovin (2019), ed. M. Škopán, Vysoké učení technické Brno, Brno, 109–112.

[4] T. Plachý et al., Monitoring of mechanical properties evolution of the cast gypsum, Procedia Engineering 48 (2012) 562–567, doi: 10.1016/j.proeng.2012.09.554.

[5] R. Černý et al., Vlastnosti modifikované sádry, Brno, CERM, 2009.

[6] M. Singh, M. Garg, Making of anhydrite cement from waste gypsum, Cement and Concrete Research 30 (2000) 571–577, doi: 10.1016/S0008-8846(00)00209-X.

[7] P. Tesárek et al., Flue Gas Desulfurization Gypsum: Study of Basic Mechanical, Hydric and Thermal Properties, CBM 21 (2007) 1500–1509, doi: 10.1016/j.conbuildmat.2006.05.009.

[8] ČSN EN 520+A1. Sádrokartonové desky - Definice, požadavky a zkušební metody. Praha: Úřad pro technickou normalizaci, metrologii a státní zkušebnictví 2010.

[9] ČSN 72 2301. Sádrová pojiva. Klasifikace. Všeobecné technické požadavky. Zkušební metody. Praha: Úřad pro technickou normalizaci, metrologii a státní zkušebnictví 1980.

Influence of Atmospheric Plasma on Mechanical Strength Fiber Reinforced Geopolymer Composite

M. Ryvolová[1,*], T. Bakalova[1]

[1] *Technical University of Liberec, Faculty of Mechanical engineering, Department of Materials, Studentská 2, 461 17 Liberec, Czech Republic*

* *martina.ryvolova@tul.cz*

Abstract: Improving the adhesion of the matrix to the reinforcement brings an increase in the quality of the composite interface and, at the same time, improves the mechanical properties of the composite system represented by the geopolymer matrix reinforced with chopped fibers. The article's topic is to monitor the effect of plasma surface treatment of mineral fibers on improving the adhesion of the geopolymer matrix to reinforcing fibers by examining the mechanical properties of the composite. The experiment was performed with glass fibers of type AR, E, R, and basalt fibers. A bending test was used to monitor changes in mechanical properties. In the research, we assumed that the plasma surface treatment of fibers should improve the overall mechanical properties of the composite system. Special attention was paid to the statistical evaluation of the results (Horn method for the analysis of small data files).

Keywords: mineral fibers; atmospheric plasma; geopolymer matrix; flexural strength; interface.

1 Introduction

A characteristic manifestation of the composite material is the synergistic effect, which arises from the combination of the continuous and dispersed components of the composite. The essence of the synergistic effect is creating an interface. It is the transition phase forms between the matrix (continuous phase) and the reinforcement (dispersion phase). The strength of the interphase determines the magnitude and character of the stress transfer from the matrix to the reinforcement; it includes the properties of the partial phases and adhesion [1].

The adhesion between the materials determines the quality of the interface in composite materials and directly contributes to the resulting composite properties. It is formed by a set of mechanisms, including absorption and wetting, electrostatic interactions, reaction bonds, and unbound interactions [2–4]. Glass and basalt fibers belong to the group of mineral fibers. Chopped fibers, continuous fibers (rovings), textiles: warp knitted (multiaxial fabrics, geogrids), braided sleeves and tapes, fabrics with plain and twill weave, hybrid fabrics, and three-dimensional woven fabrics are used as reinforcement for composites. Glass and basalt fibers have very similar densities, strengths, melting points and the technical application in the production of composites [5].

The surface treatment of fibers is a commercially applied functional coating (sizing), which simultaneously meets the requirements for protection and fibers processability and the formation of a chemical bond between the fiber and usually the polymer matrix. Organosilanes are used for the surface treatment of glass fibers [6]. These are compounds that contain silicon in their structure and function as adhesives. They have excellent reactivity with silicon-containing substances, e.g., aluminosilicates [7].

Plasma is an electrically conductive medium that contains positively and negatively charged particles, neutral atoms, and molecules. The high concentration of charged particles created by ionization, with extremely high energy compared to liquid, gaseous and solid states, forms the so-called fourth state of matter – plasma [8, 9]. Cold atmospheric plasma is a type of plasma that is often used in practice. Its advantage lies in the very high electron temperature of the plasma itself and the relatively low temperature of the gas. As a result, charged particles present in the plasma can interact with the surface of the substrate or energetically activate reactive gas molecules [8, 10].

The presented research aims to characterize the properties of composite materials after surface modification of glass and basalt fibers using atmospheric plasma. A partial goal of the presented study is to process, evaluate and compare the achieved results statistically.

2 Experiment

2.1 Used materials

The following were used to produce composite samples: BAUCIS LK cement (Czech Republic) activated with potassium water glass; silica sand was used as a filler. The components were mixed in a 1:1:0.9 ratio; was added 2 wt.% fibers:

- chopped fibers type AR (ARGF): Cem-Fil, 2400 tex, density 2.68 g/cm^3, fiber diameter 14 μm, length 12 mm;
- chopped fibers type E (EGF): manufacturer Johns Manville, PR440 2400 907, 2400 tex, density 2.6 g/cm^3, fiber diameter 16 μm, length 12 mm;
- chopped fibers type R (RGF): 2400 tex, density 2.6 g/cm^3, fiber diameter 12 μm, length 12 mm;
- chopped strands (BF): manufacturer Deutsche Basalt Faser GmbH. 2400 tex, density 2.6–2.8 g/cm^3, fiber diameter 13 μm, length 12 mm.

Samples with dimensions $10\times10\times100$ mm^3 were prepared for the experiments. Four sets of fiber-reinforced composite samples were prepared as reference samples (REF), each set containing ten pieces of samples. The surface of the mineral fibers was modified with atmospheric plasma for about 3 minutes, and then the fibers were used to prepare composite samples. Each set of samples contained 30 samples. The time required for the maturation of the geopolymer composite materials is from 28 to 30 days.

2.2 Used methods

A Piezobrush®PZ2-i from Relyon-plasma GmbH was used to modify the surface of the fibers. The purpose of the modification is to activate the surface of the fiber. Improved interface quality ensures better adhesion of the geopolymer matrix to the fiber and strengthens the bond between the composite components.

To evaluate the achieved properties of the composite. The flexural strength test was performed according to the standard ČSN EN 14617-2 Artificial stone – Test methods – Part 2: Determination of flexural strength.

The change in the composite's properties was observed on the reference samples at various time intervals – after 30 days of maturation (P30), after three months (P120), and after half a year (P210) from the end of the maturing period. The measurement results were used to determine the degree of plasma treatment of the surface of fiber reinforcement and assess the improvement/deterioration of the interface quality.

The flexural strength was calculated according to Eq. 1:

$$\sigma_F = \frac{3}{2}\frac{F_{\max} \cdot L}{bh^2} \text{ [MPa]}, \tag{1}$$

where: σ_F – flexural strength [MPa]; $F_{\max}$ – maximal force [N], L – support span [m], b – width [mm], h – thickness [mm].

The retention ratio of flexural strength was evaluated according to Eq. 2. Calculations prescribe a Standard CSN EN 14617-2 Artificial stone - Test methods - Part 2: Determination of flexural strength [11].

$$\Delta R_F = \frac{R_{F0} - R_F}{R_{F0}} 100\%, \tag{2}$$

where: ΔR_F – retention ratio [%], R_{F0} – reference value of strength [MPa], R_F – strength value after time exposure [MPa].

3 Results

The obtained results, including statistical data evaluation, are shown in Fig. 1. The research involves groups of samples with a small number of measurements, so the Horn method for the analysis of small data sets was used to evaluate the results. The result obtained from AR-type glass fiber composites after 30 days of maturation shows an increase in flexural strength of 80%. After 90 days (P120), there was approximately 48% flexural strength value. After the next 90 days (P210), flexural strength increased by 33%. The percentage increase was calculated according to Eq 2. The results, including statistical evaluation, show that plasma modification's positive effect on improving the mechanical properties of other types of fibers has not been demonstrated. The different time intervals (30, 90, and 180 days) did not affect the investigated mechanical properties.

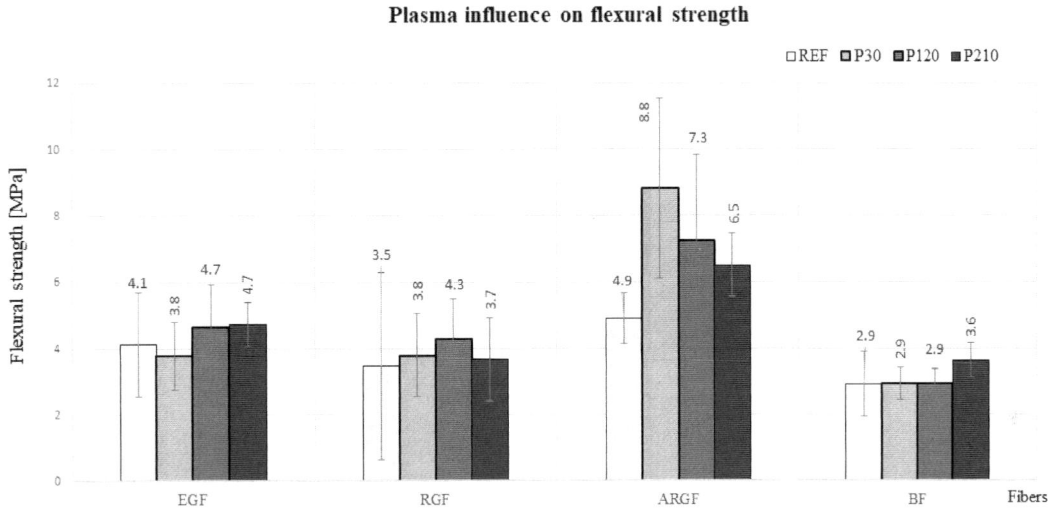

Fig. 1: Comparison of flexural strength results for reference samples and plasma-modified reinforcement.

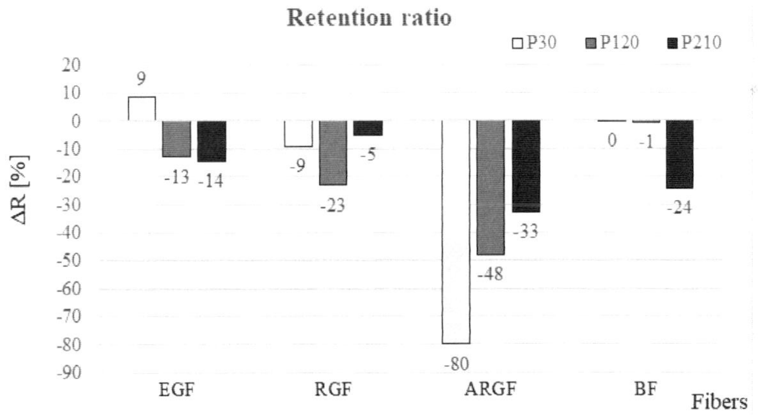

Fig. 2: Retention ratios for all type of mineral fibers after six months from preparation of samples.

The standard prescribes the evaluation of the bending test results by calculating the retention ratio (Eq. 2). The results are presented in the graph in Fig. 2. The calculated values show an increase in flexural strength of ARGF by 80% after a maturation period of 30 days against the reference value. After 90 days (P120), there was a decrease of approximately 18% in the flexural strength value against the value measured after 30 days (P30). After the next 90 days (P210), flexural strength decreased 26% against value P30. Flexural strength of BF increased after 210 days by 24%. The values of the retention ratios for exposure times P30 (0%) and P120 (1%) show no changes.

Fig. 3 and 4 show the amount of geopolymer matrix connected to the reinforcing fibers' surface. Despite plasma treatment, basalt fibers (Fig. 3) have little adhesion to the GP matrix. The surface of the fibers is clean and smooth, and the amount of matrix on the surface of the fibers is minimal to negligible. The exposure time (three and six months) had no significant influence on the adhesion of the fibers to the matrix. In the case of AR-type glass fibers (Fig. 4), an increased amount of geopolymeric material can be observed on the surface of the fibers. It indicates an excellent adhesion to the matrix even after a time delay. The results of microscopy

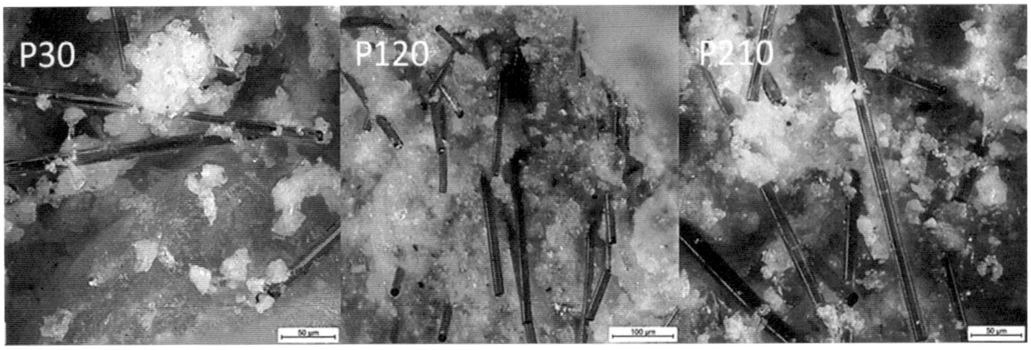

Fig. 3: Representation of basalt fiber surfaces after plasma modification (P30), after 3 (P 120) and 6 months (P210).

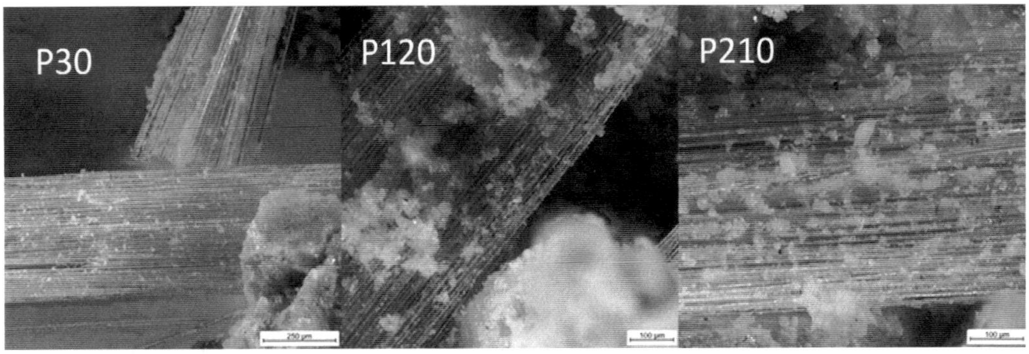

Fig. 4: Representation of AR-glass fiber surfaces after plasma modification (P30), after 3 (P 120) and 6 months (P210).

observation correspond to the results obtained in the flexural strength test; plasma treatment has been shown to affect the strength of the AR-type glass fiber reinforced geopolymer composite.

4 Conclusion

Atmospheric plasma surface treatment was used for mineral-type fibers, and a flexural strength test was performed after the maturation period. Plasma can be expected to roughen the fiber surfaces, which increases the contact areas and would result in increased adhesion between materials. The evaluation of the results showed a significant improvement in the flexural strength only of the geopolymer reinforced with glass fibers of the AR-type (with plasma surface treatment). For the other types of fibers, the improvement in the average flexural strength value between untreated and modified fiber reinforcement is statistically inconclusive. Therefore, it can be said that the atmospheric plasma surface treatment of glass (E-type and R-type) and basalt fibers did not lead to an increase in the flexural strength of the geopolymer composite system.

Acknowledgement

This work was supported by the Student Grant Competition of the Technical University of Liberec under the project No. SGS-2020-5019.

References

[1] V. Čech, New Progress in Composite Interphases: A Use of Plasma Technologies. Proc. FRC 2000, Newcastle: Woodhead Publ., 2000, pp. 246–252. ISBN 1-85573-550-4.

[2] D. Hull, T.W. Clyne, An Introduction to Composite Materials. Cambridge University Press, Cambridge 1996. ISBN 0-521-38855-4.

[3] J.K. Kim, Y.W. Mai, Engineerd interface in fiber reinforced composites. Elsevier Science Ltd, 1998, 1st ed. ISBN 0-08-042695-6.

[4] P. Buerš, Design of a methodology for determining the adhesion properties of roads, BP, University of Pardubice, 2015.

[5] T. Deák, T. Czigány, Chemical Composition and Mechanical Properties of Basalt and Glass Fibers: A Comparison, Textile Research Journal Vol 79(7): 645–651, doi: 10.1177/0040517508095597.

[6] A. Knob, Surface treatment of glass fibers for polymer composites, Dissertation thesis, VUTBR, 2016.

[7] M. Drobilík, Organosilanes and their application possibilities. University of T. Bata, Zlin 2015, 42 pp.

[8] N. Inagaki, Plasma Surface Modification and Plasma Polymerization. Lancaster: Technomic Publishing Company, 1996. ISBN 1-56676-337-1.

[9] R.L. Li, L. Ye, Y.W. Mai, Application of plasma technologies in fibre-reinforced polymer composites. Composites part, 1997, vol. 28, part A, pp. 73–86.

[10] G.M. WU, Oxygen plasma treatment of high performance fibers for composites. Materials Chemistry and Physics, 2004, vol. 85, pp. 81–87.

[11] ČSN EN 14617-2 Artificial stone - Test methods - Part 2: Determination of flexural strength.

Evaluation of Residual Stresses using ASTM E837-20

B. Řeha[1], R. Musilová[2,*], J. Václavík[2]

[1] *Bonatrans Group a.s., Revoluční 1234, 735 94 Bohumín, Czech Republic*

[2] *Výzkumný a zkušební ústav Plzeň s.r.o., Tylova 1581/46, 301 00 Pilsen, Czech Republic*

* *musilova@vzuplzen.cz*

Abstract: Procedures given in new issue of ASTM E837-20 standard for measurement of hole-drilling method including the comparison with the older standard versions and discussion of new issue benefits are described in this article. Comparison is also made on experimentally obtained residual stress data from railway axle.

Keywords: residual stresses; hole-drilling method; ASTM-E837.

1 Introduction

The new revision of ASTM E837-20 standard for residual stress measurement using hole-drilling method issued in the year 2020 with successive improving of evaluating procedures dated from first issue from the year 1981 (one step measurement), followed with multiple power series method (2001) and ending with the integral through profile method [2] (2008, 2013). The residual stresses are evaluated here from strains relieved by hole-drilling using large triangular matrices $\bar{\boldsymbol{a}}$, $\bar{\boldsymbol{b}}$. In the last standard version [1], the matrices were expressed for 20 equal drilling steps $0.01D$ up to the practical depth $0.2D$ (rosette type A and B, D is the mean rosette type diameter) i.e. only for specific hole diameter and hole depths with lower testing convenience [4].

2 Theory

In the new standard version the generalization of matrix coefficients are made with fifth order polynomial interpolation using two sets of calibration constants C_i. Using these constants first three cumulative calibration matrices $\hat{\boldsymbol{a}}$, $\hat{\boldsymbol{b}}$ have to be calculated for given three hole diameters, used drilling depths and workpiece thickness W. The reason for this is that the cumulative functions on which the integral method is based [2] are more suitable for coefficients interpolations. Then the mean cumulative matrix is calculated from these three matrixes for both coefficients. The final calibration matrices $\bar{\boldsymbol{a}}$, $\bar{\boldsymbol{b}}$ are obtained with numerical differentiation of these cumulative functions as differences of matrix values in each row. This access makes it possible to choose arbitrary hole diameters and drilling steps. The number of steps can be even in the range of 15 to 40 for rosette type A and B, the depth steps may not to be equal. The maximum drilling depth is specified as the minimum of $0.2D$ or $0.6W$, thus the deepest residual stress profile is 2 mm using the largest 1/8 in rosette.

Another benefit of the new standard is the possibility to measure the residual stresses in the intermediate workpiece thickness going from $0.25W$ to $0.6W$, which was not defined in the older standard issues. New set of constants were also defined for homogenous stress field.

The demand for the accuracy of the drilling set-up is approximately the same in both standard issues.

The evaluation procedure for residual stress profile from measured relieved strains using calculated matrices $\bar{\boldsymbol{a}}$, $\bar{\boldsymbol{b}}$ is performed as for the integral as for the power series and as for throw hole methods in the same way for both last and new standard issues.

The new standard is based on the work, published in [3]. The correction of drilling hole misalignment and the correction to elastic-plastic stress state are again not included to the standard.

3 Implementation

One set of measurement on surface hardened axle was performed using apparatus SINT RESTAN (Fig. 1), 1/8 inch rosette HBM RY61-3.2/120S, and the electric motor head with n = 20 000 rpm) in 22 drilling steps from 0.1 mm up to 2.2 mm (Fig. 2). Data seems to be very smooth.

Fig. 1: Hole-drilling device HBM Restan with electric motor head.

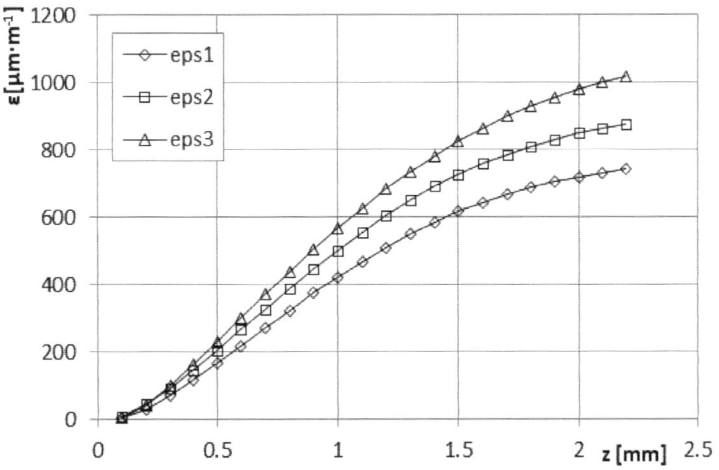

Fig. 2: Measured raw strains without interpolation.

The evaluation was made according to both last issues of standard ASTM E837 and also with constants from original integral method [2]; differential Kockelmann's and SINT HDM method were also added for comparison. Results of both principal stresses including the angle of the maximum principal stress are given in Fig. 3. Both ASTM -13a, ASTM-20 and integral method give comparable stress depth histories. HDM method does not provide so detailed distribution of residual stress with the depth but also the maximum principal values are comparable. Substantially different lower stresses were obtained with the differential method, which we do not recommend for this type of measurement.

The absolute values of stresses evaluated with ASTM E837-20 in comparison with the older issue were at the most of 6% higher.

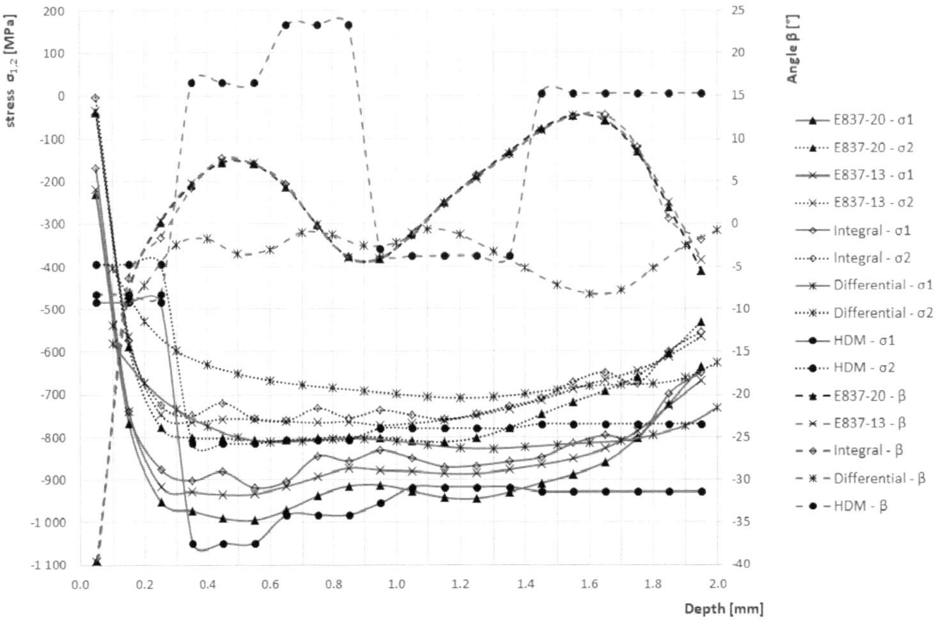

Fig. 3: Comparison of residual stress evaluation with various methods using the software SINT Eval including both ASTM E837-13 and E837-20 issues.

The difference between evaluated principal stresses according ASTM E837-13a and E837-20 is given in Fig. 4. It changes from negative to positive values and may be a mistake of polynomial interpolation in the new standard.

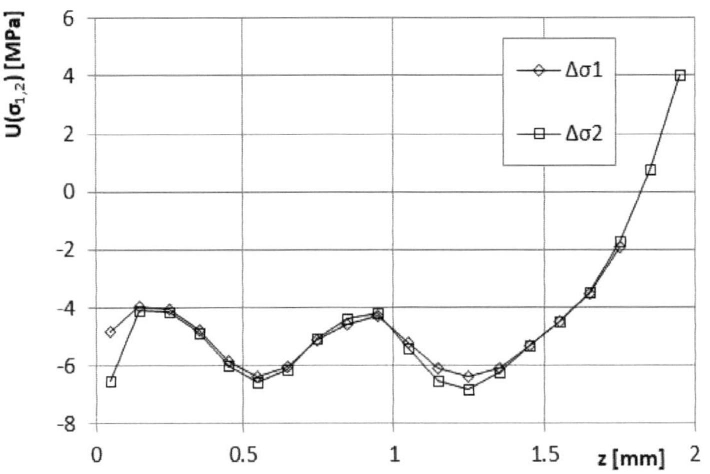

Fig. 4: The difference of evaluated principal residual stresses using ASTM E837-13 related to the new standard: older stresses are lower at the surface and higher in the maximal depth.

The residual stress evaluation from relieved strains was performed using software SINT Eval. It includes very good estimation of the uncertainty for several hole-drilling methods used for non-homogenous residual stress profile evaluation. The evaluation of residual stress uncertainty was made for both ASTM -13a and ASTM-20 methods. The results are given in Fig. 5 in form of upper tolerance intervals for both minimum and maximum principal stresses.

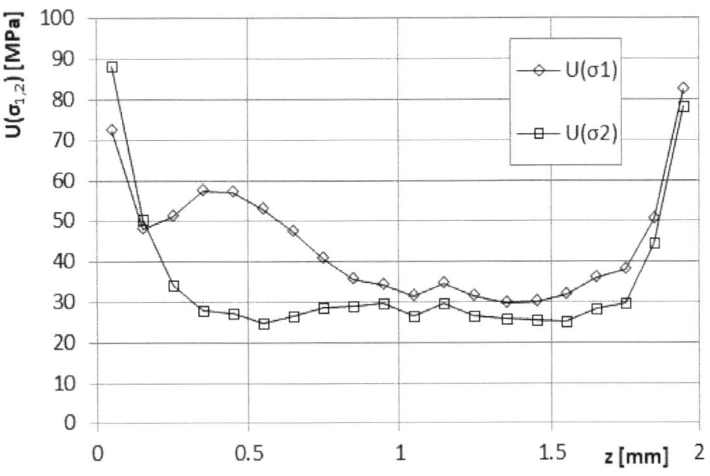

Fig. 5: Distribution of absolute uncertainties of calculated residual principal stresses through the depth is the same for both ASTM E837-13a and E837-20 standards.

4 Conclusion

The matrices in ASTM E837-13a standard issue were obtained from the old FEA model whereas new given constants are based on much more detailed finite element mesh; new calibration constants are about 2–3% higher (i.e. calculated stresses are lower). On the other hand, there rises some small error in computed stresses using interpolation procedures specified in the new standard. The effect of these differences can lead to much more higher differences in computed residual stresses especially in the case of noisy data up to 10 %, which is a generally valid criterion.

The article has originated in the framework of the institutional support for the long-term of conceptual development of the research organization.

References

[1] ASTM Determining residual stresses by the hole-drilling strain-gage method. Standard test method E837-13a. American Society for Testing and Materials, West Conshohocken, PA (2013).

[2] G. S. Schajer, Measurement of non-uniform residual stresses using the hole-drilling method. Part I J Eng Mater Technol 110(4) (1998) 338–343.

[3] G. S. Schajer, Compact Calibration Data for Hole-Drilling Residual Stress Measurements in Finite-Thickness Specimens, Experimental mechanics 60 (2020) 665–678, doi: 10.1007/s11340-020-00587-4.

[4] B. Řeha, J. Václavík, T. Návrat, D. Halabuk, Residual stresses in railway axles, Proceedings of 10th European Conference on Residual Stresses ECRS10, Leuven, Belgium, September 2018. Edited by Marc Seefeldt KU Leuven, department of Materials Engineering. Part of Material Research Proceedings Vol.6 (2018), pp. 95-100, ISBN 978-1-94529188-3 (Print).

Gyroid Structures for Use in Implants, their Characteristics and Mesh Generation

L. Řehounek[1,*], A. Jíra[1], J. Vorel[1]

[1]Czech Technical University in Prague, Faculty of Civil Engineering, Thákurova 7, 166 29 Prague 6, Czech Republic

* lubos.rehounek@fsv.cvut.cz

Abstract: This contribution presents the results of a study on gyroid structures for use in implants. The recent advent of additive manufacturing technologies enables rapid prototyping of specimens and experimental design of novel structures to achieve osseointegration. Sample gyroid structure specimens were 3D-printed via selective laser sintering (SLS) using the PA-12 polymer powder and uniaxially tested for peak stress and global elastic modulus. A great benefit is the shapeability of the structure and its predictability by FEM. Generation of this structure can either be done via CAD environments or direct modelling with the gyroid equation. Finite element method (FEM) simulations confirm the values of Young's moduli between the experiment and simulation are within ca. 15 % conformity. This approach will enable more custom designs in the future without the need for testing of every variant.

Keywords: gyroid; porous; FEM; 3D printing; osseointegration.

1 Introduction

Removal of typical stress from the bone by introducing a stiff metal implant often leads to reduction of bone mass in the adjacent bone [1]. The typical values of Young's modulus E for human bone is ca. 17.1±3.5 GPa [2] as opposed to ca. 100–120 GPa for titanium and its alloys. One of the main driving forces of novel design in the field of implantology is the motion to decrease the number of revision surgeries and reduce the risk of aseptic loosening of the implant [3]. The principal means of achieving good osseointegration of implants are usually various forms of surface treatment to increase the surface roughness (and therefore the contact area of bone) or manipulate the chemical properties of the surface [4]. Among those, the most notable are [5]:

- Blasting with particles
- Chemical etching
- Plasma-sprayed surfaces
- Ion-sputtering coating
- Anodized surface
- HA coating
- Porous surfaces

The approach to apply a porous surface stems from the need to increase the surface area at the bone-implant interface and also provide a localized area where bone trabeculae can grow *inside* the body of the implant itself. The gyroid structure has the potential to create a bone-implant composite material that can reduce (or eliminate) the effects of stress shielding [6] and create a continuous gradient of material properties in the radial implant-bone directions. The fact that bone trabeculae can effectively grow inside the implant and create an interconnected mechanism has been investigated in one of the authors' previous works [7] in in vivo tests on laboratory pigs with use of a different porous structure.

2 Gyroid equation modelling

The modelling of the structure can be done in two ways; either via CAD or directly with equations. We found that with all 3D printing applications, the former approach is usable and models have good quality for metal or plastic SLS printing. However, when FEM is considered, this approach falls short on precision as the surface mesh is intersecting and tesselation quality is bad (Fig. 1), preventing the enclosure of a solid. Rather, modelling with the TPMS (triply-periodic minimal surface) single gyroid equation (Eq. (1) can be adopted:

$$\sin\left(\frac{2\pi x}{a}\right)\cos\left(\frac{2\pi y}{a}\right)+\sin(\frac{2\pi y}{a})\cos\left(\frac{2\pi z}{a}\right)+\sin\left(\frac{2\pi z}{a}\right)\cos\left(\frac{2\pi x}{a}\right)-t=0, \tag{1}$$

where a represents the length of a circumsribed cube and x, y and z are spatial coordinates. The parameter t influences the curvature of the gyroid structure and by its manipulation we can transfer between the *sheet* and *tubular* variants of the structure.

Following the creation of the TPMS gyroid, a volumetric gyroid structure can be created by intersecting the generated surfaces with a homogeneous body (either the sheet or the tubular phase of the gyroid).

(a) Sample sheet-based single gyroid structure

(b) CAD-generated with bad tesselation

(c) Directly generated with good tesselation

Fig. 1: A typical sheet gyroid structure and details of individual means of generation with corresponding quality of tesselation. The cutouts (b) shows various errors such as overlapping faces and out-of-boundary vertices; while usually suitable for 3D printing, this generation is unaccetable for numerical analyses. Generated with MSLattice [8].

3 Mechanical tests and FEM simulations

To verify the reliability of FEM numerical analyses and compare the various variants of porous structures in regard to their mechanical properties, we performed uniaxial compression tests on sample gyroid structure specimens with relative theoretical density of $n = 0.75$ (corresponding value $t = 0.39$ for the sheet gyroid variant and $t = 0.78$ for the tubular gyroid, see Eq. (1). These specimens had dimensions of $25.12 \times 25.12 \times 25.12$ mm and had two homogeneous load-distributing plates fitted on the top and bottom. The test was carried out in the mode or controlled displacement of 1 mm/min using the electromechanic MTS Alliance RT30kN and RT50kN loading machines. The specimens were made and subsequently processed using the PA-12 polymer material on the Sinterit Lisa Pro 3D printer. Also, one Dode-Thick trabecular specimen with the same parameter of porosity was included. The uniaxial compression tests were controlled by constant velocity of displacement. The specimens are shown on Fig. 2. The test was carried out to determine the values of *global* Young's moduli E of whole individual structures in regard to comparison of viability of various geometrical variants with the same porosity under the same conditions.

4 Results and conclusions

The results of uniaxial compression tests are shown together with the results of X-FEM simulations using a strain-based periodic numerical homogenization method in Tab. 1. The proposed homogenization method was performed in a Matlab-based software the reader can learn more about in [10]. As expected, the sheet gyroid variant shows the most stiff response combined with a less localized damage behavior (Fig. 3), which

(a) Dode-Thick trabecular (b) Single sheet gyroid (c) Tubular single gyroid

Fig. 2: Three PA-12 3D-printed specimens for uniaxial compressive tests. The variants all have the same porosity $n = 0.75$.

is beneficial to implant-oriented applications as the risk of chipping and wear debris polluting the surrounding tissue of the patient is a concern for osteolysis [9].

Tab. 1: The results of uniaxial compression tests and numerical analyses (X-FEM) of the porous structures in regard to the *global* Young's modulus.

Structure variant	Compression test E [MPa]	X-FEM Homogenization E [MPa]
Dode-Thick	27.3 ± 3.8	26.4
Sheet gyroid	28.7 ± 1.8	33.2
Tubular gyroid	72.4 ± 6.1	82.9

(a) Dode-Thick trabecular (b) Single sheet gyroid (c) Tubular single gyroid

Fig. 3: All destroyed specimens after failure caused by the uniaxial compression test. Note the total dislocation and shear character of failure of (a) and (c) compared to the sheet gyroid variant (b).

Porous structures offer a solution to increasing the surface area of an implant, their system of interconnected pores allows for good fluid transport and the sheet character of their geometry does not allow for a local failure. Therefore, it seems to be worthwhile to explore the gyroid variants (single, double) in regard to optimal geometry for implant applications. Experience showed us that CAD generation is not a viable approach as the tesselation quality is bad. With direct modelling, we can achieve a good quality mesh for the purpose of optimizing the geometry via FEM analyses; mainly in regard to matching the global Young's modulus E of the structure to human bone. To further the understanding of the topic, the authors plan to launch experiments on metal SLS 3D-printed specimens to determine values of E in regard to the porosity and other geometrical parameters of the structure with the final goal of implementing it onto a surface of a functional implant, moving from polymer SLS into the area of fully SLS 3D-printed metal implants with surface porous structures.

The design should always include requirements from both mechanical and biological standpoints as strictly matching the elastic modulus of bone might not always be the first priority. While porous structures lend themselves to design of structures with much lower global moduli E (benefitial for prevention of stress-shielding), strength and prevention of localized falure can potentially take precedence in design. If experiments show good quantity of bone ingrowth and an interlocking mechanism with a gradient of material properties is achieved, many other factors such as pore size for optimal bone ingrowth and fluid transport must be considered.

Acknowledgement

The authors would gratefully like to acknowledge the financial support provided by the Faculty of Civil Engineering, CTU, Prague, project n. SGS20/155/OHK1/3T/11.

References

[1] J.E. Bidaux, Metal injection moulding of low modulus Ti–Nb alloys for biomedical applications, Powder Metall 56 (2013) 263–266, doi: 10.1179/0032589913Z.000000000118.

[2] D.T. Reilly, A.H. Burstein, V.H. Frankel, The elastic modulus for bone, Acta Polytech. J. Biomech. 7 (1974) 271–275.

[3] L. Řehounek, F. Denk, A. Jíra, Geometrical optimization of dental implants with regard to osseointegration, Acta Polytech. CTU Proc. 13 (2017) 97–101, doi: 10.14311/APP.2017.13.0097.

[4] J.C. Chen, et al., In vivo studies of titanium implant surface treatment by sandblasted, acid-etched and further anchored with ceramic of tetracalcium phosphate on osseointegration, J Aust Ceram Soc 55 (2019) 799–806, doi: 10.1007/s41779-018-00292-5.

[5] A. Gupta, M. Dhanraj, G. Sivagami, Status of surface treatment in endosseous implant: A literary overview, Indian J Dent Res 21 (2010) 433–438, doi: 10.4103/0970-9290.70805.

[6] R. Ummethala ,et al., Selective laser melting of high-strength, low-modulus Ti–35Nb–7Zr–5Ta alloy, Materialia 14 (2020) p.100941, doi: 10.1016/j.mtla.2020.100941.

[7] L. Řehounek, A. Jíra, Experimental and numerical anylses of a 3D-printed titanium trabecular dental implant, Acta Polytech. 57 (2017) 218–228, doi: 10.14311/AP.2017.57.0218.

[8] O. Al-Ketan, R.K. Abu Al-Rub, MSLattice: A free software for generating uniform and graded lattices based on triply periodic minimal surfaces, MDPC (2020) p.e205, doi: 10.1002/mdp2.205.

[9] H.C. Amstutz, et al., Mechanism and clinical significance of wear debris-induced osteolysis, Clin. Orthop. Relat. Res. (1992) 7–18.

[10] A. Jíra, et al., Mechanical Properties of Porous Structures for Dental Implants: Experimental Study and Computational Homogenization. Materials, 14(16) (2021), p.4592, doi: 10.3390/ma14164592.

The Fatigue Prediction of Devices for Long-bone Osteosynthesis

R. Sedláček[1,*], V. Drátovská[1], Z. Padovec[1]

[1] *Czech Technical University in Prague, Faculty of Mechanical Engineering, Department of Mechanics, Biomechanics and Mechatronics, Technická 4, 160 00 Prague 6, Czech Republic*

* *radek.sedlacek@fs.cvut.cz*

Abstract: This article deals with fatigue life prediction of metallic bone plates using SIMULIA fe-safe software and compare predicted results with measured data from experiments. Two materials commonly used for surgical implants were considered: stainless steel for implants 1.4441 (ASTM F138) and titanium alloy Ti6Al4V ELI (ASTM F136). Four-point bending fatigue test was carried out to obtain material parameters of both materials. The fatigue analysis based on FEA results from Abaqus was performed using the obtained material parameters. The results of the fe-safe fatigue analysis were verified by comparing them with data from measurements.

Keywords: prediction; fatigue; osteosynthesis; implants.

1 Introduction

The bone plates are in their application in human body subject to relatively high stress, that may lead to exceeding the ultimate strength of the material or more likely to the fatigue failure [1, 2]. Failure of the bone plate may cause an additional injury and further complications for the patient and leads to excessive costs related to the reoperation [3, 4]. Therefore, every new product needs to be tested to examine its fatigue endurance. However, fatigue experiments are quite expensive and time consuming and in the development process of new product there are usually several prototypes with different geometry, which needs to be tested to compare their fatigue properties. So, it would be interesting and useful to investigate, if there is a possibility to reliably predict the fatigue life of the bone plates [5, 6].

2 Methods and materials

Methodology of the four-point bending test was based on ASTM F382 standard, which describes static and cyclic testing of metallic bone plates. First, the static four-point bending test was performed to determine suitable loading levels for the fatigue four-point bending test. Afterwards, the load-controlled high-cycle fatigue test on several loading levels was done. The applied force had sinusoidal course with R ratio equals to 0.1. To obtain material properties the four-point bending fatigue test with flat specimens with rectangular cross section was performed and the S-N curve were got.

The experiments were carried out with two generations of bone plates for proximal humerus fractures and two materials (see Fig. 1). The stainless steel for implants 1.4441 (ASTM F138) and titanium alloy Ti6Al4V ELI (ASTM F136) were used. Both materials are commonly used for orthopaedic implants. The data evaluated from experiments were used to define material parameters in SIMULIA fe-safe based on elastic FEA (Finite Element Analysis) [7]. Models of bone plates were made and loaded in four-point bending mode in the same configuration as in the experiments. To perform the analysis two stress-based fatigue algorithms were used: von Mises algorithm and Normal Stress algorithm.

The fatigue analysis based on the elastic FEA results was applied in the fe-safe software. The static analysis required to gather these results was performed in Abaqus software. Models of three generations of bone plates (see Fig. 2) were created and loaded in the four-point bending mode in the same configuration as in the experiments. Each bone plate was loaded up to 1,000 N in Abaqus. The results of the Abaqus analysis (in *.odb file format) were used as input data for the fe-safe processing.

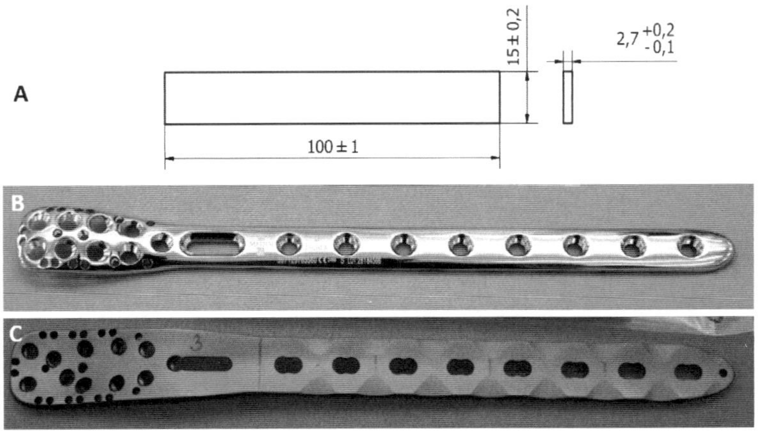

Fig. 1: (A) Dimensions of the flat sample for the four-point bending test, (B) stainless steel bone plate of the first generation, (C) titanium alloy bone plate of the second generation.

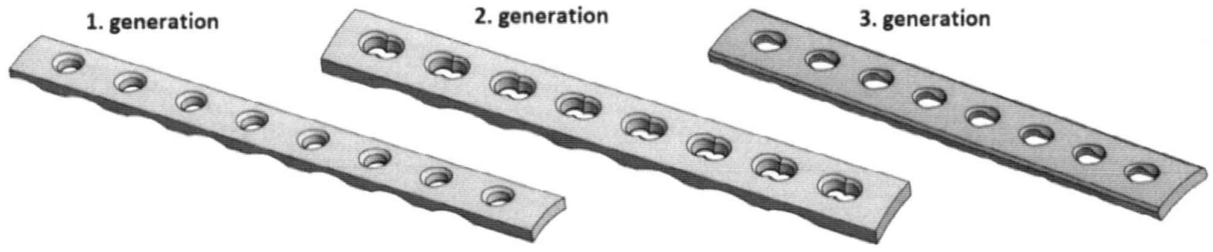

Fig. 2: Models of three generations of bone plates.

The output of the fe-safe analysis comprise the number of cycles calculated for the highest loaded node in the model, in other words the lowest lifetime calculated for the whole part. It was also possible to export the results to the *.odb file format and to display them in Abaqus as a contour plot, which showed the calculated fatigue life for the whole of the model (see Fig. 3).

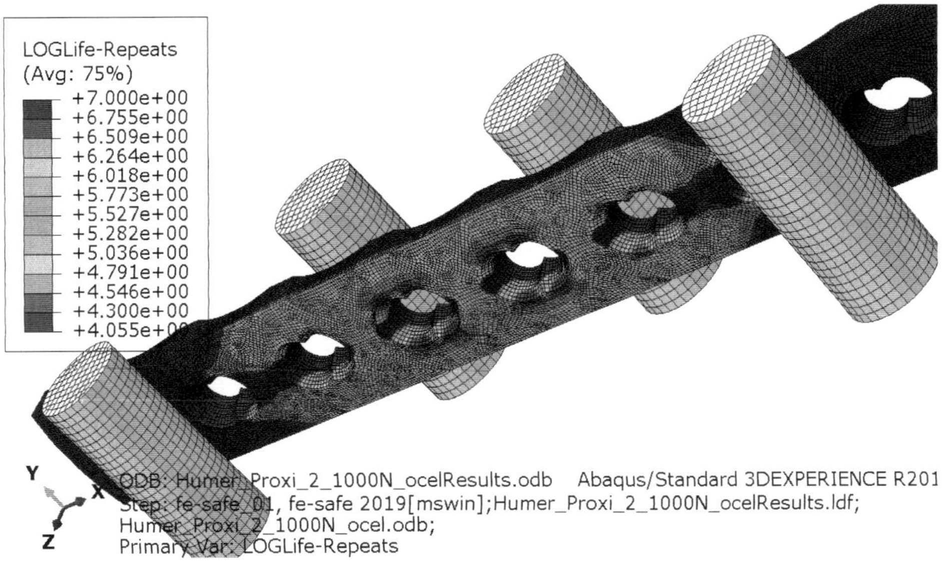

Fig. 3: Results of the fe-safe analysis for the second generation of steel bone plates displayed as a contour plot in Abaqus.

3 Results and discussion

The fe-safe analysis was run for three generations of bone plates in two material variants. The prediction results compared with experiments plotted in force amplitude are shown in the chart (see Fig. 4). The experiments are displayed as thick lines and the predictions as thinner ones, the graph contains results of both used algorithms: von Mises (VM) and Normal Stress (NS). The prediction was made for three generations, but the experiments only for the first and the second generation, because the third generation of bone plates has not been manufactured yet. Prediction for the third generation lies between the first and the second generation, this corresponds with its geometry, which is a compromise between the first and the second generation.

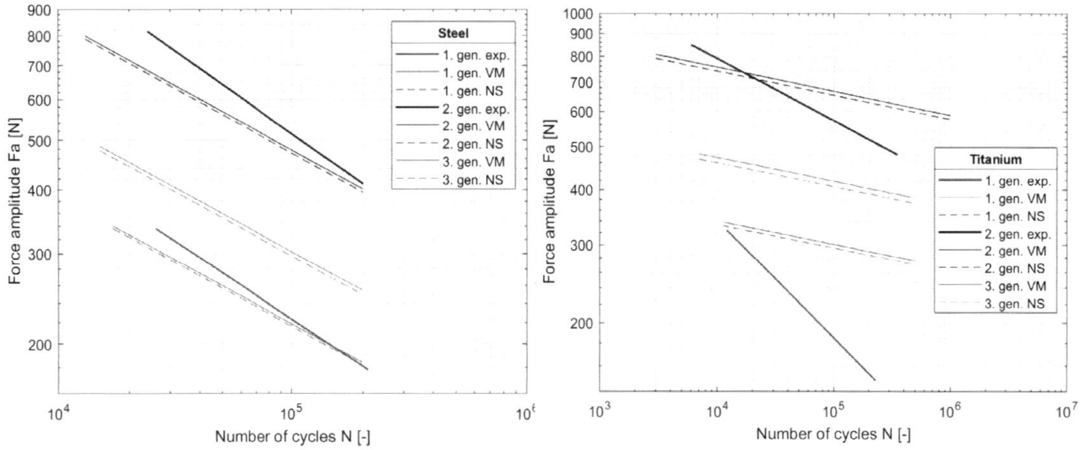

(a) steel (b) titanium

Fig. 4: Comparation of experiments and fe-safe results for all generations in stress amplitude.

The fe-safe prediction for steel is quite accurate although it predicts lower fatigue life, especially for higher stress amplitudes. However, the prediction for third generation in comparation with predictions and experiments for the first and the second generation shows us, where we could expect the results of experiment with the third generation. Results of the prediction for titanium do not really correspond with the experiments. Worse results might be caused by the input material parameters, which were for titanium computed with lower coefficient of determination or it may be caused by various quality of used titanium material.

4 Conclusion

SIMULIA fe-safe software was used for fatigue life prediction of bone plates in two material options. The prediction was based on FEA results from Abaqus and material parameters obtained from experiments. The fe-safe analysis results were compared with the measured data from experiments. Satisfying correspondence between the prediction and the experiments for the titanium alloy was not achieved. However, the prediction for the stainless steel shows better results and could be used to estimate fatigue life of bone plates with different geometry.

Acknowledgement

This study was supported by the grant project awarded by the Technology Agency of the Czech Republic No. TACR TH03010418.

References

[1] B. Wan, et al., On fatigue failure prediction of prosthetic devices through XFEM analysis. International Journal of Fatigue. 2021, 147, doi: 10.1016/j.ijfatigue.2021.106160.

[2] G. Logroscino, et al., Early failure of a locked titanium plate in a proximal humeral fracture: Case report and metallurgic analysis. Trauma Case Reports. 2018, 17, 18–22, doi: 10.1016/j.tcr.2018.09.005.

[3] N.E. Dowling, C.A. Calhoun, A. Arcari, Mean stress effects in stress-life fatigue and the Walker equation. Fatigue & Fracture of Engineering Materials & Structures, 2009, 32(3), 163–179, ISSN 8756758X, doi: 10.1111/j.1460-2695.2008.01322.x.

[4] Strain Based Fatigue Analysis. Autodesk help [online], http://help.autodesk.com/view/ASMECH/2014/ENU/?guid=GUID-ECDCD9AB-BE65-43A3-9ECB-D25C38606A3E.

[5] P. Chao, B. Conrad, D. Lewis, M. Horodyski, A. Pozzi, Effect of plate working length on plate stiffness and cyclic fatigue life in a cadaveric femoral fracture gap model stabilized with a 12-hole 2.4 mm locking compression plate, BMC Veterinary Research, 2013, **9**(1), ISSN 1746-6148, doi: 10.1186/1746-6148-9-125.

[6] D. Anitha, S. Das De, K. Sun, H. Doshi, T. Lee, Improving stability of locking compression plates through a design modification: a computational investigation. Computer Methods in Biomechanics and Biomedical Engineering, 2013, 18(2), 153–161, ISSN 1025-5842, doi: 10.1080/10255842.2013.785536.

[7] N. Mavrodontis, Fatigue assessment of a bicycle frame done with Abaqus and Fe-Safe, SIMULIA Abaqus Software, Training & FEA Consultancy, https://info.simuleon.com/blog/fatigue-assessment-of-a-bicycle-frame-withabaqus-and-fe-safe.

Effect of the Arrangement of Carbon Fiber Sensor in the Composite Layup in Regard to the SHM

N. Schmidová[1,*], R. Sedláček[1], K. Doubrava[1], M. Růžička[1]

[1] *Department of Mechanics, Biomechanics and Mechatronics, Faculty of Mechanical Engineering, Czech Technical University in Prague, Technická 4, 160 00 Prague 6, Czech Republic*

* *Nikola.Schmidova@fs.cvut.cz*

Abstract: This paper reports part of the complex study of the aspects, which can influence the possibility to detect impact damage of long fiber composite parts using integrated Carbon Fibre Sensors (CFSs). CFSs were prepared from carbon fiber tow and integrated in three different positions in relation to the composite lay-up. It was shown that the location of the CFS in the composite layup influences the possibility of the impact detection. The influence of the mechanical loading to the signal of the integrated CFSs is small compared to the effect of impact damage.

Keywords: carbon fibers; electrical properties; cyclic loading; composites; impact behavior.

1 Introduction

Structure Health Monitoring (SHM) of composites is a widely discussed and investigated topic. Impact damage detection of composite structures is one of the important tasks of SHM. in aerospace, automotive and other industries. One of the possible sensors for damage monitoring are Carbon Fiber Sensors (CFSs). CFSs are prepared from carbon fiber tow. It was reported that CFS can be used for strain monitoring [1–4], detection of microcracks [5,6] and also detection of impacts [7]. Cyclic loading of the CFS can also influence the measured signal [8]. In this paper part of a broader study of the effect of mechanical loading and impact loading on sensors will be presented. Possibility of impact damage detection using integrated CFSs is investigated. Influence of the arrangement of the CFS in the composite layup will be discussed (close to the top, close to the bottom of the specimen, in the middle of the lay-up). Relationship between number of loading cycles and electrical resistance change of the CFS integrated in different positions in the layup is also presented.

2 Experimental procedure

2.1 Specimen preparation

For the specimen preparation laminate lay-up [+45/0/-45/90]sym was used. Prepreg material made of woven fabric with epoxy resin HexPly 1454 GM/50%/1035 was used. The carbon fiber sensors were placed between the firstand second layer, between the fourth and fifth layer and between the seventh and eighth layer. Overview of the specimens is given in Tab. 1.

The following curing process was used: 125 °C for 90 min, 6 bar. All specimens which were investigated were manufactured using the peel ply – the thickness of the specimens was 1.3 mm.

2.2 Carbon Fiber Sensors (CFSs)

The CFSs used in the presented investigation were manufactured from the carbon fiber tow type PAN, producer Toray, label T300 1000-50A. CFSs were prepared in the length of 70 mm. The preparation of CFS has been described in [1], and the same procedure was used here. A nickel electrolyte coating was applied to the ends of each roving. Then electrical contacts made of thin copper wire were soldered to the ends.

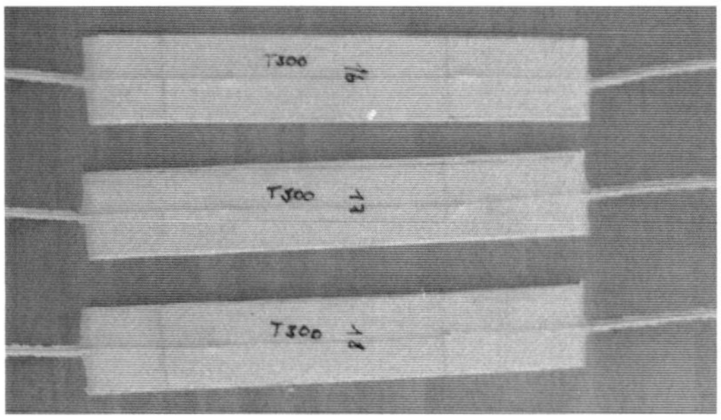

Fig. 1: Specimens with integrated CFSs.

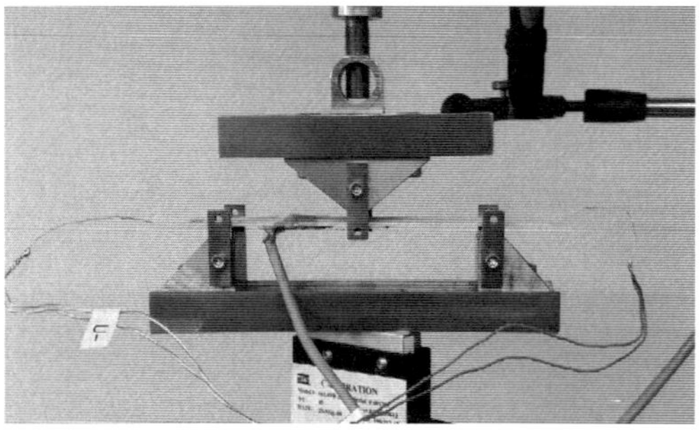

Fig. 2: Configuration of 3-point-bending test.

Tab. 1: Overview of the manufactured specimens.

Speciment Nr.	Placing of CFS between layers		Impact energy
30	1	2	2 J
31	1	2	2 J
32	1	2	2 × 2 J (first impact outside the area of the CFS)
33	1	2	2 × 2 J (first impact outside the area of the CFS)
34	4	5	Not impacted
35	4	5	Not impacted
36	4	5	2 × 2 J (first impact outside the area of the CFS)
37	4	5	2 × 2 J (first impact outside the area of the CFS)
16	7	8	2 J
17	7	8	2 J
18	7	8	2 J

2.3 Loading and damage preparation

All specimens were exposed to cyclic flexural loading. The configuration of the three-point bending (3PB) test was chosen in order to prevent damage in the area of electrical contacts. Hydraulic testing system MTS Mini Bionix (MTS, USA) was used for the load controlled procedure. The configuration of the 3PB test is shown in Fig. 2. Specimens were loaded with a compressive force in the range 1.5–15 N at a frequency of 0.1 Hz. The maximal compressive force 15 N was determined so that the longitudinal strain in the middle of the specimen on the bottom surface was 3,200 μm/m. The mechanical strain in the area of electrical contacts is then 960 μm/m. The applied strain was determined using strain gauge on one of the specimens. Total 200 load cycles were applied for each specimen to investigate the change of el. resistance depending on the mechanical loading before impact loading.

Drop weight impact test was used to test the ability of the CFSs for impact damage detection. Configuration of the impact test is depicted in Fig. 3. For the impact test, we used hemispherical impactor with the diameter of 16 mm with the weight of 410 g. The height for the drop impact test was 0.5 m, so the impact energy was 2 J. After the 2 J impact, there were no signs of damage on the impacted side of the specimen (mold side of the specimen was impacted). There were signs of damage on the opposite side observed on some specimens.

2.4 Electrical resistance measurements

Measurements of the changes in electrical resistance were performed using 344401A Agilent multimeter. The 4-wire ohms method was used in order to eliminate test lead resistances and contact resistances. The relative change in the electrical resistance of each specimen was determined according to Eq. 1. The value was used in the subsequent evaluation, and is marked $\frac{\Delta R}{R}$.

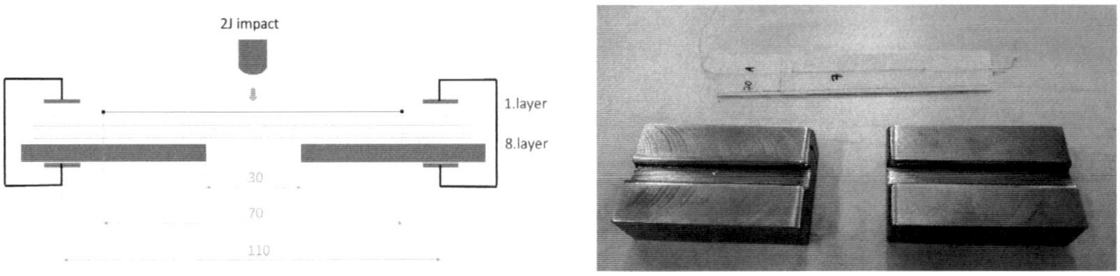

Fig. 3: Configuration of the drop weight impact test.

$$Relative\ change\ in\ electrical\ resistance = \frac{\Delta R}{R}[\%] = \frac{R - R_0}{R_0} \cdot 100. \quad (1)$$

3 Results and discussion

For the purposes of SHM and impact damage detection the following two factors are of great interest:

- the change in the electrical resistivity of the CFS due to the mechanical loading,
- the change in the electrical resistivity of the CFS due to impact loading.

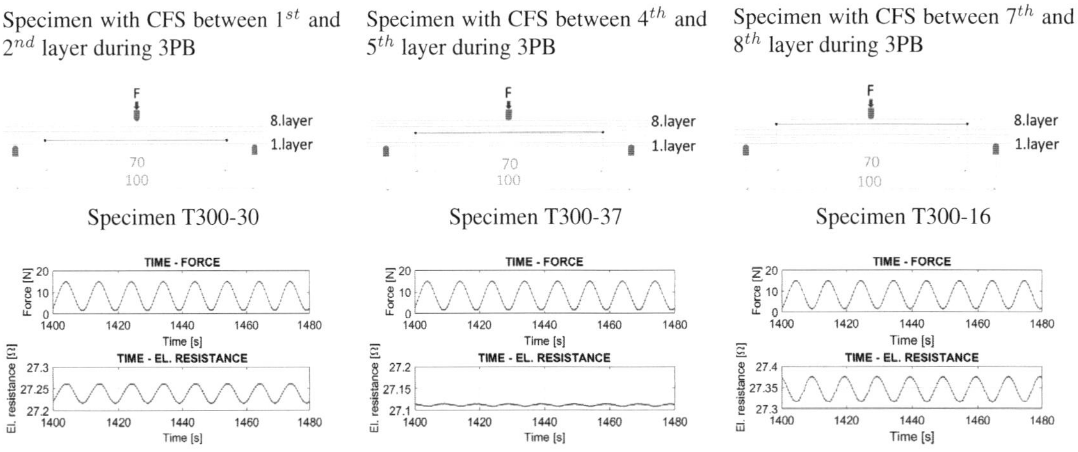

Fig. 4: Configuration of the 3PB test and measured signal from the integrated CFS.

Smooth dependence of the measured relative resistance of the fiber on the loading was measured, see Fig. 4. In order to determine the influence of the number of cycles to the change of electrical resistance the range value of the electrical resistance for the first and the 200th cycle of the mechanical loading was calculated. The results are shown in Fig. 5.

The influence of the placing CFS in different positions regarding the lay-up is shown on measured data in Fig. 4. For compressive loading the change in electrical resistivity is opposite to the change for tension loading. Small changes in electrical resistivity due to the mechanical loading was observed also for the sensors, which were placed in the middle of the lay-up.

The influence of impact loading to the measured electrical resistivity is shown in Fig. 6. In Fig. 5 is given peak-to peak relative electrical resistance change of measured signal in first and 200th cycle. It can be seen that the influence of number of cycles to measured electrical resistance is relatively small, compared to change in electrical resistance caused by the mechanical loading. According to measured data the sensors show stable behavior regarding the mechanical loading. In Fig. 6 is compared relative electrical resistance measured for maximal mechanical strain befor impact during 200th cycle and relative electrical resistance measured after impact. It can be seen that the change in measured el. resistance caused by the impact is much higher than the change in el. resistance caused by the mechanical loading. The measured change in electrical resistance between specimens with sensors placed in the same configurations differs a little bit. A big difference in measured electrical resistance change can be seen for specimen 33 (CFS placed between first and second layer same as for specimens 30,31,32). There was probably problem during the impact loading. More experiments are needed to clarify this behavior.

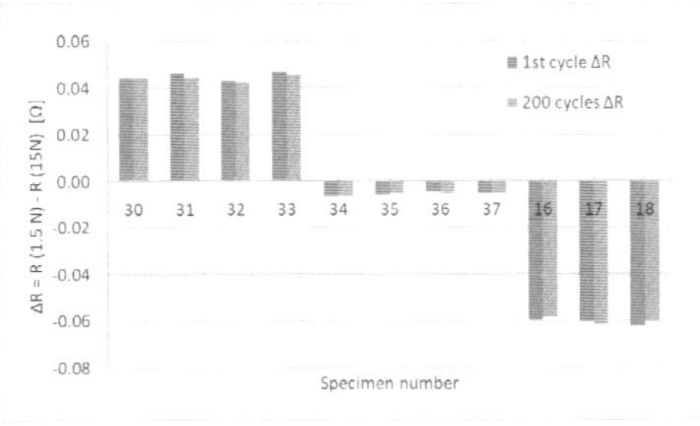

Fig. 5: Influence of the cyclic loading on the measured signal from CFSs.

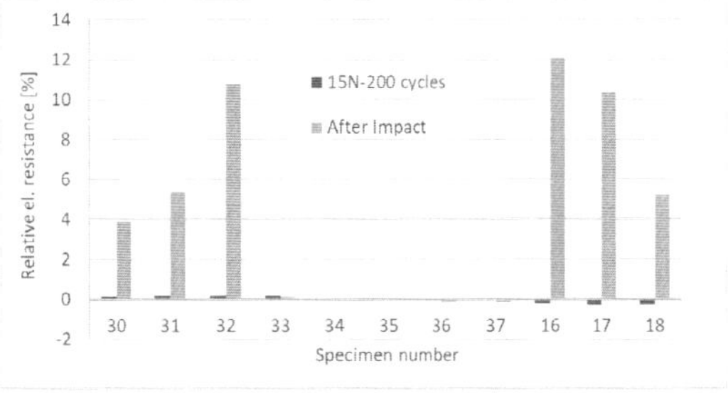

Fig. 6: Influence of the cyclic loading and impact damage on the measured el. resistance of the CFSs.

4 Conclusion

All tested CFSs show consistent dependency between applied load and measured electrical resistance. Investigated type of fiber tow used for integrated CFS shows stable behavior regarding cyclic loading. The influence of the mechanical loading to the change of electrical resistance of the integrated CFS is small compared to the effect of impact damage. For impact damage detection, it seems promising to place the CFS sensors close to the top or bottom of the composite lay-up.

Acknowledgement

This work has been supported by project No. BTHA-AP-2019-16 of the Bayerische-Tschechische Hochschulagentur.

References

[1] A. Horoschenkoff, T. Muller and A. Kroell, On the charakterization of the piezoresistivity of embedded carbon fibres, Proceedings of ICCM 17, Edinburgh, United Kingdom, 2009.

[2] X. Wang, X. Fu, D.D.L. Chung, Strain sensing using carbon fiber. J. Mater. Res. 1999, 14, 790–802.

[3] P. Scholle, M. Sinapius, A review on the usage of continuous carbon fibers for piezoresistive self strain sensing fiber reinforced plastics. J. Compos. Sci. 2021, 5, 96.

[4] E. Haentzsche, A. Nocke, T, Onggar, R.-D. Hund, Ch. Cherif, O. Weissernborn, S. Geller and N. Modler, In-situ load-monitoring of CFRP components using integrated carbon rovings as strain sensors, Proceedings of ECCM18, Athens, Greece, 2018.

[5] A. Horoschenkoff, C. Christner, Carbon fibre sensor, theory and application, Composites and their applications, Ning Hu (Ed.), ISBN: 978-953-51-0706-4, InTech, 2012.

[6] T. Müller, A. Horoschenkoff, H. Rapp, Carbon fibre sensor for crack monitoring of composite materials, Proceedings of ICCM 19, Montreal, Canada, 2013.

[7] N. Schmidová, A. Horoschenkoff and M. Růžička, Investigation of the Electrical Resistivity of Damaged Carbon Fibers Sensors With Regard to SHM, Proceeding of ECCM18,Athens,Greece, 2018.

[8] X. Wang, S.Wang and D.D.L. Chung, Sensing Damage in carbon fiber and its polymer-matrix and carbon-matrix composites by electrical resistance measurement, Journal of Materials Science, 1999, 34, 2703-2713.

New Structure Concept of the Bus Construction with the Addition Carbon Fiber Layers and Filling Foams

S. Špirk[1,*], J. Václavík[2]

[1] University of West Bohemia, Regional Technology Institute, Univerzitní 8, 306 14 Pilsen, Czech Republic
[2] Research and testing institute Pilsen, Tylova 1581/46, 301 00 Pilsen, Czech Republic
* spirks@rti.zcu.cz

Abstract: The construction of buses is a complex and complicated field with different requirements. The mass of the body frame is a very important property. The lower mass can lead to higher occupancy, but it usually means lower stiffness of the construction. Lower mass (while maintaining rigidity) can be achieved by optimizing the geometry or using new advanced materials. In this paper the new structure concept of the bus construction with the addition carbon fibre layers and filling foams, used as local reinforcement of the whole structure, is described. The main goal is the investigation of the feasibility of these hybrid joints in the bus construction (hybrid beams and structural nodes).

Keywords: bus construction; composite; foam; explicit; crash.

1 Introduction

The most significant load on the bus skeleton falls into the area of fatigue and crash. The raw materials' cost in the steel design is cheaper than in the composite one, but the number of hours employed to manufacture the steel design is much higher than in the composite one [1]. This paper is mainly focused on the crash field (especially roll-over Ro.66). Here the composite materials and manufacturing technologies reduce the weight of the bus and the tram vehicle [2]. The effort to lighten the construction of the bus (with same or better stiffness) has two levels. First, it is necessary to create a numerical model (based on available material data) as the first support for experimental work. Second, it is necessary to validate the numerical model using data from those experimental measurements. It is a good idea to validate individual materials separately and subsequently validate more complex units (such as a combination of a steel beam with glued composite strip and foam). The main work is connected with a simple beam model, but the general results will be applied in the bus construction. The commercial solver Pam-Crash is used for simulations.

2 The theory of explicit numerical simulation

The beginnings of the development of explicit solvers date back to the 1960s. At that time, such development took place mainly at universities. The HEMP program, whose code was freely accessible, was the very beginning of today's software packages. Explicit time integration is suitable for simulating processes which involve large strains and changes in shape. It offers better representation of the nonlinear behavior of materials, and failures. Explicit solvers are generally better suited for problems with complex contact. The essence of an explicit code is Newton's second law of motion. It is an equation of motion in the matrix Eq. 1. This equation is defined for the given time. In order to maintain equilibrium between dynamic forces, the relationships below must be met [4].

$$\{a_t\} = [M]^{-1}\left(\{F_t^{ext}\} - \{F_t^{\mathrm{int}}\}\right) \tag{1}$$

Here, $\{a_t\}$ denotes the acceleration vector (at time instant t), $[M]$ is the mass matrix, F_{text} stands for the vector of external forces acting on the body, and F_{tint} is the vector of internal forces.

Once the internal forces have been defined and some fundamental elements have been added, an equation for the numerical solution can be obtained in the following Eq. 2. The element $\{F_{houg}\}$ was added to prevent the hourglassing effect, and $\{F_{cnt}\}$ is a vector of contact forces. Furthermore, $\{\sigma_n\}$ is an internal stress matrix, and $[B]$ is a strain matrix.

$$\{\mathrm{F}_{\mathrm{t}}^{\mathrm{int}}\} = \sum \left(\int_{\Omega} [B]^T \{\sigma_n\} \, d\Omega + \left\{F^{houg}\right\} \right) + \{F^{cnt}\} \tag{2}$$

Solvers that use the explicit code are conditionally stable. This means that they are only stable under certain conditions, referring mainly to the time step size. This, in turn, is related to the propagation of stress waves through the material (see the following Eq. 3). Here, c is the velocity of the wave propagating through the material, l is the characteristic size of an element, E is the modulus of elasticity, and ρ the density of the material.

$$t_{comp} \leq t_{krit} = \frac{l}{c} = l\sqrt{\frac{\rho}{E}} \tag{3}$$

A great advantage of the explicit method is usage of elements with a single integration point. A downside is the reduced stability of computation. If an element deforms symmetrically, no corresponding change in internal energy takes place. Eventually, the computation leads to an imbalance between the kinetic and the internal energy of the system. This numerical error is known as hourglassing. The total energy must be controlled in dynamic calculations. The recognized critical threshold is an increase in the hourglassing energy above 5 % of the total energy of the system. In extreme cases with a significant hourglassing effect, the simulation run may even crash. Various methods are available to control hourglassing.

3 Numerical model

For calculations in the field of passive safety, the project use the FEM solver Pam-Crash from the VPS (Virtual Performance Solution) software package. This solver allows to calculate the events of structural dynamics through an explicit integration scheme and is suitable for problems with significant nonlinearities (geometric, material, contact). To compile the calculation, it is necessary to create a FEM mesh with the necessary entities (welded joints, glued joints, contacts,. . .) and above all it is necessary to create relevant material models. Innovative design requires the usage of models that can be divided into categories:

- Standard steel materials;
- Composite materials;
- Foam;
- Glue.

Because some materials and their uses are relatively unique, this task requires a considerable amount of experimentation, and conventional material testing (like tensile test only) cannot be used. Three-point and four-point bending tests seem to be relatively suitable (quite similar approach to literature [3]), as they can be performed on samples corresponding to the real construction of the bus and their deformation mode is close to the deformation in the rollover type of test. Material models are compiled gradually. Steel parts of the structure can be defined on the basis of routine tensile tests. The beam test, from which the bus structure is formed (and which most influences the behavior of the rollover test), can be considered a suitable validation of the FEM material model (see Fig. 1). If the deformation characteristics match, the material model of the steel can be considered sufficient. The measure of the numerical reliability of the FEM model is the energy balance, where the total energy change must be less than 10 %.

4 Possible approaches for glue joint idealization

4.1 Link with basic material and additional rupture model

The most widely used link (in Pam-Crash) is based on penalty algorithm (with node to segment coupling). Nonlinear contact force can be computed by Eq. 4 where A and B are stiffness parameters and is the distance. The more significant disadvantage of this approach is, than external rapture criterion is the same for all-possible modes of rupture (even same in shear and normal direction).

$$FNL = A \cdot \delta + B \cdot \delta^3 \tag{4}$$

4.2 Link with advanced material model

The more sophisticated solution is to use the material model with failure criterion calculated for two modes of rupture. It is useful to have different criterion for glue joint than for shear and normal direction. The modulus in normal direction E_0 and modulus in shear direction G_0 depend on the energy absorbed in the connection after the rupture starts progress (see Eq. 5).

$$\begin{pmatrix} \sigma_n \\ \sigma_t \end{pmatrix} = \begin{pmatrix} E_0 & 0 \\ 0 & G_0 \end{pmatrix} \begin{pmatrix} \varepsilon_n \\ \varepsilon_t \end{pmatrix} \tag{5}$$

4.3 Cohesive element and link with advanced material model

This approach is the most accurate one. The model is based on a "traction-separation" law with rate dependent trilinear elastic plastic cohesive zone "mixed" mode. It is using the Benzeggagh and Kename criteria or the power criteria for cohesive elements. The theory is described in literature [5, 6]. In the numerical model, the cohesive element does not influence the time step (in Eq. 3). This is a very important feature of a thin layer of glue (typically elements with characteristic length smaller than 0.1 mm).

5 First results of numerical simulations

The very first results of numerical simulations (based on adhesive rupture model) are finished. If the deformation characteristics is in coincidence with experiment, the material model can be considered sufficient. The measure of the numerical reliability of the FEM model is the energy balance, where the total energy change must be less than 10 %.

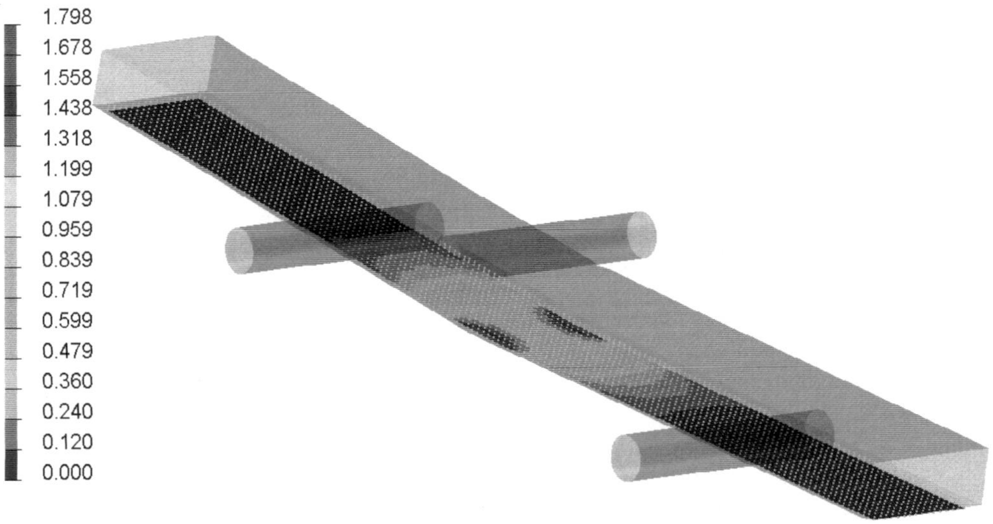

Fig. 1: The calculation of three-point bending of a component reinforced with CFRP tape (VM stress [GPa]).

6 Conclusion

The paper shows some very first results and approaches for numerical simulations of a hybrid beam structure. The partially validated model of a steel hollow profile with glued composite strip and filled with foam gives some observations. The good cooperation of the composite and the steel has better results for composites

Fig. 2: The calculation of a three-point inside view of the foam equivalent plastic strain [-].

with young modulus close to young modulus of steel. The glue connection will be probably the key point of the good working part of construction. The three-point bending test with its simulation can help to validate the material models, but the numerical simulation and the full-scale experiment of the bus segment is recommended. This type of experiment will be performed in the near future.

Acknowledgement

This work has been supported by the project TH71020003 "Composite reinforcement in a light stainless steel bus structure" provided by the Ministry of Industry and Trade of Czech Republic.

References

[1] E. Larodé, et al., A new concept of a bus structure made of composite materials by using continuous transversal frames. Composite Structures 32 (1995), 345–356.

[2] K. Hee-Young, et al., A study on the crashworthiness and rollover characteristics of low-floor bus made of sandwich composites. Journal of Mechanical Science and Technology 23 (2009), 2686–2693.

[3] M. Kriescher, et al., Possibilities for the use of metal-hybrid-structures for vehicle crash load cases. Proceedings o 1. International Conference Euro Hybrid Materials and Structures 2014, 10–11. April 2014, Stade, Germany.

[4] T. Belytschko, J. L. Lin, C. S. Tsay, Explicit algorithms for the nonlinear dynamics of shells, Computer Methods in Applied Mechanics and Engineering, July 1983.

[5] P.P. Camanho, C.G. Davila, M.F. de Moura, Numerical Simulation of Mixed-Mode Progressive Delamination in Composite Materials, Journal of Composite Materials, 2003, 37(16), 1415–1438.

[6] G.I. Barenblatt, The Mathematical Theory of Equilibrium Cracks in Brittle Fracture, Advances in Applied Mechanics, Elsevier, Volume 7, 1962, Pages 55–129, ISSN 0065-2156, ISBN 9780120020072.

Real Structure and Residual Stresses of Ground Laser Cladded H13 Tool Steel

K. Trojan[1,*], J. Čapek[1], J. Čech[2], N. Ganev[1], V. Ocelík[3], K. Kolařík[1]

[1] *Department of Solid State Engineering, Faculty of Nuclear Sciences and Physical Engineering, Czech Technical University in Prague, Trojanova 13, 120 00 Prague, Czech Republic*

[2] *Department of Materials, Faculty of Nuclear Sciences and Physical Engineering, Czech Technical University in Prague, Trojanova 13, 120 00 Prague, Czech Republic*

[3] *Department of Applied Physics, Zernike Institute for Advanced Materials, Materials Science group, University of Groningen, Nijenborgh 4, 9747 AG, Groningen, The Netherlands*

[*] *karel.trojan@fjfi.cvut.cz*

Abstract: Since laser cladding does not achieve such accuracy, after cladding the material, it is always necessary to machine the surface to the required final shape. In this article, the real structure, the surface state of the residual stresses and the hardness of the ground surface are studied. The values of these parameters significantly affect the properties of the surface and thus the service life of the repaired part. Compressive residual stresses on the ground surface were observed, as well as a hardness difference of up to 300 HV.

Keywords: laser cladding; X-ray diffraction; AISI H13 tool steel; real structure; residual stresses.

1 Introduction

AISI H13 hot working tool steel is one of the most common die materials used in metal and casting industries. Dies suffer damage due to wear and thermo-dynamic stresses during their lifetime [1]. Therefore, various methods have been developed for their repair, which is cheaper than manufacturing new ones. A great benefit of laser cladding in this field is high productivity with minimal influence of surrounding material due to low heat input. However, when cladding multiple layers, the previous layers are thermally affected, which can significantly modify their microstructure, real structure, and hardness. For this reason, not only the cooling rate but also the temperature history during the cladding affects the resulting microstructure [2].

Since laser cladding achieves less accuracy than, for example, the selective laser melting (SLM) process, which uses a powder bed, after cladding the material, it is always necessary to machine the surface to the required final shape [3]. It is crucial to carefully determine, whether by machining the part into the desired shape, an area with poorer mechanical properties does not reach the surface. This would lead to lower service life and consequently unprofitability of the laser cladding. Such a study has not been described in the available literature; therefore, it is necessary to research the state of the real structure and other mechanical properties on the machined surface.

2 Experiment

Laser cladding was carried using an *IPG 3kW Yt:YAG* fibre laser. Laser power density of 90 J/mm^2 was applied to form clads in multilayers. A five-layer sample was formed from six and seven overlapping beads on the substrate made from AISI H11 tool steel. The surface of the clad was ground using an oscillating surface grinder, when 1.7 mm from the contact has been removed.

Surface macroscopic residual stresses on the ground surface were determined by X-ray diffraction. The values of residual stresses were calculated from lattice deformations determined on the basis of experimental dependencies of $2\theta(\sin^2\psi)$ assuming biaxial state of residual stress (θ is the diffraction angle, ψ the angle between the sample surface and the diffracting lattice planes). The diffraction angle was determined as the centre of gravity of the Cr$K\alpha_1\alpha_2$ doublet diffracted by the lattice planes $\{211\}$ of the α-Fe phase. The X-ray elastic constants $\frac{1}{2}s_2 = 5.76\ TPa^{-1}$, $s_1 = -1.25\ TPa^{-1}$ were used for the stress calculation. The experimental error is the standard deviation according to the "$\sin^2 \psi$" residual stress calculation algorithm.

In order to determine the real structure, diffraction patterns were obtained in classical Bragg–Brentano focusing configuration with cobalt X-ray tube. Quantitative analysis was evaluated using the Rietveld analysis in the MStruct software.

Hardness was measured by the instrumented indentation technique. Tests were carried out on MHT micro-hardness tester with diamond Vickers indenter. Indentation cycle consisted of loading to a maximum force of 9.81 N, holding at maximum load, and unloading for 30 s, 10 s, and 30 s, respectively. Data were evaluated by the Oliver-Pharr method.

3 Results

Figs. 1 and 2 describes the surface macroscopic residual stresses of the ground surface in the L direction, i.e. in the cladding and grinding direction, and in the T direction, i.e. perpendicular, respectively. The average error of the residual stress calculation is 26 MPa for both directions. The maps consist of 33 values in three rows with a spacing of 4 mm and the data were linearly interpolated.

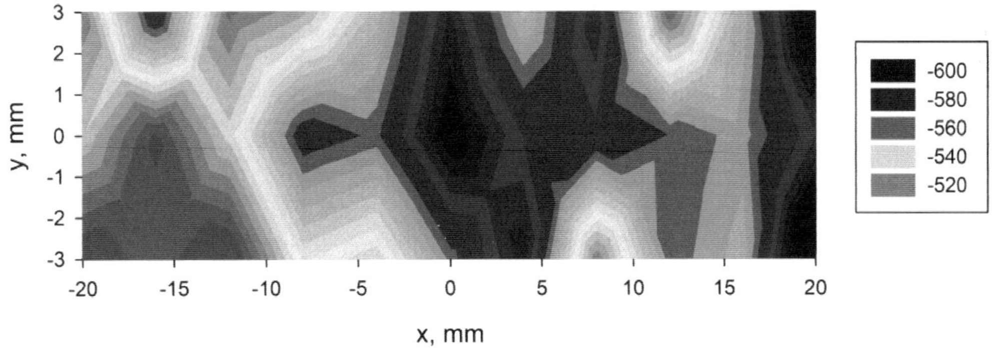

Fig. 1: Map of residual stresses [MPa] on the ground surface of the clad in the L direction.

From the point of view of the development of residual stresses during grinding, the direction of grinding is unfavourable, i.e. in the direction of the grinding lines (in this case the L direction). The material is strongly plastically deformed in this direction during grinding, which can cause tensile residual stresses, especially if the depth of cut per pass is large. However, only compressive residual stresses were detected in both directions. In the L direction, as expected, the compressive residual stresses reached smaller values (average value –551 MPa vs. –867 MPa in the T direction), but due to the small depth of cut per pass and sufficient cooling, tensile residual stresses did not occur.

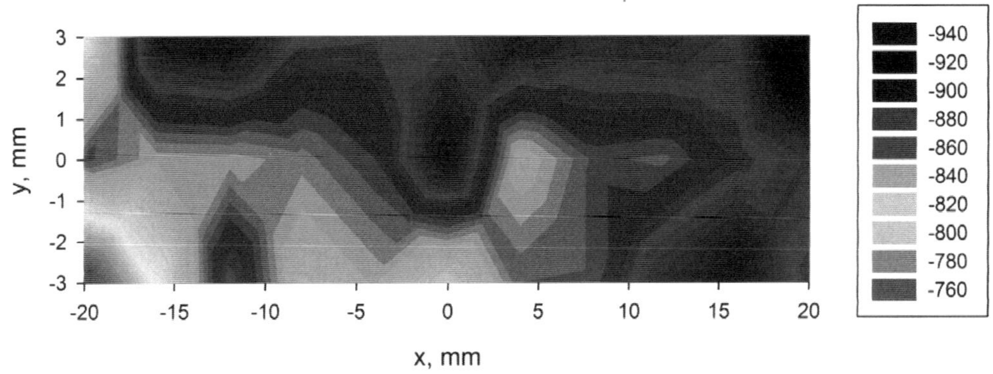

Fig. 2: Map of residual stresses [MPa] on the ground surface of the clad in the T direction.

From the figures describing the residual stresses of the ground surface, it can be seen that higher values of compressive stresses were found in the right part of the clad for both directions. This dependence was most likely caused by the cladding process, when the beads from which the clad was made end in the right part. Therefore, a different cooling rate and temperature history probably occurred in this area.

The real structure was determined by Rietveld refinement, where Fig. 3 denotes the crystallites size and Fig. 4 the microdeformation. The average crystallite size error is 0.8 nm and for microdeformations 1×10^{-4}. Slightly higher values of crystallite size were determined in the left half of the ground surface and, conversely, microdeformation was higher in the right. However, the obtained values show a low standard deviation – 1.2 nm and 1.8×10^{-4} which is almost comparable to the average error. Therefore, in terms of real structure the ground surface appears homogeneous.

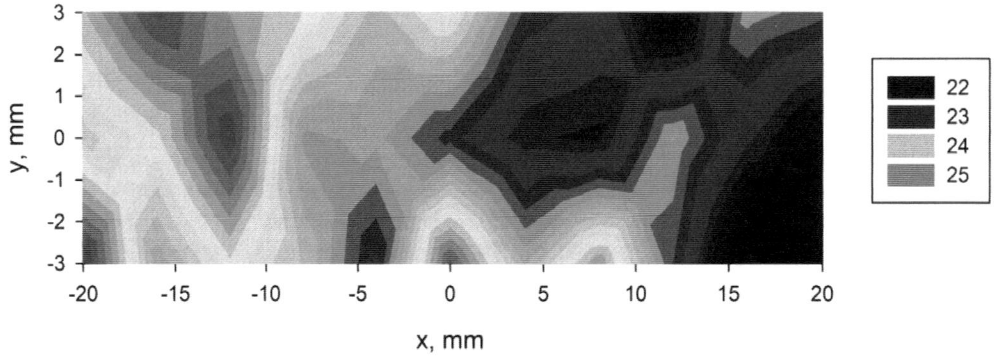

Fig. 3: Map of crystallite size [nm] of ferrite phase on the ground surface of the clad.

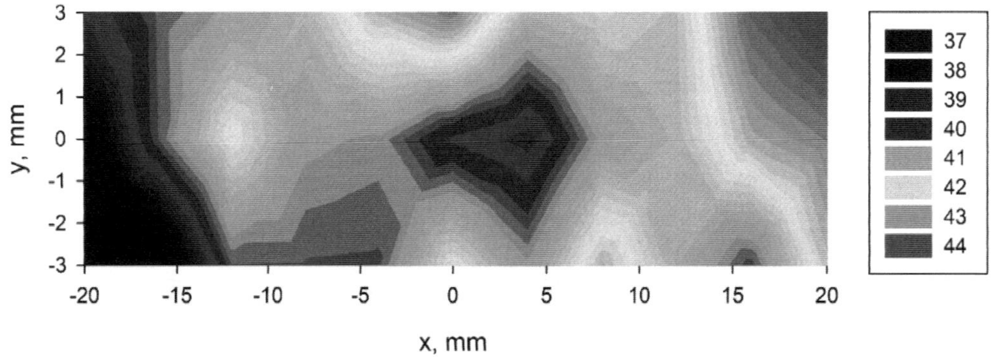

Fig. 4: Map of microdeformation $[10^{-4}]$ of ferrite phase on the ground surface of the clad.

Fig. 5 defines the hardness $HV_{IT}1$ of the ground clad. The average hardness value is 681 $HV_{IT}1$ with a standard deviation of 50 $HV_{IT}1$. The map consists of 111 values in three rows and the data were linearly interpolated. At a distance of approx. $x = -15$ mm on the upper and lower side of the clad, the hardness reached only 535 $HV_{IT}1$. The occurrence of an area with a lower hardness value on the surface of the repaired part is unfavourable in terms of its service life.

It is also clear from the Fig. 5 that the higher hardness values are in the right part of the clad. Although the differences between microdeformation values and crystallite size are small, it is possible to observe that area with higher hardness values correlate with areas with smaller crystallite size and higher microdeformation. This is due to the concentration of dislocations and boundaries of crystallites, which in turn leads to higher hardness. In the area with higher hardness, the compressive residual stresses also reach higher values in both directions.

When optimizing the parameters of laser cladding, it would be appropriate from the point of view of service life to achieve the parameters that occur in the right part of the clad, where the beads always ended during the cladding process.

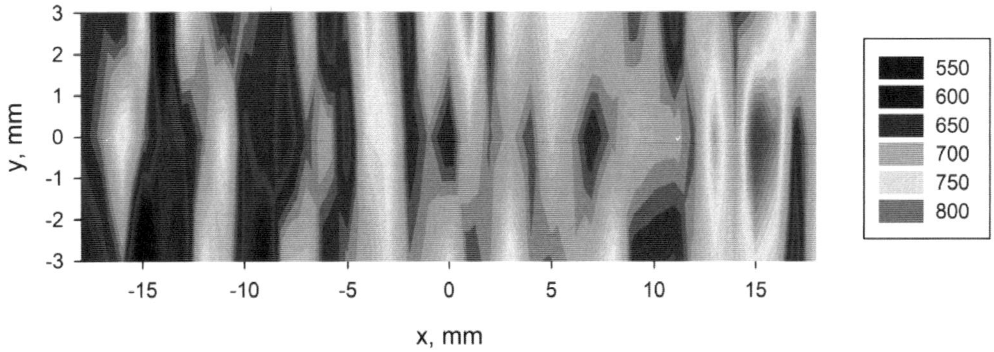

Fig. 5: Hardness [$HV_{IT}1$] map of the ground surface of the clad.

4 Conclusion

Laser deposition of the H13 tool steel showed great application potential. The ground surface showed compressive residual stresses in both directions, which are convenient from the point of view of component life. Even the real structure did not indicate large differences within the analysed ground surface. However, greater variance was observed at the hardness, when the difference between the highest value and the lowest was 300 HV. It is very important that areas with lower hardness did not cover a large part of the surface after machining. When optimizing the parameters of laser cladding, it would be appropriate from the point of view of service life to achieve the microstructure that occurs in the right part of the clad, where the beads always ended during the cladding process. This fact would be appropriate to investigate in further research using numerical methods.

Acknowledgement

Authors thanks to the project CZ.02.1.01/0.0/0.0/16_019/0000778 "Centre of advanced applied science" within the Operational Program Research, Development and Education supervised by the Ministry of Education, Youth and Sports of the Czech Republic. Laser cladding was performed at The University of Groningen, The Netherlands. K.T.'s work was supported by the Grant Agency of the Czech Technical University in Prague, grant No. SGS19/190/OHK4/3T/14.

References

[1] R.G. Telasang, et al., Microstructure and Mechanical Properties of Laser Clad and Post-cladding Tempered AISI H13 Tool Steel, Metall. Mater. Trans. A.46A (2015) 2309–2321, doi: 10.1007/s11661-015-2757-z.

[2] G. Roberts, et al., Tool Steels. Materials Park: A S M International, 1998.

[3] K. Salonitis, L. D'Alvise, B. Schoinochoritis. Additive manufacturing and post-processing simulation: laser cladding followed by high speed machining, Int. J. Adv. Manuf. Technol. 85 (2015) 2401–2411, doi: 10.1007/s00170-015-7989-y.

Load Carrying Capacity of the Hybrid Beam

J. Václavík[1,*], L. Dopierała[2], B. Walczak[2], M. Hejman[3]

[1] *Výzkumný a zkušební ústav Plzeň s.r.o., Czech Republic*
[2] *Solaris Bus & Coach sp. z o.o., Poland*
[3] *Fatigue Analysis RI s.r.o., Czech Republic*
[*] *vaclavik@vzuplzen.cz*

Abstract: Hybrid beams were designed for the local reinforcement of the bus structure. They are formed from hollow profile, CFRP plate and the foam. To verify the load-bearing capacity of the beam and the analytical and FEM calculations, several static and dynamic tests were performed. Tests on basic materials are described in this article. The goal is to determine the allowable structural stress ensuring that the plate will not be deboned during bus performance.

Keywords: bus structure; CFRP; foam; hybrid beam; adhesive joint.

1 Introduction

Carbon fibre-reinforced polymers (CFRP) are often used for strengthening of concrete and more recently steel structures in civil engineering. The use of CFRP materials was spread out also to the vehicle industry. Nowadays vehicle producers must achieve the lowest possible weight of the vehicle with a view to reducing the energy consumption and environmental impact. High interest is to use hybrid materials, which do not require modification of the vehicle production process and lead to strengthen the steel structure [1].

CFRP materials can produce a significant improvement of the local structural response through a significant increment of the local stiffness and strength. This is particular appealing for flexural strengthening at the ultimate limit state, fatigue reinforcement, strengthening against local buckling and confinement of hollow profiles. In particular, for flexural strengthening in the elastoplastic regime, the bonding of CFRP materials produces a significant increment of the plastic moment of the reinforced steel section leading to a considerable increment of the failure load.

In [3], results of some test on samples with bonded pultruded CFRP strips Sika CarboDur M514 bonded with thixotropic epoxy resin Sikadur 30 are presented. Static tests after artificial aging had shown that even if the stiffness and strength capability of the bond is not affected, the ductility of the reinforcement drops to 50%. The study also covered the surface treatment techniques, adhesive curing, and support condition under cyclic loading including fatigue performance, crack propagation, and failure modes with FE simulation. The application of CFRP strengthening composites not only delays the initial crack, reduces the crack growth rate, and extends the fatigue life, but also decreases the stiffness decay with residual deflection.

The influence of the CFRP plate end form was studied in [4]. To help maximize the capacity of bonded joints, all plate ends should implement a reverse tapered detail. Nevertheless the failure by debonding typically occurred when the peak shear stress in the adhesive reached a limiting value of.

Our effort is to improve the vehicle structure with hybrid beams, which are stiffened with CFRP plates and foam inside the hollow profiles. To verify the load-bearing capacity of the beam and the analytical and FEM calculations, several static and dynamic tests were performed. Tests on basic materials are described in this article.

2 Theory

2.1 Analytical solution

The dominant failure mode which can prevent the full load bearing capacity of the hybrid bus structure is the debonding of the CFRP plate. During the debonding process the part of the steel plate not covered by the CFRP plate undergoes plastic deformation while the other part, covered by the CFRP strips remains elastic.

An analytical formulation for prediction the failure load of steel/CFRP joints including plastic deformation of the steel was derived in [2]. Using stress based approach; the shear stress distribution close to the reinforcement ends is given by

$$\tau_{a1} = -\frac{P}{2b_a}\frac{\lambda}{(\delta+1)}e^{-\lambda x} = \tau_{a1,\max}\cdot e^{-\lambda x}, \tag{1}$$

where

$$\delta = \frac{E_s A_s}{2E_f A_f} \text{ and } \lambda = \sqrt{\frac{f_2}{f_1}} = \sqrt{\frac{G_a b_a}{t_a}\left(\frac{1}{E_f A_f} + \frac{2}{E_s A_s}\right)}, \tag{2}$$

where P is the loading force, b_a, t_a are CFRP width and thicknes, x is the distance of reinforced end, E_s, A_s, E_f, A_f are steel plate end CFRP plate Young's modulus and section area, G_a is adhesive sheer modulus.

The sheer stress level is the indicator of unbonding the CFRP layer. The debonding failure load can be estimated as follows

$$P_{lim} = (1 - e^{-\lambda l})\frac{-2b_a\tau_{a2,\max}\cdot(\delta+1)}{\lambda\cdot\delta}. \tag{3}$$

An estimation of the required length l, of the adhesive joint in order to achieve a 99% of the maximum strength is $l_{\min} \geq \frac{5}{\lambda}$, which for our sample under investigations was $l_{\min}$ = 39 mm.

Based on this theory the distribution of sheer stress and peeling stresses in steel adhesive beginning at the CFRP end for our hybrid sample were calculated (Fig. 1).

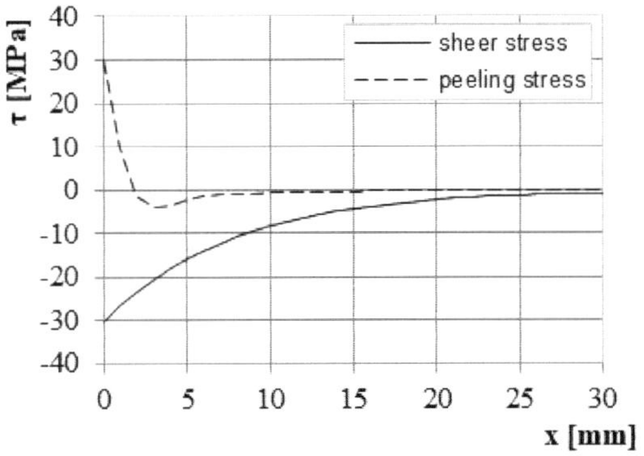

Fig. 1: Shear and peeling stress distribution in the DSR adhesive sample (P = 60 kN), corresponding to tensile stress σ = 400 MPa.

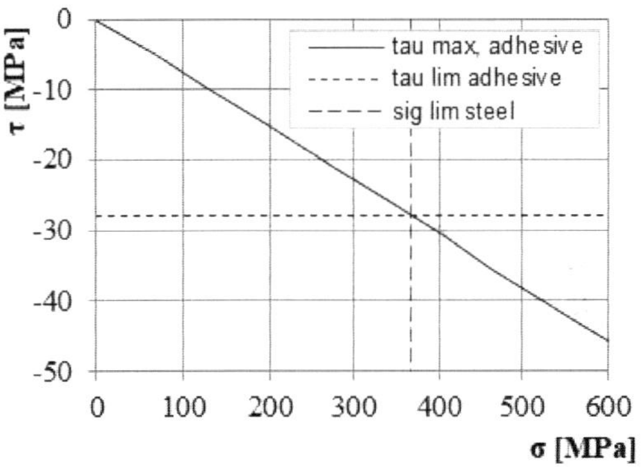

Fig. 2: Relation between sheer stress in the adhesive and the tensile stress with allowable steel and adhesive stress limits.

According the linear relation between the structure tensile stress and sheer stress in adhesive at the reinforced end presented in Fig. 2 it is obvious, that the maximum allowable tensile stress at the CFRP end without interface delamination can reach 370 MPa; this value is higher than the yield point of the rolled plate but lower than that of cold-drawn hollow profile used for the bus structure. The plastic joints are formed in front of the reinforcement in first case but the second case means that the CFRP reinforcement is able to accommodate the development of plastic deformation in the steel. In any case the debonding process to start at the CFRP plate ends after yielding of the steel [3].

3 Implementation

The reinforcement was made with unidirectional pultruded CFRP plates SIKA Carbodur S512 and M512, thickness 1.2 mm and length of 140 mm, E = 165,000 MPa and R_m = 2,900 MPa. CFRP plates were bonded to the stainless steel plate quality 1.4003 by an adhesive SikaPower 1277 with the thickness 0.3 mm E = 2,000 MPa and R_m = 28 MPa). Before performed tests the basic mechanical quantities of steel, CFRP plate, adhesive and foam were tested and compared with guaranteed values, given from the producer. No significant variation was found. The special sample for tensile tests of the adhesive is given in Fig. 3. Sheer test on DSL joints were also performed. The breaking sample after this test is visible in Fig. 4. The variation of measured sheer strength was between 18 and 24 MPa.

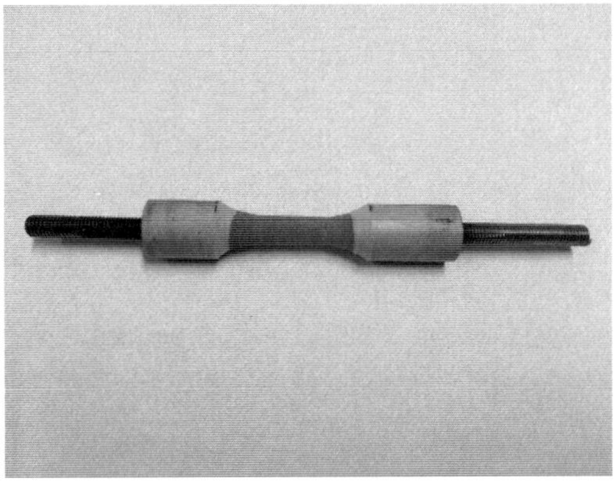

Fig. 3: Tensile sample of the adhesive, casted to the tube and subsequently turned.

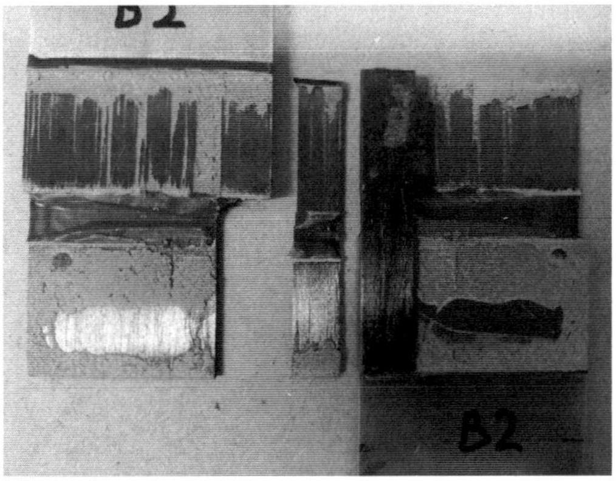

Fig. 4: Separation of double side lapped joint with Carbodur plate.

3.1 Static tensile tests of reinforced specimens

In order to validate analytical relations, the one (OSR) and double-side reinforcement (DSR) steel plates were tested under tensile tests (Fig. 5) and one-side steel plate under 3- and 4-point bending. The reinforcement for the DSR specimen was realized by two unidirectional pultruded CFRP plates Sikadur S512 and M512, with the length equal to 140 mm. All samples were made from the stainless steel quality 1.4003 as it is used for the bus structure. Flat samples made from rolled plates $R_{p0.2}$ = 350 MPa) as well as cut from hollow profile plates $R_{p0.2}$ = 430 MPa) were used for the tests. Measured maximum CFRP strains together with the maximum reached loading stress till unbonding in the distance x from the end of CFRP plate during tensile test are given in Fig. 6. Here SO corresponds to CFRP S512 bonded from one side, SB from both sides and MO CFRP M512 glued from one side.

No debonding was observed in the elastic region of the steel during tensile tests. With large stress the steel undergoes large plastic deformation outside the CFRP (tear about 20 mm of the CFRP end) with small CFRP debonding of some millimeters.

Fig. 5: DSR specimen under static tensile test.

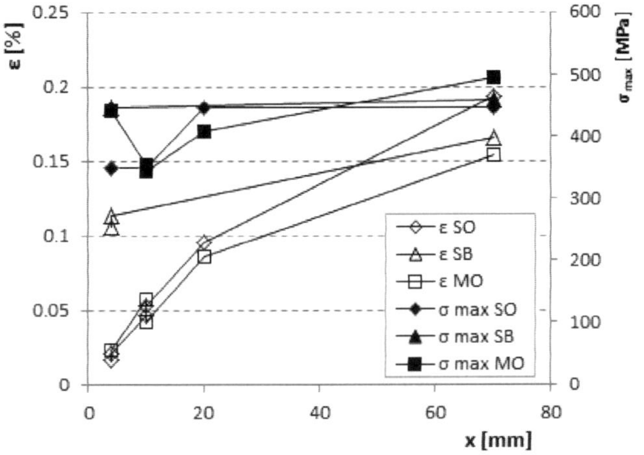

Fig. 6: Maximum CFRP strains and corresponding steel stresses at the moment of CFRP plate unbonding as a function of the distance of the CFRP end.

The comparison of stresses in the middle part of the CFRP and the middle part of steel, opposite to the CFRP for SO specimen is provided in the Fig. 7. It is clearly visible, that the stress transferring by CFRP is much lower, than the stress in the steel (from 35% at beginning of the loading to 65% after the yield point). This is not caused only with the different elastic moduli of steel and CFRP. The measured strains at CFRP and steel are not the same which is caused due to the bending of the sample (bending is worst in the elastic part of the loading curve).

From this point of view it can be concluded, that steel stress is nearly the same as the total stress (Fig. 7, green and black curves) and that the CFRP does not lower the stress substantially. However the advantage of the CFRP plate is obvious – keeps of the high plastic deformation of the sample in the area, where it is glued. The theoretical values of stress in the CFRP and steel were calculated using the total test force, elastic moduli and cross sections of the sample and the CFRP plate and are also given in Fig. 7. These stresses differ not so much, as measured elastic modulus of the steel was low and no bending was considered.

CFRP debonding occurs at 0.38% steel strain, which corresponds to the apparent tensile stress 800 MPa.

The distribution of the longitudinal strain in the CFRP plate from the beginning of the plate to the plate center is shown in Fig. 8. The different curves characterize the different time sections of the loading history up to reaching the yield point. Each cut is specified with the reached testing stress. The curves are normalized to the strain in the middle of the sample. It is interesting, that at the beginning of the loading the stress maximum is not in the middle of the CFRP but closer to the end of the CFRP (here at the distance of 20 mm). The stress does not transferee whole the CFRP plate. During increasing the force the stress is distributed in the CFRP plate according the theoretical assumptions.

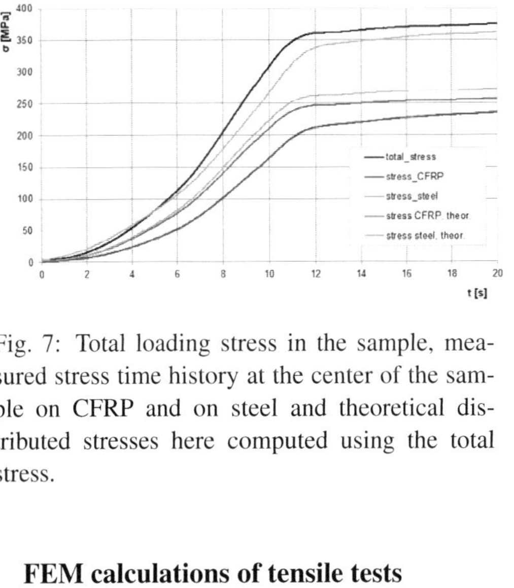

Fig. 7: Total loading stress in the sample, measured stress time history at the center of the sample on CFRP and on steel and theoretical distributed stresses here computed using the total stress.

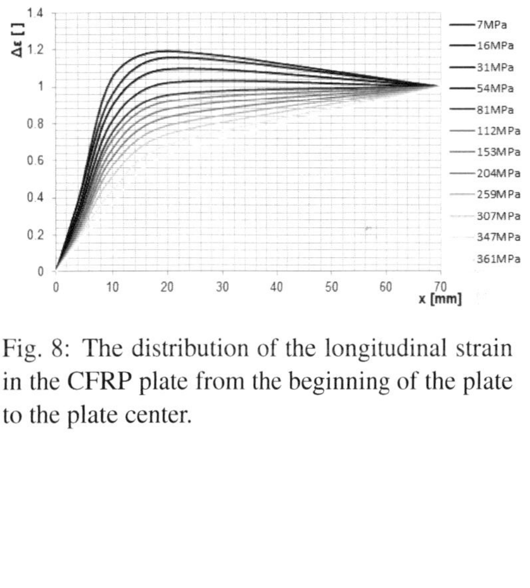

Fig. 8: The distribution of the longitudinal strain in the CFRP plate from the beginning of the plate to the plate center.

3.2 FEM calculations of tensile tests

FEM calculations of the reinforcement specimen was performed in software ANSYS. The linear isotropic material properties were considered according to chapter 3. Quadratic hexagonal elements were used for steel plate, glue and CFRP plate. Three test specimens were modeled. Steel plates with glued CFRP plate Sikadur S512 on one side SO, on both side SB and CFRP plate Sikadur M512 on one side MO. The model was loaded by force 1 kN and scaled for loading force at the moment of CSRP plate unbonding, see Fig. 9. The force corresponding to the total stress see Tab. 1.

Tab. 1: Loading Force at the moment of CSRP plate unbonding.

	Sigma [MPa]	Force [N]
SO	350	52,500
SB	455	68,250
M0	345	51,750

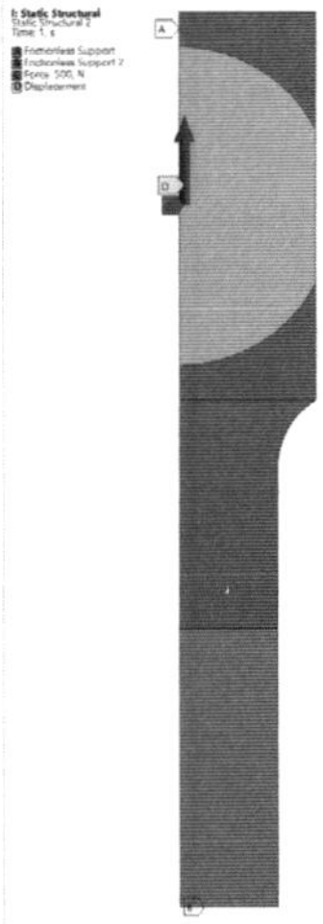

Fig. 9: Loading and boundary conditions of the ¼ test specimen FEM model.

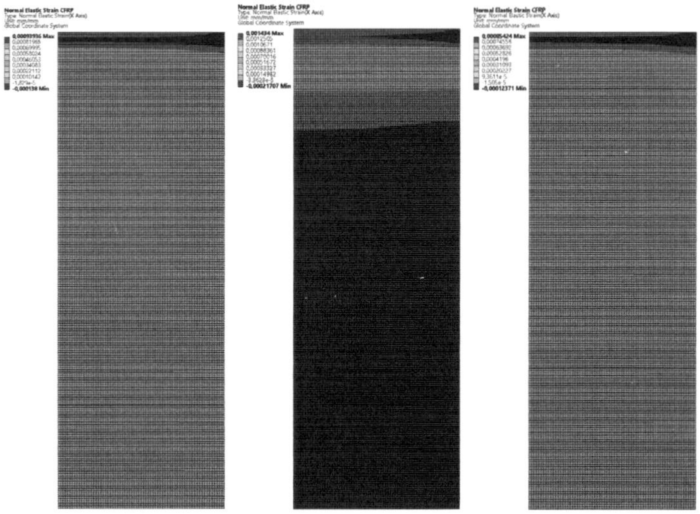

Fig. 10: Field of strain ε [%] in CFRP strap of SO, SB and MO test specimens at the moment of CSRP plate unbonding.

It is possible to compare evaluated strain in Fig. 11 with result of measurement in Fig. 6. The test specimens SO and MO have approximately three time higher strain then in calculations. The reason can be the effect of debonding of CFRP strap or simplified FEM modeling.

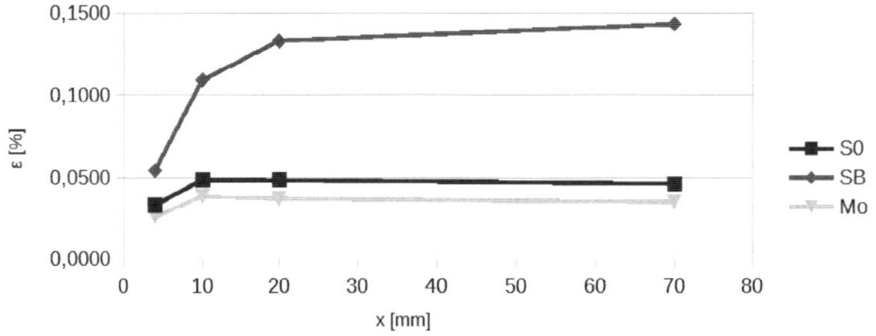

Fig. 11: Strain ε [%] of SO, SB and MO test specimens at the moment of CSRP plate unbonding as a function of the distance of the CFRP end.

To achieve a suitable approximation of SO and MO tests, it is necessary to ensure better measurement of the characteristics of the materials used and implement them in the FEM model.

3.3 Static bending tests of reinforced specimens

3-point and 4-point bending test was realized with the rectangular steel 1.4003 beam with glued CFRP plate of the length 180 mm from the bottom.

During increasing of the deflection, plastic joints were formed in the steel in the area behind the end of the CFRP strips with where very high strain. Plastic strains were also measured on the upper side of the steel above the glued CFRP plate between the applied force pair, but no visual deflection was observed here. The separation of the strip occurred at a steel strain at the end of the strip at $\varepsilon = 3{,}000\ \mu\text{m}\cdot\text{m}^{-1}$, which corresponds to an apparent stress of about 600 MPa.

In the elastic state the sample stiffness with CFRP plate has increased of about 65 %. Due to the shift of the beam neutral axis the strain on the beam upper part drops of about 14 %. However in the elastic plastic state the highest strains were comparable.

Fig. 12: Four-point bending test of steel plate with glued CFRP plate. Plastic joints on both reinforced ends and visible debonding of the CFRP plate.

Fig. 13: Detail of debonding of the reinforce end after plastic deformation of the beam.

4 Conclusion

Results of static tests of hybrid beam individual components were in agreement with producer datasheets. In elastic state the improvement of sample stiffness was substantially improved. After reaching the yield point the plastic deformation occurs in areas not covered with CFRP plate. The conclusion is that the reinforcement ends have not end in the area with high stresses. However it was proved that the CFRP reinforcement is able to accommodate the development of plastic deformation in the steel.

Acknowledgement

The article has originated in the framework of M-ERA.NET call 2019 and was supported with Technological Agency of Czech Republic under the No. TH71020003.

References

[1] J. Korta, T. Uhl, Multi-material design optimization of a bus body structure, J KONES Powertrain Transp 2013, 139–146.

[2] M. Bocciarelli, P. Colombi, G. Fava, C. Poggi, Prediction of debonding strength of tensile steel/CFRP joints using fracture mechanics and stress based criteria, Engineering Fracture Mechanics 76 (2009) 299–313.

[3] M. Bocciarelli, P. Colombi, G. Fava, G. & C. Poggi, Some issues on the strengthening of steel structures with fibre-reinforced polymer materials, Australian Journal of Structural Engineering, Vol. 15, No. 4, October 2014, 337–354.

[4] M. Dawood, S. Rizskala, Bond and splice behavior of CFRP laminates for strengthening steel beams, https://www.researchgate.net/publication/252419022.

Shear Analysis of Perforated Steel Strip

P. Vanova[1,*], D. Dubecky[1], V. Kvocak[1]

[1] *Technical University of Košice, Faculty of Civil Engineering, Institute of Structural Engineering, Vysokoskolska 4, 042 00 Kosice, Slovakia*

[*] *patricia.vanova@tuke.sk*

Abstract: The paper presented contains of analytical, experimental as well as numerical study of perforated steel strip – often referred to as perfobond connector developed at Technical University of Kosice. New geometrical option for such type of shear connector is presented and analyzed, with focus on shear connection – and therefore the push-out tests are the main topic of the article below.

Keywords: composite bridges; shear connector; steel strip; perfobond; push-out test.

1 Introduction

One of the first studies on of the perfobond strips done was performed by Oguejiofor and Husain. They used the steel strip, 127 mm of height, which included one row of 50 mm in diameter holes, located 45 mm axially from the top of the wall. By comparing this shear connector to the shear stud, which was and still is used as main provider of the shear connection in steel concrete composite bridges, the suitability of such a connection was proven [1, 2].

Vianna et al. did a deep study of layout of holes in perfobond strips. They used strips with zero, two and four holes, which also included two options of either passing reinforcement or no reinforcement. With this study they checked the relevancy of the formula used in Eurocode, which concluded positively [3].

Ahn et. al. compared a use of single or twin perfobond connector and development of the cracks regarding its use. They proved the twin connector to be 1.5 to 2 times as strong and it had a ratio of 0.60 of crack to ultimate load [4].

2 Geometrical attributes

The steel strip contains two rows of holes – which are its main geometrical attribute. After filled with concrete, they serve as concrete studs – and therefore strengthen the shear resistance of the strip. As depicted in the Fig. 1, the second row of holes was cut open. Both rows of holes were of 16 mm radius, with the top (open) row 10 mm axial distance from the top of the wall and the second (closed) row 25 mm distant. Both rows were shifted about 15 mm from each other in horizontal direction (vertical in Fig. 1 as for the push-out tests).

For the push out tests, the steel strip was made out of I-beam (specifically IPE200), which was cut in longitudinal direction into two strips.

Through several holes, reinforcement bars were put in order to secure the concrete blocks of the specimen in push-out test layout. In total, 8 reinforcing bars were used, each of 12 mm diameter.

3 Analytical study

Before conducting the experimental study, the presumed results were counted. Using formula (see Eq. 1) originally developed by Oguejiofor and Husain [2, 5], the expected results from experimental study were to be 735 kN. However, this equation does not consider the special characteristic of this continuous shear connector – the fact that the second, outside, row of the holes is cut open. For this reason the formula was adjusted into the form of Eq. 2, which substitute the diameter of the holes of the outside rows in the way that the areas equal.

$$P_{U,teor} = 4.50htf_c + 3.31nd^2\sqrt{f_c} + 0.91A_{tr}f_{yr}, \quad (1)$$

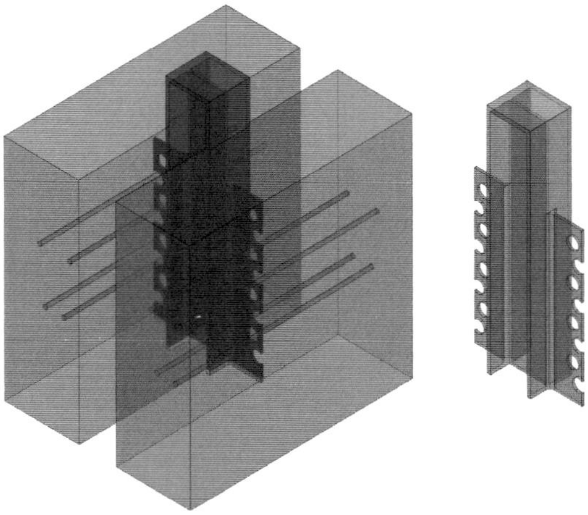

Figure 1: The layout of the perforated steel strip with two rows of holes in the push-out tests.

$$P_{U,teor} = 4.50htf_c + 3.31(n_1d_1^2 + n_2d_2^2)\sqrt{f_c} + 0.91A_{tr}f_{yr}, \tag{2}$$

where *P* represents the expected shear resistance of the specimens; *f* represents strength of each material (with indexes *c* and *yr* representing concrete and reinforced steel, respectively); *A* represents the area of the reinforcing bars that cross the connectors holes; *h* and *t* represent height and thickness of the steel wall, respectively; and *n* and *d* represent number of holes as well as their diameter, respectively.

With the second, adjusted, equation (see Eq. 2), the expected shear resistance was lowered to 725 kN.

4 Experimental study

The normative push-out tests were performed. Two 600 mm long strips were welded vertically onto 100 mm wide middle steel part that served for transforming the pushing force from the hydraulic press into the steel strips, which were lying onto the 10 mm thick polystyrene, in order not to push directly to the concrete – as they were poured into the 600×600×200 mm blocks each. That way, only the shear resistance of the perforated steel strip tested would be measured [6].

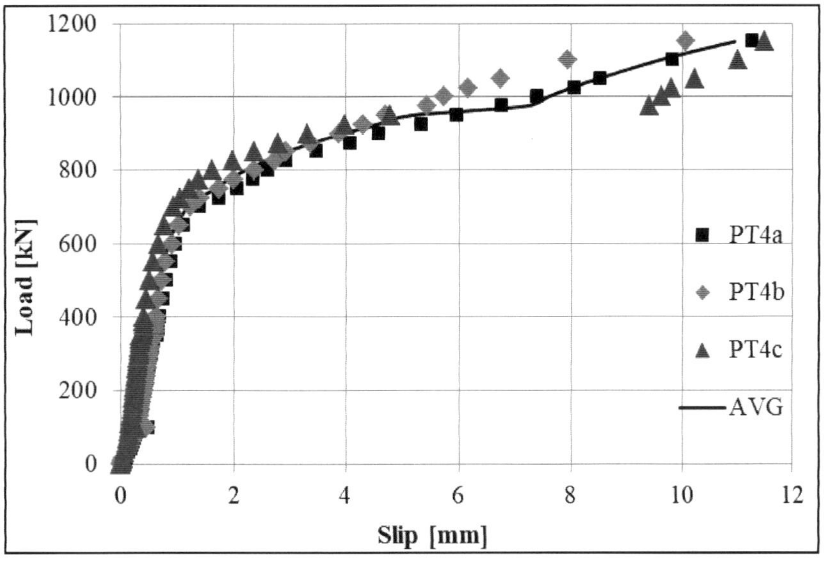

Figure 2: The push-out tests results.

All the specimens were loaded exactly the same by loading cycles. The maxim reached loading cycle included the force of 1150 kN from hydraulic press after the 10 mm polystyrene was compressed and the specimens developed plastic fractures.

The shear resistance of this type perforated steel strip was set at the point where the specimens stopped acting elastically – at the point of the graph (see Fig. 2) where the vertical part turns into horizontal part – which happened after the loading cycle equal to 700 kN.

5 Numerical study

Finite element model was created using Abaqus/CAE software. The model created was the replica of the entire tested specimens, with its idealized geometrical measurements. It consist of around 1 million elements [7].

Material properties set in the Abaqus software were based on the experimental results of materials used. Compressive and tensile strength of concrete at the time of the push-out tests was 58 MPa and 7 MPa, respectively. Its Young Modulus was approximately 33.5 GPa. Yield strength of steel was measured by Faculty of Metalurgy, Technical University of Kosice, who determined it to be 315.3 MPa. Young Modulus of steel was not measured and therefore the normative value of 210 GPa was considered in the numerical study. Reinforcing bars used in the specimens were not tested, therefore only the normative values of their material characteristics were put into the software [7–11].

Dynamic, explicit step performed the analysis in the time of 0.1 s. Nonlinear geometry option was set on ON to take into consideration all the possible geometrical nonlinearities of the specimens during push-out tests [7].

One of the most important aspects of analysis of the push-out tests is the interaction between the two main materials – concrete and steel. In the Interaction Module the General Contact option chosen. To specify the connection, in this option the additional characteristic was added – the fraction coefficient between the touching surfaces of concrete and steel. Its value was 0.25 [7].

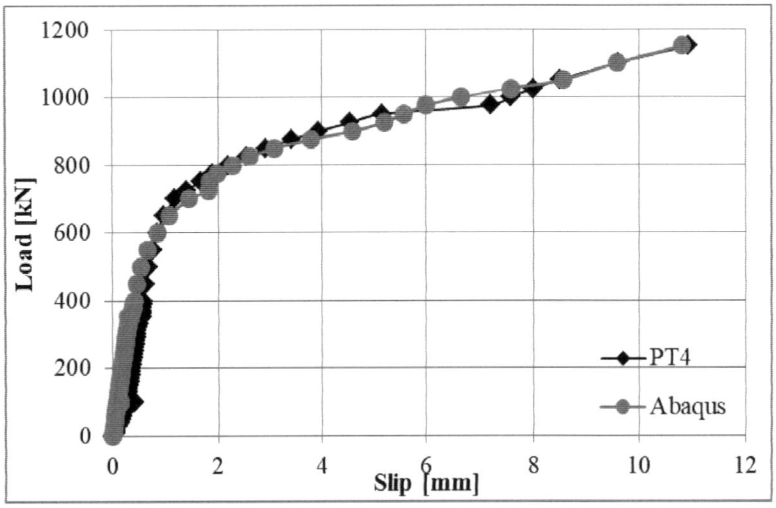

(a) the comparison between software analysis results and experimental results

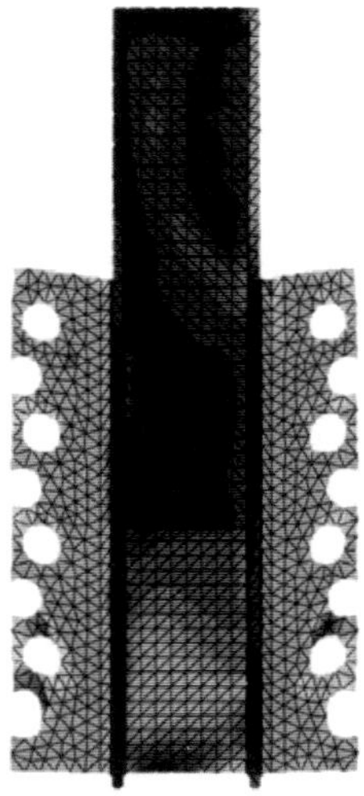

(b) the deformated stressed perforated steel strip during the maximum loading cycle

Figure 3: The software analysis of the specimens during push-out tests.

As visible in the Fig. 3a the curve of the graph representing the numerical study from Abaqus software copied the curve representing the average experimental results sufficiently. The end of the linear section of the software curve was equal to the end of the linear section of the experimental curve with its value of 700 kN.

6 Conclusion

In conclusion, the value of 700 kN for shear resistance of the perforated steel strip is sufficient to secure full shear connection for future use in bridges. However, for use in bridges, flexural tests as well as fatigue tests need to be analyzed.

By adjusting the formula, the ratio between experimental study versus the analytical study was risen from 0.95 to 0.97, however for future use new regression analysis would be required.

The numerical study proved to be satisfactory and therefore can be used for future parametric study.

Acknowledgement

The paper has been supported by The projects: VEGA 1/0172/20 "Stress and deformation analysis of load bearing components made out of steel, glass and composite materials" of the Scientific Grant Agency of the Ministry of Education, science, research and sport of the Slovak Republic and the Slovak Academy of Sciences and by the project Slovak Research and Development Agency under the contract No. APVV-15-0486 "Analysis of Shear Connection Influence in Bridges with Encased Beams".

References

[1] E.C. Oguejiofor and M.U. Hosain, Behaviour of perfobond rib shear connectors in composite beams: full-size tests, Canadian Journal of Civil Engineering 19 (1992) 224–235.

[2] E.C. Oguejiofor and M.U. Hosain, Numerical analysis of push-out specimens with perfobond rib connectors, Computers and Structures 62 (1997) 617–624.

[3] J.d.C. Vianna, S.A.L. de Andrade, P.C.G.d.S. Vellasco and L.F. Costa-Neves, Experimental study of perfobond shear connectors in composite construction, Journal of Constructional Steel Research 81 (2013) 62–75.

[4] J.-H. Ahn, C.-G. Lee, J.-H. Won and S.-H. Kim, Shear resistance of the perfobond-rib shear connector depending on concrete strength and rib arrangement, Journal of Constructional Steel Research 66 (2010) 1295–1307.

[5] M. Rovňák, A. Ďuricová, K. Kundrát and Ľ. Naď, Spriahnuté oceľovobetónové mosty, Elfa s. r. o. (2006).

[6] STN EN 1994-1-1, Eurokód 4: Navrhovanie spriahnutých oceľovobetónových konštrukcií. časť 1-1: Všeobecné pravidlá pre budovy, Technical report (2006), SÚTN Bratislava.

[7] Dassault Systèmes, Abaqus 6.13 online documentation (2013), http://dsk.ippt.pan.pl/docs/abaqus/v6.13/books/usi/default.htm.

[8] STN EN 10002-1, Skúška ťahom. časť 1: Skúška ťahom pri teplote okolia, Technical report (2002), SÚTN Bratislava.

[9] STN EN 12390-13, Skúšanie zatvrdnutého betónu. časť 13: Stanovenie sečnicového modulu pružnosti v tlaku, Technical report (2014), SÚTN Bratislava.

[10] STN EN 12390-3, Skúšanie zatvrdnutého betónu. časť 3: Pevnosť v tlaku skúšobných telies, Technical report (2010), SÚTN Bratislava.

[11] STN EN 12390-5, Skúšanie zatvrdnutého betónu. časť 5: Pevnosť v ťahu pri ohybe skúšobných telies, Technical report (2011), SÚTN Bratislava.

Mechanical Response of Explanted Collagen-nanofiber Composites Intended for Pulmonary Artery Banding

M. Vltavský[1], L. Horný[1,*], J. Kronek[1], T. Suchý[2], L. Vištejnová[3], P. Klein[3],

E. Kuželová Košťáková[4], K. Havlíčková[4], H. Chlup[1], R. Sedláček[1]

[1] *Czech Technical University in Prague, Faculty of Mechanical Engineering, Technická 4, 160 00 Prague 6, Czech Republic*

[2] *Institute of Rock Structure and Mechanics, Czech Academy of Sciences, V Holešovičkách 94/41, 182 09 Prague 8, Czech Republic*

[3] *Charles University, Faculty of Medicine in Pilsen, Husova 3, 301 00 Pilsen, Czech Republic*

[4] *Technical University of Liberec, Faculty of Science, Humanities and Education, Univerzitní náměstí 1410/1, 461 17 Liberec, Czech Republic*

* *lukas.horny@fs.cvut.cz*

Abstract: Pulmonary artery banding (PAB) is an intervention used in neonates suffering from certain congenital heart diseases to reduce pulmonary over-circulation. Our project deals with the research and development of a new banding material based on collagen-polycaprolactone (COLL-PCL) and collagen-poly[L-lactide/caprolactone] (COLL-PLCL) composites. The composites are biodegradable and give the advantage of PAB as a one-step procedure. However, the rate of degradation of the composites in a real body environment is unknown but can be determined by means of biomechanical analysis. COLL-PCL and COLL-PLCL materials were prepared by electrospinning, followed by the impregnation of the layers with collagen (type I) dispersion and implanted into the abdominal walls of adult Wistar rats. The composites were explanted after 4 weeks and their mechanical properties were evaluated by means of the uniaxial tensile test. Both COLL-PCL and COLL-PLCL were found to maintain their integrity up to explantation, and the results of the tensile experiments confirmed sufficient ultimate stress (0.51 ± 0.13 MPa).

Keywords: caprolactone; L-lactide; nanofiber composite; pulmonary artery banding; tensile test.

1 Introduction

There are several procedures in the field of cardiovascular surgery which are based on the mechanical interaction of the arterial wall and the external support. A typical example is Pulmonary Artery Banding (PAB) in infant patients [1–3]. PAB is a palliative procedure that reduces pulmonary over-circulation in neonates suffering from certain congenital heart diseases and constitutes the first stage of intervention prior to the complete repair of cardiac defects. The principle of PAB consists in the implantation of a band around the pulmonary artery, which results in the reduction of blood flow.

Our project deals with the research and development of a new banding material based on collagen-nanofiber composites, which are biodegradable and give the advantage of PAB as a one-step procedure. The rate of degradation of the composites in a real body environment is unknown. However, PAB materials have to maintain their integrity for a sufficiently long time to be capable of fulfilling their hemodynamic function. Thus the present study is designed to determine whether PAB materials based on biodegradable collagen-nanofiber composites will preserve their integrity after a 4-week exposition to *in vivo* conditions.

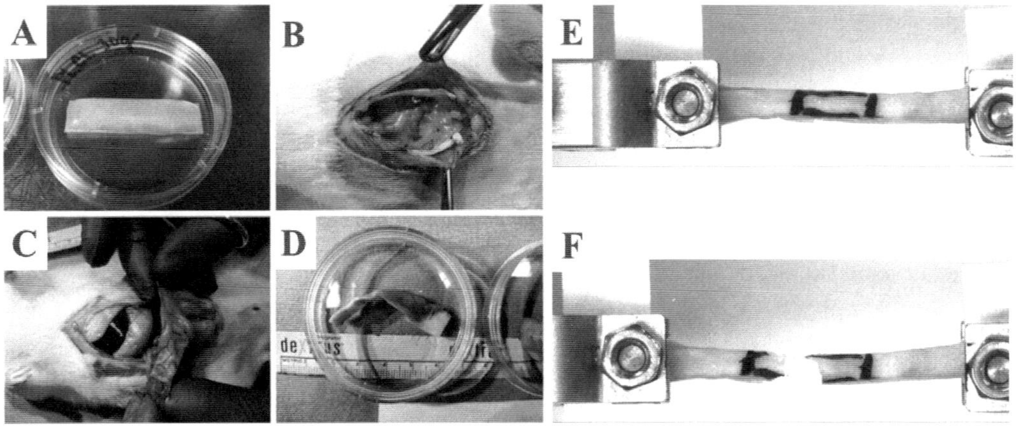

Fig. 1: Representative example of PAB composite. (A) Banding before implantation moistened for 1 hour by physiological solution; (B) Two bandings implanted and fixed by suture on the abdominal wall; (C) Banding 4 weeks after implantation; (D) Explanted banding; (E) A sample mounted in the testing machine; (F) The resulting state after tensile test.

2 Methods

COLL-PCL and COLL-PLCL composite bandings were prepared from nanofibrous layers based on L-lactide/caprolactone copolymer (PLCL; 70/30 mol/mol; PLC 7015, Corbion, The Netherlands) or polycaprolactone (PCL; Merc, USA). The PLCL nanofibrous layers were prepared by needleless electrospinning (NS 1WS500U, Elmarco, Czech Republic) under a direct current from a 10 wt% solution of chloroform and ethanol (Penta, Czech Republic) in 8:2 weight ratio. PLCL layers were prepared with a surface density of 15 (PLCL 15) and 30 (PLCL 30) g/m^2, both with an average fiber diameter of 50–1,000 nm. PCL nanofibrous layers were prepared by needleless electrospinning (NS 1WS500U, Elmarco, Czech Republic) under a direct current from a 16 wt% solution of chloroform and ethanol (Penta, Czech Republic) in 8:2 weight ratio. PCL layers were prepared with the surface density of 30 (PCL 30) g/m^2, with an average fiber diameter of 50–700 nm. The matrix of the composite was composed of type I collagen (COLL, calf skin, VUP Medical, Czech Republic). The volume fraction of all fibrous reinforcement in the composite was 70 $\pm$ 10 vol%. The composite bandings were prepared using the following procedure. An aqueous collagen dispersion (5 wt%) was prepared by means of the swelling of collagen in deionized water (37 °C for 12 hours), double homogenized using a disintegrator (5.000–15.000 rpm, 1–20 min with a delay of 20 minutes). Following homogenization, five separate layers of nanofibrous reinforcement were impregnated with this dispersion, placed in form and left for 36 hours at room temperature. After achieving the constant weight of composites, the stability of the collagen component was enhanced by cross-linking with a 95 wt% ethanol solution containing EDC (N-(3-dimethylaminopropyl)-N-ethylcarbodiimide hydrochloride) and NHS (N-hydroxysuccinimide) (4/1 w/w, Sigma Aldrich, USA). Following a reaction period of 2 hours at 37 °C, the composites were washed in 0.1 M Na_2HPO_4 (for at least 2$\times$ 20 min), followed by rinsing with deionized water (for at least 2$\times$ 20 min) and left in a form for 36 hours at room temperature until the constant weight was reached. The final stage involved cutting the samples into 40 mm $\times$ 6 mm strips and sterilization by gamma irradiation (25 kGy, BIOSTER, a.s., Czech Republic). Fig. 1A documents the final composite in the hydrated state ready to be implanted into the experimental animal.

Experimental implantation was approved by the Animal Welfare Advisory Committee of the Ministry of Education, Youth and Sports of the Czech Republic (approval ID MSMT-19760-2020-3). Six male Wistar rats aged 41 $\pm$ 2 weeks (Masaryk University Brno, Czech Republic) were divided into 3 groups; a group with implanted COLL-PLCL 15g, COLL-PLCL 30g and COLL-PCL 30g. The surgical procedure was performed under general anesthesia. The anesthetized animals were placed on a tempered operating table and continually supplied with oxygen through a face mask and monitored by pulse oximetry. Then an approximately 30–40 mm incision was made on the *linea alba* and on the abdominal wall, where 2 composite bands were inserted into the peritoneum (that is $n = 4$ for COLL-PLCL 15g group, $n = 4$ for COLL-PLCL 30g and COLL-PCL 30g groups). Bandings were fixed at the implantation site with a non-absorbable suture. The wound was closed in 2 layers

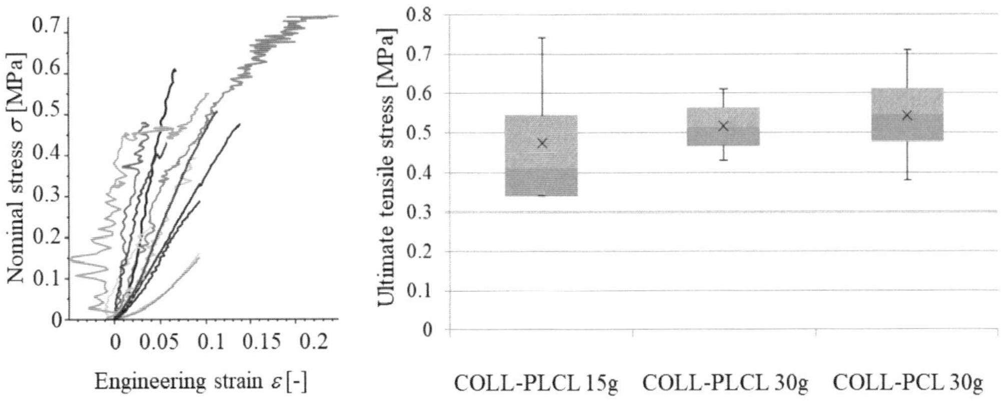

Fig. 2: Results of the uniaxial testing. Mechanical response in the first loading cycle (left). Box-plot of the nominal strength for all investigated groups (n = 4 in each group). There is not a statistically significant difference amongst the medians at the 95% confidence level.

with an absorbable suture and treated with a liquid bandage. Following surgery, the animals were provided with analgesia for 3 days and the animals were kept under conventional conditions (12/12 dark/light cycle) for the duration of the experiment according to EU directive 2010/63/EU in sterile polycarbonate microisolators (Bioscape, Castrop-Rauxel, Germany) with bedding (Lignocel Select fine, JRS, Rosenberg, Germany) and free access to water and pelletized feed (Sniff, Germany). The animals were sacrificed after 4 weeks, the bandings were explanted, immersed in physiological solution, and transported for the biomechanical analysis. Panels B–D of the Fig. 1 document above described phases of the experiment.

The mechanical integrity of explanted PAB materials was evaluated by means of a uniaxial tensile test. Mechanical testing was carried out with a multipurpose electromechanical tensile testing machine Zwick/Roell with built-in videoextensometer and U9C HBM load cells (±25N). Tensile tests were carried out either as a simple monotonous loading up to failure, or as a cyclic loading-unloading procedure, where the 6th cycle was conducted up to the failure. The reference dimensions of the samples were determined by means of an image analysis of photographs of the samples taken before testing. Quantities determined during the experiment were the ultimate tensile stress σ_{ult}, the strain at failure ε_{ult}, and the initial Young modulus E_{ini}. Ultimate tensile stress was defined as the ratio of the force at failure to the reference cross-section area. The strain at failure was calculated from the distance of marks made on the surface of the sample, which was measured by the videoextensometr, by employing $\varepsilon = \Delta L/L$, where L is the reference distance between the marks. Finally, E_{ini} was determined as the slope of the tangent made to the stress-strain curve in the first loading cycle considering $\varepsilon \in [0, 0.02]$.

3 Results and discussion

It was found that both the COLL-PCL and COLL-PLCL materials kept their integrity until explantation and the samples were capable of withstanding normal manipulation (see Fig. 1D,E). Prior to the tensile testing, explanted samples were halved in their width, thus the number of conducted experiments increased from 6 to 12, and the average dimensions of a sample were 40 mm × 3 mm. Seven samples were subjected to a simple monotonous tensile test to a failure, and five were tested in the cyclic procedure ending with ultimate loading.

Fig. 2 displays the stress-strain relationships recorded in the first loading cycle of all tested samples (left panel). Experimental curves exhibit a certain degree of longitudinal oscillations which is an experimental artefact induced by light reflections which disturb the auto-tracking function of the videoextensometr. The imperfect geometrical shape of the marks, which were made manually on the surfaces of the samples, also contributes to these artefacts (see Fig. 1E, F). It is worth noting that the explanted samples exhibit a rather irregular geometry. This is a result of an interaction with the biological environment (see Fig. 1A, D, E).

The specific values of the ultimate tensile stress, the strain at failure, and the initial Young modulus determined in the first cycle are presented in Tab. 1. The strength of materials is also presented in the form of the

Tab. 1: Results of the uniaxial tensile tests. Each group consisted of four samples.

	COLL-PLCL 15g				COLL-PLCL 30g				COLL-PCL 30g			
σ_{ult} [MPa]	0.48	0.74	0.34	0.34	0.61	0.48	0.43	0.55	0.58	0.71	0.51	0.38
mean ± SD	0.48 ± 0.19				0.52 ± 0.079				0.55 ± 0.14			
ε_{ult} [-]	0.14	0.23	0.28	0.25	0.066	0.043	0.057	0.11	0.26	0.16	0.11	0.13
mean ± SD	0.23 ± 0.060				0.069 ± 0.029				0.17 ± 0.067			
E_{ini} [MPa]	2.9	6.4	2.5	2.0	2.5	6.5	9.9	7.0	5.0	3.3	3.4	4.0
mean ± SD	3.5 ± 2.0				6.5 ± 3.0				3.9 ± 0.78			

box-plot in Fig. 2 (right panel). The data shows that all materials retained a certain elasticity during four weeks of *in vivo* degradation. Statistical comparison based on the Kruskal-Wallis test suggests that the composites do not differ from each other with respect to the ultimate stress ($p = 0.63$), however, one should not overrate this result based on a small number of observations. On the other hand, Tab. 1 shows that significant differences could exist in the deformability (here represented by the strain at failure), which is clearly higher for COLL-PLCL 15g and COLL-PCL 30g compared to COLL-PLCL 30g. An inverse result is obtained with regard to the Young modulus, which is a clear consequence of its definition. Thus, from the strength point of view, all composites seem to be interchangeable. However, when deformability is taken into account, COLL-PCL 30g could represent a suitable candidate for a PAB material. Nevertheless, this is a preliminary analysis which will be followed by a more detailed study, involving a larger number of samples and a longer period of *in vivo* degradation.

4 Conclusions

Tensile experiments confirmed a sufficient strength of all variants tested (0.51 ± 0.13 MPa, total pooling). Thus, these preliminary results suggest that bands based on COLL-PCL and COLL-PLCL should preserve their mechanical function under *in vivo* conditions for at least one month. When both strength and deformability are taken into account, COLL-PCL 30g could be a suitable candidate for a PAB material. However, the small number of experiments prevents us from reaching a final conclusion. Further study will also involve other methods of biomedical research such as microscopy, histological investigation, and FTIR.

Acknowledgement

This study has been supported by a grant project awarded by the Ministry of Health of the Czech Republic under project no. NU20-02-00368 “Biomechanically defined absorbable materials for cardiovascular surgery”.

References

[1] E. Buratto, et al., Long-Term Outcome After Pulmonary Artery Banding in Children With Atrioventricular Septal Defects, Annals of Thoracic Surgery 106 (2018) 138–144, doi: 10.1016/j.athoracsur.2018.02.086.

[2] B. Cools, et al., Pulmonary artery banding as 'open end' palliation of systemic right ventricles: An interim analysis, European Journal of Cardio-Thoracic Surgery 41 (2012) 913–918, doi: 10.1093/ejcts/ezr078.

[3] J. Kobr, et al., Right Ventricular Pressure Overload and Pathophysiology of Growing Porcine Biomodel, Pediatric Cardiology 37 (2016) 1498–1506, doi: 10.1007/s00246-016-1463-y.

Inherent Variability of Lattice Discrete Particle Models Caused by Particle Placement Strategies

J. Vorel[1,*], M. Marcon[2], J. Podroužek[3] and R. Wan-Wendner[4]

[1] *Department of Mechanics, Faculty of Civil Engineering, Czech Technical University in Prague, Thákurova 7, 166 29 Praha 6, Czech Republic*

[2] *Christian Doppler Laboratory, University of Natural Resources and Life Sciences Vienna, Peter-Jordan-Strasse 82, 1190 Vienna, Austria*

[3] *Institute of Computer Aided Engineering and Computer Science, Faculty of Civil Engineering, Brno University of Technology, Veveří 331/95, 602 00 Brno, Czech Republic*

[4] *Magnel Laboratory for Concrete Research, Ghent University, Tech Lane Ghent Science Park - Campus A, Technologiepark-Zwijnaarde 60, 9052 Gent, Belgium*

[*] *jan.vorel@fsv.cvut.cz*

Abstract: The paper aims at characterizing the influence of particle placement and clustering in lattice discrete particle models (LDPM) on structural response. More specifically, the meso-structural features are mimicked by the proposed particle placement schemes for LDPM, which are no longer independent and random but are correlated to prescribed fields. The study is based on high-dimensional Monte Carlo (MC) LDPM simulations of three classical concrete tests in which the inherent variability and production process are represented by the proposed particle placement schemes with varying parameters, and constant material and composition properties.

Keywords: Lattice Discrete Particle Model; particle placement; Monte Carlo; structural response.

1 Introduction

Many materials used in civil engineering are heterogeneous and characterized by several length scales, e.g. concrete, fibre reinforced, or particulate filled composites. Various numerical models have been developed to simulate miscellaneous material behaviour. Providing the quasi-brittle materials are in focus, two different groups of approaches have been formulated and developed throughout the years: (1) the discrete fracture formulation, such as cohesive (fictitious) crack model and finite elements (FE) with discontinuity (see [1, 2]), that are not considered in this study; (2) the models with distributed cracking/damage using continuum finite elements and discrete elements such as random lattice or particle model. In the latter approach, many constitutive models have been developed over the years to describe the behaviour of concrete: based on the concepts of plasticity [3], damage mechanics [4], a combination of both, or fracture mechanics [5]. They are typically formulated in tensorial (classical continuum-based theory) or vectorial form (e.g., microplane theory, discrete particle models) [6].

In this paper, we put our attention on the discrete particle model, more specifically Lattice Discrete Particle Model (LDPM) introduced by Cusatis et al. [7]. The LDPM material model response is dependent on particle distribution, and thus multiple simulations are needed to provide credible results [8]. This variability is often considered similar to experimental scatter. However, the numerical model scatter is usually much smaller than the experimentally observed. Therefore, the particle placement influence and its affect on the response scatter is investigated in the paper.

2 Lattice Discrete Particle Model (LDPM)

A well-established member of the discrete framework, the lattice discrete particle model (LDPM), has been extensively calibrated and validated. It has shown superior capabilities in reproducing and predicting concrete behaviour in many practically relevant applications [8, 9, 10]. It simulates the mesostructure of concrete by a three-dimensional (3D) assemblage of particles generated randomly according to a given grain size distribution (Fuller curve). After the particles are randomly placed in the concrete domain from the biggest to the smallest, the lattice mesh is generated, connecting the centres of the spheres. The topology of the interaction among particles is obtained by the Delaunay tetrahedralisation and a tessellation of the domain; for more details see [7]. Displacements and rotations of such adjacent particles form the discrete compatibility equations in terms of rigid body kinematics. At each cell facet, the mesoscale constitutive law is formulated such that it simulates cohesive fracture, compaction due to pore collapse, frictional slip and rate effect. For every single particle, equilibrium equations are finally formulated.

3 Particle Placement

The numerical models of classical concrete experiments are introduced, i.e. cubes and cylinders loaded in compression and unnotched beams loaded in three-point bending. As already mentioned above, the internal structure of the material is modelled through discrete elements to capture the fundamental aspects of heterogeneity. Essential inputs for the models are the maximum and minimum aggregate sizes which are used for the particle size generation based on the classical Fuller curve. The higher bound of the sieve curve is defined by the maximum aggregate size (d_a) while the minimum aggregate size (d_0) defines its arbitrary lower cut-off, i.e. the diameter under which no particles are discretely generated and placed. Thus, the minimum aggregate size affects the refinement of the discrete mesh and consequently also the computational cost. In the original formulation, a statistically isotropic random mesostructure is utilized to create the numerical models. In the present paper new placement procedure is introduced to mimic the segregation or clustering of large particles in specimens caused by the casting process. More specifically, gradient-based generation is utilized in the paper. The response scatter of these approaches are compared, and conclusions are drawn. In the paper, the random and gradient field-based generations are studied (Fig. 1).

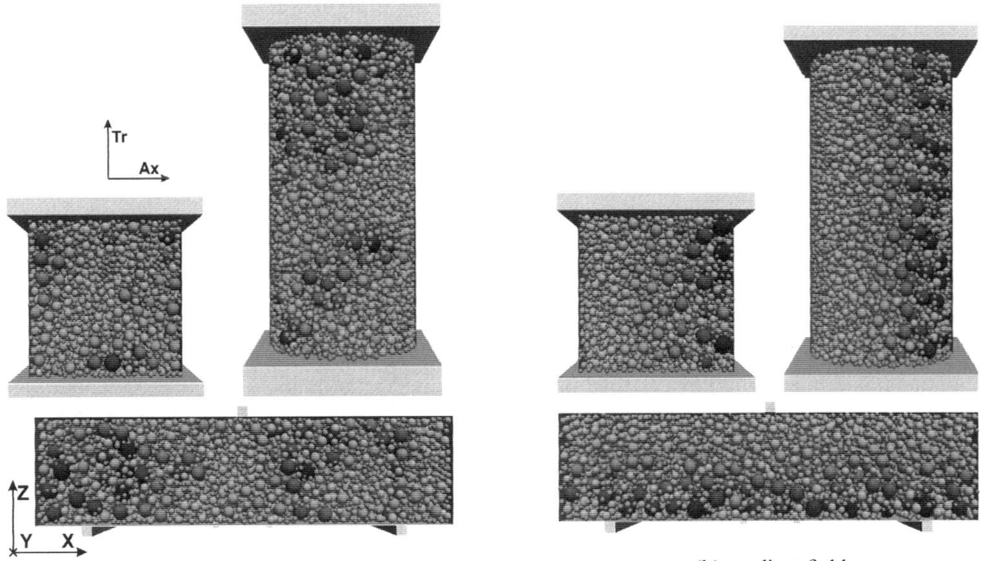

(a) random field

(b) gradient field

Fig. 1: Visual representation of particle placement for compression and three-point bending tests.

Particle generation governed by a field is a modified version of a standard geometrical characterization of the concrete mesostructure presented in [7]. In the present study, the generated mesostructure has to follow both the particle distribution curve and the distribution of a given field, e.g., random, directional. In the first step, particles represented by spheres are generated following the defined concrete granulometric distribution. The main difference between the standard and the new procedure lies in the particle placement strategy. The particle centres are positioned throughout the specimen volume one by one (from the largest to the smallest). Assuming

Tab. 1: Results for the three-point bending tests. The COV is expressed in $\%$.

Property	IRPP	PGGF		
		Z^-	Z^-	Y^+
L@P [MPa]	11.1±2.8%	11.4±2.0%	11.0±1.7%	11.2±1.8%
D@P [-]	0.039±4.5%	0.040±3.3%	0.038±2.7%	0.039±2.9%

Tab. 2: Results for the cylinder compression tests. The COV is expressed in $\%$.

Property	IRPP	PGGF	
		A_x	T_r
L@P [kN]	21.5±0.4%	20.8±0.7%	21.2±0.5%
D@P [μm]	1.2±1.2%	1.1±1.0%	1.2±1.7%

that N_0 particles have to be placed, N_0 random particle positions are generated, and the intensity for each of them is evaluated based on the prescribed field. The positions are then ordered following the given intensity (from the highest to the lowest), and the position with the highest intensity is assigned to the largest particle. The largest particle is then placed at this position (assuming that it does not cross the domain's border), and both the particle and the position are deleted from their lists. Next, the new position with the highest intensity is utilized to position the new largest particle (previously second in the particle list). If there is no conflict with the previously placed particle(s) and the domain's boundary, the particle is placed and again deleted from the list. However, suppose it exceeds the domain boundary or overlaps with the previously placed particle(s). In that case, this position is discarded, a new random position is generated, and its intensity is evaluated. Then the positions are again ordered based on the given intensity, and the particle placing procedure continues as described before. To minimize the geometrical bias of the discretization, a minimum distance between two adjacent particles is defined as $\delta_s\,(r_1 + r_2)$, where $r_{1,2}$ stand for the radii of the particles and $\delta_s \geq 0$ is the non-dimensional scaling parameter. The utilized minimum distance rule allows a smaller distance between small and large particles compared to the distance between two large particles. $\delta_s = 0.1$ is utilized in the current study.

4 Results

Along with simulations in which the particle generation is governed by a field (PGGF), the independent and random particle placement (IRPP) simulations were run for direct comparison. In all cases, 20 repetitions per specimen configuration were run. The results used in the comparison are: (a) the mean stress or the mean force at peak (mean L@P); (b) the mean strain or the mean displacement at peak (mean D@P) for beams and compression specimens, respectively. Also, their coefficients of variations were computed for the comparison.

Tables 1-3 show the IRPP and PGGF results for the three geometries and also different orientations of gradient fields. The results reveal only a slight change in the results and scatter compared to IRPP if the PGGF based generation is utilized.

Tab. 3: Results for the cube compression tests. The COV is expressed in $\%$.

Property	IRPP	PGGF	
		A_x	T_r
L@P [kN]	26.2±0.6%	25.8±0.6%	26.0±0.7%
D@P [μm]	1.7±3.6%	2.9±1.0%	3.5±1.7%

5 Conclusion

A spatial variability package for the LDPM has been presented, including two new abstraction levels for the discrete framework where an initial random field governs material characterization and particle generation. The presented work is the first step of a more extensive investigation. Modelling concepts for different sources of spatial variability in materials (concrete) are being investigated, including spatially variable material property fields. In order to study the pure effects of the particle generation process governed by random or gradient-based fields, the material properties have been kept constant for all of the presented analyses.

Based on the presented results, the following conclusions can be drawn:

- Directional effects, mimicking production processes (concrete casting) and represented by gradient-based fields, may slightly affect both the mean values of force at peak, displacement at peak, and their respective coefficients of variation of the response;
- Correlated spatial variability models (random fields) governing the particle generation process moderately influence the COV of the response compared to the independent and random generation of particles;
- The investigated particle placement schemes with constant material and composition properties enhance the realism of the simulations but are insufficient to reproduce the experimental scatter.

Acknowledgement

The financial support provided by the GAČR grant No. 21-28525S is gratefully acknowledged.

References

[1] T. Belytschko, T. Black, Elastic crack growth in finite elements with minimal remeshing, International journal for numerical methods in engineering 45 (1999) 601–620, doi: 10.1002/(SICI)1097-0207(19990620)45:5¡601::AID-NME598¿3.0.CO;2-S.

[2] M. Jirásek, Comparative study on finite elements with embedded discontinuities, Computer methods in applied mechanics and engineering 188 (2000), 307–330, doi: 10.1016/S0045-7825(99)00154-1.

[3] M. Jirásek, Z. Bažant, Inelastic analysis of structures, Chichester, John Wiley & Sons, 2002.

[4] Z.P. Bažant, L. Cedolin, Stability of Structures: Elastic, Inelastic, Fracture and Damage Theories, 2nd ed., New York, Oxford University Press, 1991.

[5] Z. Bittnar, J. Šejnoha, Numerical methods in structural mechanics, New York, ASCE Press, 1996.

[6] G.I. Taylor, Plastic strain in metals, J. Inst. Metals 62 (1938) 307–324.

[7] G. Cusatis, D. Pelessone, A. Mencarelli, Lattice discrete particle model (LDPM) for failure behavior of concrete. I: Theory, Cement and Concrete Composites 33 (2011) 881–890, doi: 10.1016/j.cemconcomp.2011.02.011.

[8] M. Marcon, J. Vorel, K. Ninčević, R. Wan-Wendner, Modeling Adhesive Anchors in a Discrete Element Framework, Materials 10 (2017), doi: 10.3390/ma10080917.

[9] G. Cusatis et al., Lattice discrete particle model (LDPM) for failure behavior of concrete. II: Calibration and validation, Cement and Concrete composites 33 (2011), 891–905, doi: 10.1016/j.cemconcomp.2011.02.010.

[10] I. Boumakis et al., Discrete element framework for modeling tertiary creep of concrete in tension and compression, Enginnering Fracture Mechanics 200 (2018), 263–282, doi: 10.1016/j.engfracmech.2018.07.006.

Index of Authors

LABORTECH je ryze česká firma, která se více než 26 let podílí na světovém trhu vývojem a výrobou zkušebních strojů, zařízení a zkušebních automatů. Díky vlastnímu inovativnímu vývoji produktů, komplexnímu portfoliu výrobků a především nadstandardními službami našich aplikačních techniků, poskytujeme našim zákazníkům komplexní služby od A až po Z – a to jak v oblasti výzkumu a vývoje, tak i při zajišťování 100% kontroly kvality ve více než 20 průmyslových odvětvích, a to vše podle standardů ISO 9001 a 14001. Naším cílem je i nadále poskytovat našim zákazníkům moderní technická a technologická řešení v oblasti zkušebnictví.